T0271017

BIOMEDICAL MICROSYSTEMS

ELLIS MENG

CRC Press
Taylor & Francis Group
Boca Raton   London   New York

CRC Press is an imprint of the
Taylor & Francis Group, an **informa** business

CRC Press
Taylor & Francis Group
6000 Broken Sound Parkway NW, Suite 300
Boca Raton, FL 33487-2742

© 2011 by Taylor and Francis Group, LLC
CRC Press is an imprint of Taylor & Francis Group, an Informa business

No claim to original U.S. Government works

10 9 8 7 6 5 4 3 2 1

International Standard Book Number: 978-1-4200-5122-3 (Hardback)

Visit the Taylor & Francis Web site at
http://www.taylorandfrancis.com

and the CRC Press Web site at
http://www.crcpress.com

Contents

Preface

The decades-long history of microelectromechanical systems (MEMS) technology has long featured silicon-based devices for telecommunications, automotive sensors, and display devices. The development of MEMS has already left a lasting impact on many fields of engineering and in our everyday lives. The benefits of miniaturization now turn to biochemical and biomedical applications of MEMS. Emerging biomedical microsystems (bioMEMS) technologies are poised to dramatically impact human health and benefit our quality of life.

BioMEMS is a highly interdisciplinary topic; thus, this text combines fundamental knowledge from the material sciences, biology, chemistry, physics, medicine, and engineering to convey the basic principles of these disciplines in a format suitable for advanced undergraduates, graduate students, industrial practitioners, and enthusiasts. The text examines bioMEMS, microfabrication, and nanotechnology, as well as practical examples of devices in use today, through original text and graphical aids. Further complementing these materials are companion homework problems (accompanying all chapters) and suggested laboratory exercises (accompanying several chapters).

The major goal of this text is to provide upper-level undergraduates, beginning graduate students, and other engineers with a limited background in MEMS and bioMEMS with a practical introduction to the technology used to make these devices, the principles that govern their operation, and examples of areas for their application. Chapter 1 begins with an introduction to the benefits of miniaturization. Chapters 2 and 3 introduce the reader to materials and fabrication technology, the necessary components of all biomedical microsystems. Fundamental principles and building blocks are then covered in Chapters 4–6. Finally, Chapters 7–9 discuss a survey of several important applications of bioMEMS.

While the text introduces many topics relevant to bioMEMS, it is not meant to be an exhaustive review of all related areas. Instead, it is intended as a starting point to understanding advanced topics and allows readers to begin to formulate their own ideas about to the design of novel bioMEMS. Many references are provided as a springboard for enthusiasts to launch their own in-depth research on topics of interest.

It is my hope that this text engenders greater interest in bioMEMS and draws fresh minds into finding new ways to harness science and engineering in order to serve humanity.

Ellis Meng

Acknowledgments

This textbook began as a project to fill an unmet need for the introductory course in biomedical microsystems that I started at the University of Southern California in the fall of 2005. I embarked on a journey to write a text to accompany my course with the encouragement and support of my editor, Michael Slaughter, and the fantastic staff at Taylor & Francis. There are many who have contributed their time to reviewing and editing the text, including my colleagues, many patient students who served as "beta testers" for the first drafts, teaching assistants, and graduate students (especially Heidi Gensler and Dr. Ronalee Lo). In particular, special thanks go to my husband, Dr. Tuan Hoang, who drew many new figures for the text and tolerated many lost weekends spent on the preparation of this manuscript. All of you have my gratitude and thanks.

Author

Dr. Ellis Meng received a BS degree in engineering and applied science and MS and PhD degrees in electrical engineering from the California Institute of Technology (Caltech), Pasadena, in 1997, 1998, and 2003, respectively. She joined the Department of Biomedical Engineering at the University of Southern California in 2004, where she currently holds a joint appointment in the Ming Hsieh Department of Electrical Engineering. She holds the Viterbi Early Career Chair in the Viterbi School of Engineering. Her research interests include bioMEMS, implantable bioMEMS, microfluidics, multimodality integrated microsystems, and packaging.

Dr. Meng was a recipient of the Intel Women in Science and Engineering Scholarship, the Caltech Alumni Association Donald S. Clark Award, the NSF CAREER Award, the Wallace H. Coulter Foundation Early Career Translational Research Award, and the Caltech Special Institute Fellowship. She was recently recognized as one of *Technology Review*'s 35 Young Innovators Under 35. She is a member of Tau Beta Pi, the American Society of Mechanical Engineering, the Biomedical Engineering Society, the Society of Women Engineers, and the American Society for Engineering Education.

1

Introduction

Microelectromechanical systems (MEMS) have their roots in the semiconductor industry and share part of their history with the development of the transistor. The similarities between MEMS and transistors are their miniaturized formats (a typical size of nanometers to millimeters) and mass production enabled by batch-based microfabrication technologies. However, unlike a purely electronic transistor, MEMS incorporate multiple modalities, including electrical, mechanical, chemical, and biological. This allows the integration of many functions in a very compact space, including sensing, computing, actuation, control, communication, and power. A large part of the driving force behind the development of MEMS technology is the promise of increased performance achieved with a simultaneous decrease in cost.

Research in the early years of MEMS development focused on silicon-based devices that targeted applications for industrial and automotive use. These devices are now readily available and are used in consumer electronics and in display and communication devices. A recent key development in MEMS is the emergence of their application in biology, chemistry, and medicine. These biomedical microdevices, or bioMEMS, seek to improve technology related to preserving human health and improving quality of life.

1.1 Evolution of MEMS

We start our exploration of bioMEMS with a brief history of their roots in both MEMS and integrated circuits. The basic building block for the transistor is the silicon substrate on which it is fabricated. Although the first transistor was demonstrated in 1947, it took two more years for the development of pure, single-crystalline silicon suitable for transistors and integrated circuits. The methods used to define these small features were already centuries old by this point in time, with the discovery of lithographic processes dating back to the 1500s. However, following the invention of the transistor, MEMS development lagged, with little progress during the integrated circuit revolution; the MEMS revolution began a few decades later. The key events throughout the history of MEMS are summarized in Figure 1.1.

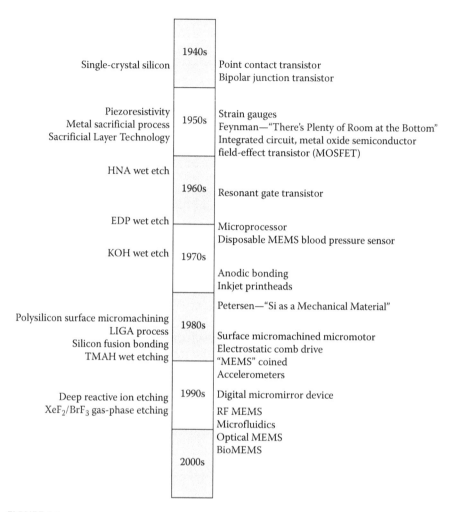

FIGURE 1.1
Major events in the evolution of microelectromechanical systems. Process developments are on the left of the timeline. Devices and other events are on the right.

The resonant gate transistor invented by Nathanson and Wickstrom in 1965 is considered one of the first examples of a MEMS device. This transistor featured a moving gate positioned by electrostatic forces. However, a patent was filed for tiny electrostatic shutter mosaic devices in 1952. The acronym "MEMS" was finally coined in the late 1980s in the United States; the term is now recognized internationally. Before the 1980s, terms such as "micromachining" and "microdynamics" were used. In other parts of the world, regional terms were used, for example, "microsystems" in Europe and "micromachines" in Japan.

A major force in the development of MEMS was an inspiring vision set forth by the Nobel Prize–winning physicist Richard Feynman. Ahead of his time, in late 1959 Feynman presented a vision for the future in a lecture to the American Physical Society, entitled "There's Plenty of Room at the Bottom" [1]. He described the development of microelectronics, micromachines, microsurgery tools, and even nanotechnology. According to Feynman, these technological advances would require new equipment and technologies that did not exist at that time.

To motivate the engineering community to pursue miniaturization, Feynman created two challenges supported by cash prizes funded from his personal funds. The first challenge was to construct an operational electrical motor no larger than 1/64 of an inch on one side. Within a mere 4 months of the speech, William McLellan created a working electrical motor with tweezers and a microscope constructed of only 13 parts and measuring only 250 μg and 3.81 mm wide. Although McLellan did claim the cash prize, he used existing technologies, which was not what Feynman had intended. However, the second challenge, which was to write the contents on a page of a book in 1/25,000 of the original space, was completed 26 years later using the new technology he had described. Tom Newman carved the first page of *A Tale of Two Cities* by Charles Dickens with an electron beam writer so that each letter measured only 50 nm wide.

Feynman's vision has inspired generations of miniaturization enthusiasts and spurred the decades-long pursuit of miniaturization. Early MEMS devices include pressure sensors, accelerometers (acceleration sensors), inertial sensors, micromotors, microengines, microgears, and microtransmissions. These were constructed of materials commonly found in integrated circuits, such as silicon, polysilicon, silicon dioxide, silicon nitride, and metals. With the emergence of biological, chemical, and medical applications, many new materials, especially polymers, have been adopted for MEMS. New fabrication processes have also accompanied the introduction of these new materials. The shift of interest to bioMEMS is attributed in part to the maturation of MEMS technologies, breakthroughs in molecular biology, and demand for improved health care.

1.2 Applications of MEMS

MEMS development has been driven by a number of commercial applications in the automotive, aerospace, telecommunications, and consumer electronics (from displays to mobile handsets to gaming devices) industries. Cars are safer with pressure sensors to monitor tire inflation and accelerometers to trigger airbag deployment. Accelerometers are now increasingly used in consumer electronics, most notably in the Apple iPhone and Nintendo Wii for motion detection. However, the road from discovery to full commercialization has been long due to the time required to develop the technology. This is evident in the timeline shown in Table 1.1.

Other MEMS products that exist commercially, in addition to those already listed, include

- Digital light projectors
- Inkjet print heads
- Gyroscopes
- Microfluidics
- Atomic-force microscope probe tips
- Wafer probers
- Variable optical attenuators
- Infrared image detectors
- Gas chromatographs
- Microphones

According to the market research of Yole Development, the future of MEMS is bright, with an anticipated growth of ~15% per year. BioMEMS are poised to lead this continued growth and perhaps push it to new highs.

TABLE 1.1

Commercialization Timetable for MEMS Products (after [2])

Product	Discovery	Product Evolution	Cost Reduction	Full Commercialization
Pressure sensors	1954–1960	1960–1975	1975–1990	1990
Accelerometers	1974–1985	1985–1990	1990–1998	1998
Gas sensors	1986–1994	1994–1998	1998–2005	2005
Valves	1980–1988	1988–1996	1996–2002	2002
Nozzles	1972–1984	1984–1990	1990–2002	2002
Photonics/displays	1980–1986	1986–1998	1998–2005	2005
Biosensors and biochemical/ chemical sensors	1980–1994	1994–2000	2000–2010	2010
Radio frequency devices	1994–1998	1998–2001	2001–2009	2009
Rate sensors	1982–1990	1990–1996	1996–2006	2006
Micro relays	1977–1993	1993–1998	1998–2010	2010
Oscillators	1965–1980	1980–1995	1995–2010	2010

1.3 BioMEMS Applications

BioMEMS include technologies that enable scientific discovery, detection, diagnostics, and therapy and span the fields of biology, chemistry, and medicine. Rapid progress by researchers worldwide has led to the development of many commercial devices, primarily for noninvasive applications. These include microfluidic devices with pumps and valves, micronozzles for sample introduction into mass spectrometry systems, flow cytometry, microreactors, electrode arrays, genetic analysis devices, biosensors, and point-of-care systems. Existing commercial medical microdevices include pressure sensors (e.g., for blood pressure monitoring, drug dispensing, and respiration monitoring), miniaturized sensors for use with catheters, accelerometers for adaptive rate pacemakers, drug-delivery devices, and components for hearing aids (e.g., MEMS microphones and speakers). Many bioMEMS devices are still being developed. These, along with existing technologies, will continue to drive growth and development in bioMEMS. Exciting possibilities that are yet to be achieved are tools for microsurgery, portable multiple diagnostic devices, and neural implants to restore lost functions. There indeed exists "plenty of room at the bottom" to drive bioMEMS development for decades to come.

1.4 MEMS Resources

Many textbooks now include the science and technology of MEMS, bioMEMS, and other subfields. Several textbooks, which focus on areas of microfabrication technology [3–5] and MEMS devices [6–10] that are beyond the scope of this text, are recommended for further reading. Current developments, however, can only be found in the proceedings of professional meetings and in archival journals. The reader is urged to peruse major journals associated with the field, including *Journal of Microelectromechanical Systems, Journal of Micromechanics and Microengineering, Sensors and Actuators A: Physical, Sensors and Actuators B: Chemical, Lab on a Chip*, and *Biomedical Microdevices*. Other journals that may include bioMEMS include *Analytical Chemistry, Biosensors and Bioelectronics, Institute of Electrical and Electronic Engineers (IEEE) Transactions on Biomedical Engineering*, and *Langmuir*.

Professional meetings of note include Micro Total Analysis Systems (MicroTAS), Microtechnologies in Medicine and Biology (MMB), Micro Electro Mechanical Systems (MEMS), International Conference on Solid-State Sensors, Actuators and Microsystems (Transducers), and Solid-State

Sensor and Actuator Workshop (Hilton Head). Other technical meetings that include bioMEMS are the IEEE Engineering in Medicine and Biology Society Conference and the Biomedical Engineering Society Annual Fall Scientific Meeting. Many of these meetings include published conference proceedings that are accessible through electronic resources affiliated with academic or corporate libraries.

Web portals may also provide valuable information, including the following:

- MEMS and Nanotechnology Clearinghouse: http://www.memsnet. org
- Small Times magazine (on small-tech research, engineering, and applications intelligence): http://www.smalltimes.com
- eMicroNano (a resource for bio-, micro-, and nanofluids and systems): http://www.emicronano.com
- mstnews (international magazine on smart systems technologies): http://www.mstnews.de/

1.5 Text Goals and Organization

The main goal of this text is to introduce readers at various levels and with differing backgrounds to the diverse and highly interdisciplinary field of biomedical microdevices, with an emphasis on bioMEMS and microtechnologies. Students of the field will learn the building blocks of biomedical devices, the methods of their construction, and the principles governing their operation and performance. The focus in this text on fundamental principles of microdevices allows the reader to appreciate both the technical challenges and opportunities that biomedical microdevices bring to medical and life sciences. The text is not intended to be an exhaustive discourse but rather an introductory reference from which the reader can delve into more advanced topics. Specifically, the text will enable the reader to appreciate the motivation and technology behind the latest developments presented in conference proceedings and journal articles.

The text introduces both fundamental concepts and their applications in biomedical engineering problems. This chapter begins with an introduction to the world of biomedical microdevices and explores the merits of miniaturization. The building blocks of all biomedical microdevices, that is, the materials and the microfabrication processes, are discussed in Chapters 2 and 3. Chapter 4 introduces some key concepts in microfluidics and basic elements for microflow control. These microfluidic principles and structures are then applied to lab-on-a-chip and micro total analysis systems in Chapter 5. Chapter 6 introduces sensing and detection principles and technology.

Chapters 7–9 focus on applications of biomedical microdevices, from technology intended for scientific discovery to devices used for clinical monitoring in hospitals. Nascent biomedical microdevices that are yet to be translated from the research laboratory to the clinic are also presented.

1.6 Miniaturization and Scaling

In MEMS, we deal with features in the micrometer to millimeter scale. Thus, we start our exploration of biomedical microdevices by considering the implications of scaling down in size. There are many physical and economical consequences of scaling; some are beneficial, whereas others are not. We can use scaling laws derived from physical laws to rule out physically unfeasible approaches. Identification of economically unfeasible approaches requires synthesis of many factors. Some considerations are as follows:

- Less consumption of material, sample, or reagent
- Disposability and recyclability
- Use of batch fabrication
- Potential for higher yield and fewer defects

There are two types of scaling laws. The first relates to geometry and its implications on physical forces. The second applies to phenomenological behavior.

1.6.1 Geometrical Scaling

The effects of scaling can be nonintuitive. For example, we typically think in a linear fashion, but many of the implications of scaling affect nonlinear parameters such as area, volume, or even forces. Let us look at a simple geometrical scaling example. The solid in Figure 1.2 has a rectangular geometry, with differing characteristic side lengths having the relationship $a > b > c$. The volume of the solid is $V = abc$ and the surface area is $S = 2(ac + bc + ab)$.

Instead of dealing with three different side lengths, we can generalize and say that l represents the linear dimension of the solid. Then, we can express volume and surface area as $V \propto l^3$ and $S \propto l^2$, respectively. The surface area-to-volume ratio then becomes

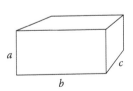

FIGURE 1.2
Solid rectangle having three unequal characteristic side lengths.

$$\frac{S}{V} \propto l^{-1} \qquad (1.1)$$

If we say that l is 1, then the surface area-to-volume ratio is 1. Therefore, a reduction in size by 10 times (or changing l from 1 to 0.1) will result in a 10^3 or 1000-time reduction in volume and a 10^2 or 100-time reduction in surface area. Therefore, as the length scale decreases, volume decreases at a more rapid rate than surface area for an equivalent reduction in the scale. In other words, the surface area-to-volume ratio increases (10:1 in this case, compared to 1:1) as the length decreases.

Let us look at another example. Consider a solid with cubic geometry and side length a. The expressions for volume and surface area are $V = a^3$ and $S = 6a^2$, respectively. This results in a surface area-to-volume ratio of

$$\frac{S}{V} = \frac{6a^2}{a^3} = \frac{6}{a} \tag{1.2}$$

Table 1.2 shows the impact of scaling on surface area, volume, and surface area-to-volume ratio for a cubic solid in relative terms. It is apparent that as the scale decreases, the surface area-to-volume ratio increases.

1.6.1.1 Scaling in Nature

As noted by Thompson in 1917 [11], scaling plays an important role in the sizes of living creatures (Table 1.3). Consider the size extremes of an elephant and a mouse; both are warm-blooded creatures. Larger animals must fight to overcome gravity in order to generate movement; this greater use of energy requires a greater food intake. One might hypothesize that smaller warm-blooded animals are more efficient as they require less food intake and energy to generate movement. The smallest warm-blooded vertebrates are bee hummingbirds, which weigh only 1.5 g. However, a practical limit does exist on the lower bounds of size for warm-blooded animals. Heat loss is proportional to surface area. The significance of this fact lies in the relationship to food intake, which is proportional to volume. This means that smaller animals must spend a significant portion of their energy budget to balance heat loss, which makes tiny warm-blooded animals impractical. (This in part explains why the tiniest of creatures are cold-blooded.) To put the surface

TABLE 1.2

Impact of Increasing and Decreasing Scale on Surface Area, Volume, and Their Ratio of a Cubic Solid

Relative Length of Side	Relative Surface Area	Relative Volume	Surface Area-to-Volume Ratio
0.01	0.0006	0.000001	600:1
0.1	0.06	0.001	60:1
1	6	1	6:1
10	600	1000	0.6:1
100	60,000	1,000,000	0.06:1

TABLE 1.3

Size Scales of Natural and Manmade Objects

Object	Size
Ant	5 mm
Pin head	1–2 mm
Dust mite	200 μm
Human hair	60–120 μm
Ragweed pollen	20 μm
Red blood cell	7–8 μm
Bacteria	2 μm
Rhinovirus	20 nm
ATP synthase	10 nm
MEMS devices	~nm-mm
DNA	2.5 nm
Carbon nanotube	1.3 nm
Carbon buckyball	1 nm
Silicon atom	0.24 nm

area-to-volume ratio of large and small animals in perspective, we note that an elephant possesses a 10^{-4}/mm ratio, whereas a dragonfly has a 10^{-1}/mm ratio.

1.6.2 Scaling of Forces

Scaling laws are particularly useful for understanding the relative strengths of different forces. Relations that would otherwise be complex can be understood with simple mathematical analyses. However, these laws can provide guidance but cannot be regarded as exact solutions.

1.6.2.1 Notation

Physical phenomena are linked to linear dimensions. Trimmer used this concept to create a useful matrix notation to quickly express different scaling relationships [12]. The following force scaling vector is useful for tracking all possible forces of interest:

$$F = \begin{bmatrix} l^1 \\ l^2 \\ l^3 \\ l^4 \end{bmatrix} \tag{1.3}$$

1.6.2.2 Implications

From the force vector, we can group physical phenomena loosely into two categories based on their power dependence on length. Forces that follow l^2 and higher power dependencies have a large power dependence on length; the opposite is true for forces with l^2 and lower power dependencies. For example, gravity follows an l^4 dependence and plays a very significant role at astronomical size scales. On the other hand, magnetic effects may range over l^2–l^4 and have a more complicated relationship with size scale. Thus, mathematically speaking, phenomena that have large power dependencies on length will have a reduced effect at smaller length scales. The opposite is true of phenomena having small power dependencies on length. Such phenomena with lower power dependencies play an important role in the design of MEMS devices in the micrometer to millimeter range.

Let us look at a practical example of how the Trimmer matrix notation can be used for the parameters of acceleration (a), transit time (t), and power density (P/V).

In physics force is expressed as $F = Ma$, and we can rewrite this in terms of acceleration to obtain

$$a = \frac{F}{M} \tag{1.4}$$

Mass, M, is proportional to l^3. Substituting this and the force vector into Equation 1.4, we obtain

$$a = [l^F][l^3]^{-1} = \begin{bmatrix} l^1 \\ l^2 \\ l^3 \\ l^4 \end{bmatrix} [l^{-3}] = \begin{bmatrix} l^{-2} \\ l^{-1} \\ l^0 \\ l^1 \end{bmatrix} \tag{1.5}$$

Transit time is related to Equation 1.4 in the following manner

$$F = Ma = \frac{2xM}{t^2} \tag{1.6}$$

Rewriting, we get

$$t = \sqrt{\frac{2xM}{F}} = \left([l^1][l^3][l^{-F}]\right)^{1/2} \tag{1.7}$$

where x is distance. After substituting the force scaling vector, we get

$$t = [l^2] \begin{bmatrix} l^1 \\ l^2 \\ l^3 \\ l^4 \end{bmatrix} = \begin{bmatrix} l^{1.5} \\ l^{1} \\ l^{0.5} \\ l^0 \end{bmatrix} \tag{1.8}$$

Power, P, is the work, W, done per unit time (or $P = W/t$), and work is force times distance ($W = Fx$). Using these relations, power density, or power per unit volume, is expressed as follows:

$$\frac{P}{V} = \frac{Fx}{tV} = \frac{[l^F][l]}{\left([l^1][l^3][l^{-F}]\right)^{1/2}[l^3]} \tag{1.9}$$

Simplifying Equation 1.9 results in

$$\frac{P}{V} = [l^F]^{1.5}[l^{-4}] = \begin{bmatrix} l^1 \\ l^2 \\ l^3 \\ l^4 \end{bmatrix}^{1.5} [l^{-4}] = \begin{bmatrix} l^{-2.5} \\ l^{-1} \\ l^{0.5} \\ l^2 \end{bmatrix} \tag{1.10}$$

Combining the results in Equations 1.5, 1.8, and 1.10, we now have a set of scaling laws for rigid-body dynamics (Table 1.4). A graphical comparison of the relative importance between force across different lengths is also provided in Figure 1.3. Examples of scaling laws and their length dependencies are summarized in Table 1.5.

TABLE 1.4

Set of Scaling Laws for Rigid-Body Dynamics

Order	Force (F)	Acceleration (a)	Transit Time (t)	Power Density (P/V)
1	l^1	l^{-2}	$l^{1.5}$	$l^{-2.5}$
2	l^2	l^{-1}	l^1	l^{-1}
3	l^3	l^0	$l^{0.5}$	$l^{0.5}$
4	l^4	l^1	l^0	l^2

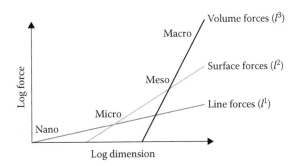

FIGURE 1.3
Graphical comparison of the relative importance of forces across length scales (after [13]).

TABLE 1.5

List of Some Scaling Laws as a Function of Length Scale (after [14,15])

Quantity	Scaling Law
Intermolecular Van der Waals force	l^{-7}
Density of Van der Waals force between interfaces	l^{-3}
Electric field	l^{-1}
Time	l^{0}
Capillary force	l^{1}
Length	l^{1}
Flow velocity	l^{1}
Hydrostatic pressure	l^{1}
Thermal power transferred by conduction	l^{1}
Area	l^{2}
Reynolds number	l^{2}
Electrostatic force	l^{2}
Diffusion time	l^{2}
Volume	l^{3}
Mass	l^{3}
Gravitation force	l^{3}
Magnetic force with exterior field	l^{3}
Magnetic force without exterior field	l^{4}
Centrifugal force	l^{4}

1.6.3 Scaling of Phenomena

1.6.3.1 Electricity

This section describes scaling laws related to electricity.

Electrical field energy, U, is given by

$$U = \frac{1}{2}\varepsilon E^2 \propto l^{-2} \qquad (1.11)$$

where ε is the permittivity of the dielectric and E is the electric field strength.

Electric resistance, R, of a block of material having length L and cross-sectional area A is expressed as

$$R = \frac{\rho L}{A} \propto l^{-1} \qquad (1.12)$$

where ρ is the resistivity.

Resistive power loss, P, through a resistor with a resistance R having an applied voltage V is expressed as

$$P = \frac{V^2}{R} \propto l^{1} \qquad (1.13)$$

Electrostatic forces developed between parallel-plate systems are often used in MEMS devices. Two charged plates separated by a dielectric are represented in Figure 1.4. The plates are of equal size and separated by a gap d. When a voltage, V, is applied to the plates, the electrical charge that develops induces a capacitance, which is expressed as

$$C = \varepsilon_0\varepsilon_r \frac{A}{d} = \varepsilon_0\varepsilon_r \frac{wl}{d} \qquad (1.14)$$

where A is the area of the plates; w and l are the width and length of the plates, respectively; ε_0 is the permittivity of vacuum; and ε_r is the relative permittivity.

The energy associated with the applied potential in this parallel plate system is

$$U = -\frac{1}{2}CV^2 = -\frac{\varepsilon_0\varepsilon_r A}{2d}V^2 \qquad (1.15)$$

Note that the negative sign indicates the loss of potential energy with increasing applied voltage. The force in the d direction is then derived as follows:

$$F_d = -\frac{\partial U}{\partial d} = -\frac{1}{2}\frac{\varepsilon_0\varepsilon_r wl}{d^2}V^2 \qquad (1.16)$$

There also exist forces in plane with the plates (i.e., in the width or length direction) if misalignment occurs. This is generally expressed as follows:

$$F_i = -\frac{\partial U}{\partial x_i} \qquad (1.17)$$

For the width direction, the corresponding force is

$$F_w = \frac{1}{2}\frac{\varepsilon_0\varepsilon_r l V^2}{d} \qquad (1.18)$$

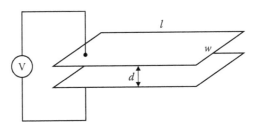

FIGURE 1.4
Two charged parallel plates.

For the length direction, the corresponding force is

$$F_l = \frac{1}{2}\frac{\varepsilon_0\varepsilon_r wV^2}{d} \tag{1.19}$$

Although low in magnitude, electrostatic forces can drive micromotors and microgrippers.

Now, let us see how these forces scale. First, let us assume that applied voltage scales proportionally to the gap ($V \propto d$ or $V \propto l^1$). (Note that this assumption holds true for small gaps on the order of 10 μm or less.) We can express the scaling of the electrostatic potential energy and the various electrostatic forces as follows:

$$U \propto \frac{l^0 l^0 l^1 l^1 \left(l^1\right)^2}{l^1} = l^3 \tag{1.20}$$

$$F_d, F_w, F_l \propto \frac{l^0 l^0 l^1 \left(l^1\right)^2}{l^1} = l^2 \tag{1.21}$$

These scaling laws imply that for a 10-time decrease in the linear dimension, the potential energy will decrease by a factor of 10^3 and the electrostatic forces by a factor of 10^2.

Now, we just need power for operation of the system. A system that uses electrostatic forces and carries its own power supply can be evaluated by examining the ratio of power loss to energy available for powering the system. Available power is a function of the volume, so, using Equation 1.13, we get

$$\frac{P}{E} = \frac{l^1}{l^3} = l^{-2} \tag{1.22}$$

Thus, the decision to scale down power supplies requires careful consideration of this scaling impact and other practical factors.

Passing a current, i, through a conductor placed in a magnetic field will result in the induction of electromagnetic forces. If the conductor possesses an inductance, L, the electromagnetic energy can be found as follows:

$$U = \frac{1}{2}Li^2 \tag{1.23}$$

An electromagnetic force is also present on the conductor. For constant current flow, the electromagnetic force is

$$F = \left.\frac{\partial U}{\partial x}\right|_{i=constant} = \frac{1}{2}i^2\frac{\partial L}{\partial x} \tag{1.24}$$

Because current depends on the cross-sectional area of the conductor ($\sim l^2$) and $\partial L/\partial x$ is dimensionless, the scaling law in this case is

$$F \propto \left(l^2\right)^2 = l^4 \tag{1.25}$$

Electromagnetic forces do not scale as favorably as electrostatic forces. For a 10-time reduction in scale, electromagnetic forces are reduced by a factor of 10^4! Even so, electromagnetic forces are utilized in certain applications, in which they are preferred over electrostatic forces in miniaturized devices.

1.6.3.2 Mechanical Systems

If we have a beam with dimensions w, t, and l, what happens to its mechanical stiffness, k, as it is scaled down? Such a beam is shown in Figure 1.5 and has an elastic modulus E. The expression for the beam's mechanical stiffness is

$$k = \frac{wt^3 E}{4l^3} \tag{1.26}$$

From Equation 1.26, we can see that $k \propto l$ (elastic modulus is a material property and is assumed not to be scale-dependent). Therefore, if we reduce the scale by a factor of 10, k also scales down by a factor of 10. The consequence of this scaling law is that small beams are incredibly strong and can withstand tremendous accelerations without breaking. Microaccelerometers are constructed of small beams and exploit this scaling effect.

1.6.3.3 Fluidic Systems

Reynolds number (Re) is the ratio of inertial to viscous forces and is defined as

$$Re = \frac{\rho UL}{\mu} \propto l^2 \tag{1.27}$$

where ρ is the density of the fluid, U is the fluid velocity, L is the characteristic length of the system, and μ is the viscosity. The Reynolds number is a dimensionless number that can be used to determine whether flow is laminar or turbulent. Typically, we say that for Re < 2000, flow is laminar, and for Re < 2000, flow is turbulent. For MEMS, the flow is in the laminar

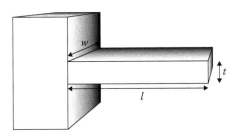

FIGURE 1.5
Simple mechanical beam.

regime and Re $\ll$ 1 is the norm. In scales that we relate on a daily basis, flow can fall into either category. The consequence of a low Reynolds number is that viscous forces dominate and inertial forces may be neglected. This also means, however, that mixing in microsystems will occur slowly and only through diffusion.

Diffusion time, τ, for a particle to travel a distance L is generally expressed as

$$\tau = \frac{L^2}{\alpha D} \tag{1.28}$$

where α is a geometrical constant and D is the diffusion constant. So, $\tau \propto l^2$.

1.6.3.4 Heat Transfer

Thermal energy is proportional to thermal mass, so it scales as l^3. Heat loss (or, more generally, heat transfer) can occur by conduction, convection, and radiation. Let us consider the case of conduction (Figure 1.6). The rate of heat conduction, Q, can be derived from Fourier's law, which describes heat flux, q. For one-dimensional heat conduction,

$$Q = qA = -kA \frac{\Delta T}{\Delta x} \tag{1.29}$$

where A is the area through which heat transfer occurs, k is the thermal conductivity of the solid, and ΔT is the temperature gradient over the distance Δx. At the meso- and microscales, heat conduction scales as

$$Q \propto l^2 l^{-1} = l^1 \tag{1.30}$$

This assumes that the thermal conductivity is a constant. At submicron ranges, this assumption does not hold and $k \propto l^1$, which leads to

$$Q \propto l^1 l^2 l^{-1} = l^2 \tag{1.31}$$

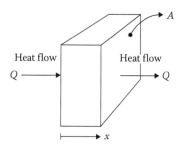

FIGURE 1.6
Conductive heat transfer through a solid slab of material.

Thus, even scaling laws need to be modified to account for differences between macro- and microscale behaviors of materials. We will examine this point further when we delve more deeply into microfluidic phenomena (Chapter 4).

1.7 Problems

1. Identify an important unmet medical need. Research, identify, and summarize the key issues related to your topic. Brainstorm possible technological solutions, including new devices or improvements to existing devices.

2. Search the Internet and find a bioMEMS product (e.g., medical pressure sensor). Obtain the product data sheet from the manufacturer. Use this information and the product's Web site information to answer the following questions: What medical need does the product fulfill and who are the target customers? What MEMS-specific advantages are touted?

3. What are the key advantages of MEMS? List at least five examples and provide brief justifications for each. At least one example should involve scaling considerations.

4. Get to know your library resources. Using a search engine for journal articles (e.g., Web of Knowledge, IEEE Xplore, Engineering Village, or PubMed), find a recent article related to biomedical microdevices. Provide the reference in the following IEEE format. Insert your text within the brackets (and then remove the brackets).

   One author:

   <Initials Author1>. <Last Name Author1>, "<Article Title>," *<Journal Title>*, vol. <Volume Number>, pp. <First page>–<Last page>, <Month> <Year>.

   Multiple authors:

   <Initials Author1>. <Last Name Author1>, <Initials Author2>. <Last Name Author2>, ... and <Initials Last Author>. <Last Name Last Author>, "<Article Title>," *<Journal Title>*, vol. <Volume Number>, pp. <First page>–<Last page>, <Month Abbreviated>. <Year>.

   Example:

   R. Lo, P. Y. Li, S. Saati, R. Agrawal, M. S. Humayun, and E. Meng, "A passive refillable intraocular MEMS drug delivery device," *Lab on a Chip*, vol. 8, pp. 1027–1030, Jul. 2008.

5. Conference proceedings are important for finding the latest developments. Using a search engine containing records of conference

proceedings (e.g., IEEE Xplore or Engineering Village), find a recent conference proceedings paper on a topic related to bioMEMS from one of the conferences listed in the text. Provide the reference in the following IEEE format. Insert your text within the brackets (and then remove the brackets).

One author:

<Initials Author1>. <Last Name Author1>, "<Paper Title>," in *<Conference Proceedings Title>*, <Conference City>, <Year>, pp. <First page>–<Last page>.

Multiple authors:

<Initials Author1>. <Last Name Author1>, <Initials Author2>. <Last Name Author2>, ... and <Initials Last Author>. <Last Name Last Author>, "<Paper Title>," in *<Conference Proceedings Title>*, <Conference Location>, <Year>, pp. <First page>–<Last page>.

Example:

P. Y. Li, R. Sheybani, J. T. W. Kuo, and E. Meng, "A Parylene bellows electrochemical actuator for intraocular drug delivery," in *Proc. 15th Int. Conf. on Solid-State Sensors, Actuators and Microsystems (Transducers)*, Denver, 2009, pp. 1461–1464.

6. Learn to access electronic journals. Select and read an article from the most recent issue of *Lab-on-a-Chip* or *Biomedical Microdevices*.

   a. Summarize and explain the unique benefits of the bioMEMS for the application described. (One paragraph)

   b. Now argue the other side. Are bioMEMS really the best solution? Could another approach provide a better solution? (One paragraph)

   c. Provide the reference (in the IEEE format) and a hardcopy of the paper you selected.

7. Richard Feynman's 1959 speech "There's Plenty of Room at the Bottom" has inspired MEMS researchers internationally. Read the speech at http://www.its.caltech.edu/~feynman/plenty.html. What do you find to be most surprising given that this speech was delivered in 1959? Try to identify predictions that have come true and match them with currently available technology. Think critically and identify predictions that are (a) possible in the near future and (b) altogether unrealistic.

8. Consider the simple cube scaling example in Section 1.6.1. Different cube sizes were examined and the surface area-to-volume ratio was determined. Assume that the cubes (made of the same material) are heated to the same high temperature and allowed to cool to the same low temperature. What behavior do you expect to observe in this system? Hint: it helps to relate heat storage and dissipation to either the surface area or volume. Then, discuss the cooling rate.

9. Let us try to understand what we mean by "small" in the context of microdevices. (a) How small is a microliter of fluid? How small is a nanoliter of fluid? Answer both questions assuming that the fluid takes on a spherical shape. Give the diameter of the sphere. (b) What about the size of a drop of water? Use reasonable estimates to size the drop of water—they should be close to a real droplet. How many microliters are in a drop of water? How many nanoliters are in a drop of water?

10. Visit the MEMS and Nanotechnology Clearinghouse at http://www.memsnet.org. Browse the job listings. What standard qualifications are employers looking for in MEMS positions? List four of them. What types of engineering/scientific backgrounds are commonly sought in potential employees? List three relevant majors.

References

1. Feynman, R. P. 1992. There's plenty of room at the bottom. *J Microelectromech Syst* 1:60.
2. Grace, R. H. 2008. MEMS: Lessons for Nano. *Mechanical Engineering Magazine*, available at http://www.memagazine.asme.org/Articles/2008/august/MEMS_lessons_nano.cfm.
3. Madou, M. J. 2002. *Fundamentals of Microfabrication: The Science of Miniaturization*. 2nd ed. Boca Raton, FL: CRC Press.
4. Bhushan, B. 2004. *Springer Handbook of Nanotechnology*. Berlin: Springer.
5. Gad-el-Hak, M. 2006. *MEMS: Design and Fabrication*. 2nd ed. Boca Raton, FL: CRC Press/Taylor & Francis.
6. Kovacs, G. T. A. 1998. *Micromachined Transducers Sourcebook*. Boston: WCB.
7. Senturia, S. D. 2001. *Microsystem Design*. Boston: Kluwer Academic Publishers.
8. Liu, C. 2006. *Foundations of MEMS*. Upper Saddle River, NJ: Pearson/Prentice Hall.
9. Gad-el-Hak, M. 2006. *MEMS: Applications*. 2nd ed. Boca Raton, FL: CRC Press/Taylor & Francis.
10. Gad-el-Hak, M. 2006. *MEMS: Introduction and Fundamentals*. 2nd ed. Boca Raton, FL: CRC Press/Taylor & Francis.
11. Thompson, D. A. W. 1917. *On Growth and Form*. Cambridge, UK: University Press.
12. Trimmer, W. S. N. 1989. Microrobots and micromechanical systems. *Sens Actuators* 19:267.
13. Judy, J. W. 2001. Microelectromechanical systems (MEMS): Fabrication, design and applications. *Smart Mater Struct* 10:1115.
14. Wautelet, M. 2001. Scaling laws in the macro-, micro- and nanoworlds. *Eur J Phys* 22:601.
15. Tabeling, P. 2005. *Introduction to Microfluidics*. Oxford, UK: Oxford Univ. Press.

2

BioMEMS Materials

Selection of appropriate materials is critical to the success of microelectrome-chanical systems (MEMS) devices (or any manmade engineering device for that purpose). Many well-known engineering failures (e.g., the space shuttle *Challenger* disaster) could have been avoided with a proper material selection and an appreciation for its properties. In particular, we are faced with incomplete knowledge of material properties at the micro- and nanoscales. For example, most material properties are only available for bulk materials, but MEMS rely on thin film versions, which, in this format, can have significant differences from bulk properties. Also, more so than in conventional bulk materials, the manner in which a material is processed impacts its final properties and behavior. With these factors in mind, we begin our study of biomedical MEMS (bioMEMS) with materials first and then continue with the fabrication of these devices in Chapter 3.

2.1 Traditional MEMS and Microelectronic Materials

Many of the materials used in MEMS have their roots in the semiconductor industry. For integrated circuits, the electrical properties of materials are the primary concern. In general, microfabrication is performed on a flat substrate. For many semiconductors, processing is performed on a flat substrate known as a "wafer," which is sliced from a larger single crystal ingot (a cylindrical piece). The thickness is typically on the order of ~300–700 μm, depending on the diameter of the wafer. However, only the top few micrometers of the wafer are used for manufacturing integrated circuits. The bulk of the substrate serves as a mechanical support and is largely left untouched. When selecting materials for MEMS, many properties, such as mechanical, optical, chemical, and biological, in addition to electrical properties, should be considered. We will first explore some materials traditionally associated with microelectronics that have transitioned successfully into MEMS materials, and then will discuss new MEMS materials and, in particular, those having biological or medical importance.

2.1.1 Classification of Electronic Materials

There are three classes of electronic materials: (1) insulators (also known as "dielectrics"), (2) semiconductors, and (3) conductors. They differ in their electrical resistivity, ρ (Ω cm), and allow a convenient material classification, as shown in Table 2.1.

Semiconductors fall between insulators and conductors in terms of their electrical resistivity. Their versatility lies in the ability to tune the resistivity through a process called *doping*, so that semiconductors can have insulating or conductive properties. Examples of commonly used semiconductors are as follows: elemental semiconductors (e.g., silicon and germanium), compounds (e.g., AlAs, GaN, GaP, GaAs, and InP), and alloys (e.g., $Al_xGa_{1-x}As$ and $GaAs_{1-x}P_x$). Many metals and insulators are commonly used in MEMS; among semiconductors, silicon is undoubtedly the most popular MEMS material.

TABLE 2.1

Classification of Materials Based on Electrical Resistivity

Classification	Range of Electrical Resistivity ($\Omega \cdot$ cm)	Examples
Insulator	$>10^8$	Oxide, glass, diamond, quartz
Semiconductor	$10^{-3} - 10^8$	Germanium, silicon, gallium arsenide
Conductor	$<10^{-3}$	Silver, copper, gold, aluminum, platinum

2.1.2 Silicon

Silicon is the second most abundant material on earth but is not found in pure form in nature and must be purified before microfabrication. Silicon is the material of choice for the fabrication of integrated circuits for many reasons, including

- Semiconducting properties can be modified
- Mechanical strength and stability
- High melting point
- Naturally occurring oxide insulation

Due to its widespread use in the semiconductor industry and the historical connection between MEMS and fabrication of integrated circuits, silicon has also gained popularity as a MEMS material.

2.1.2.1 Crystal Structure

Silicon is a crystalline material, although amorphous forms also exist. *Crystalline materials* possess a predictable, long-range atomic or molecular arrangement (i.e., position and stacking sequence; Figure 2.1). *Amorphous materials* do not exhibit a long-range order and do not possess periodic arrangements of atoms or molecules. *Polycrystalline materials* possess regions that exhibit a short-range atomic or molecular order. It is this microstructure that governs many of the material properties of bulk materials.

In crystallography, the basic unit of a crystal is referred to as the *unit cell*. Atoms in a unit cell are located at the fixed locations on a structure known as a *lattice*. The lattice is a three-dimensional arrangement of points in space; these points are referred to as lattice points. Lattice points are associated with a single atom or group of atoms (Figure 2.2). One might think of the lattice model as tennis balls (representing the atoms) connected with sticks (representing the bonds). In close-packed models, the sticks are discarded leaving only an arrangement of spherical objects (e.g., tennis or ping-pong balls; Figure 2.3). Each unit cell has predictable dimensions. For instance, in the simplest of unit cells, a cubic crystal, the lattice has equal dimensions on all

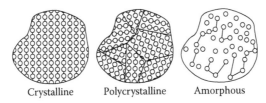

Crystalline   Polycrystalline   Amorphous

FIGURE 2.1
Solids classified according to atomic arrangement. The circles represent atoms or molecules and the connecting lines represent the presence of chemical bonds.

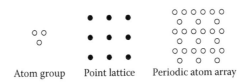

Atom group   Point lattice   Periodic atom array

FIGURE 2.2
Example of a periodic atom array assembled using the point lattice and basic atom group.

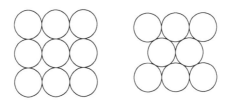

FIGURE 2.3
Examples of two-dimensional close-packed atom arrangements.

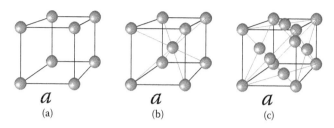

FIGURE 2.4

Three types of cubic lattice unit cells each with lattice constant a: (a) cubic, (b) body-centered cubic, (c) face-centered cubic.

TABLE 2.2

Lattice Constants for Materials Having Cubic-Type Crystal Structure

Material	Crystal Structure	Lattice Constant, a (Å)
Silicon	Diamond	5.43
Germanium	Diamond	5.65
Carbon (diamond)	Diamond	3.57
Gallium arsenide (GaAs)	Zincblende	5.65
Gallium phosphide (GaP)	Zincblende	5.45

sides (Figure 2.4). This defining dimension is called the lattice constant. The lattice constant, a, of silicon is 0.543 nm. Although there are crystal lattices having noncubic structures and thus nonidentical lattice constants, we will not explore them here because silicon takes on a cubic structure (Table 2.2).

Silicon is based on a face-centered cubic (FCC) unit cell. The FCC crystal derives its name from the fact that the atoms reside at the center of each face of the cubic structure (there is also a body-centered cubic cell with an atom located in the center of the cube; Figure 2.4). One additional feature of the silicon crystal is that it is derived from two interpenetrating FCC unit cells offset by one-fourth of the lattice constant in each of the x, y, and z directions. Another way to think of this is to consider an FCC unit cell with an extra atom placed at one-fourth of the lattice constant from each of the FCC atoms. This adds four additional atoms to the silicon unit cell compared to a normal FCC crystal; this structure is referred to as a *diamond* lattice (Figures 2.5 and 2.6). In total, a silicon unit cell has 18 atoms: (1) eight atoms in corners, (2) six atoms on faces, and (3) four interior atoms. Note that only a portion of the atoms located at the corners and faces are associated with each unit cell. At the corners, the lattice points are shared by eight unit cells, and at the faces, the lattice points are shared by two unit cells. In the corner of a diamond crystal, there is a tetrahedron formed by the four nearest neighbor atoms. These atoms possess the closest spacing (0.235 nm) in a diamond lattice.

Note that the overall structure of the silicon crystal is asymmetric, and unequal distances exist between the constituent atoms. These structural details result in the anisotropic nature of silicon's thermophysical and

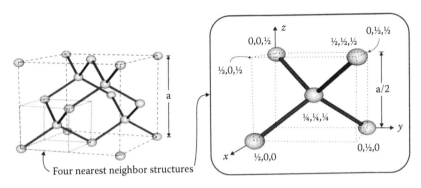

FIGURE 2.5
Diamond lattice unit cell (left) with the four nearest-neighbor structures shown in detail (right).

mechanical properties. This also impacts the micromachining of silicon, which will be explored in Chapter 3.

2.1.2.1.1 Crystallographic Notations

It is useful to devise a uniform nomenclature for designating crystal planes and orientations. We start with a crystal unit cell that has three crystal axes *a*, *b*, and *c*. To specify the direction indicated as OP in Figure 2.7, we resolve the components of OP into three axes *u*, *v*, and *w*, in which the unit distances are *a*, *b*, and *c*. Therefore, the *uvw* components for OP are 1, 1, and ½, respectively. To simplify the notation, we remove the fractions and express the numerals as the smallest set of integers with the same ratio. These numbers, placed in square brackets, denote the direction OP as [221]. If we have negative components, as in Figure 2.8, we indicate this by placing a bar over the corresponding number (e.g., [22$\bar{1}$]). Other direction examples are shown in Figure 2.9. We can have crystallographically equivalent directions such as [100], [010], [001], [$\bar{1}$00], [0$\bar{1}$0], and [00$\bar{1}$]; this family of equivalent directions is conveniently expressed as $\langle 100 \rangle$.

Miller indices provide a standardized system for specifying crystal planes. In Figure 2.10, there is a plane labeled ABCD that has

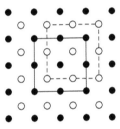

FIGURE 2.6
Top view of diamond lattice structure as viewed along any <100> direction in which the black and white circles represent two different face-centered cubic sublattices in which the white circle lattice is the interpenetrating face-centered cubic lattice.

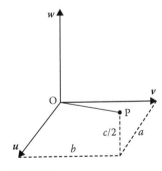

FIGURE 2.7
The crystallographic direction [221] denoted by the line OP.

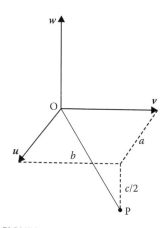

FIGURE 2.8
The crystallographic direction [22$\bar{1}$] denoted by the line OP.

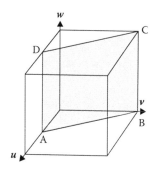

FIGURE 2.10
Plane ABCD with intercepts at ½, 1, and ∞.

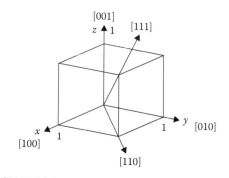

FIGURE 2.9
Several important crystal directions in a cubic unit cell.

intercepts at ½, 1, and ∞. We take the reciprocals of the intercepts and remove fractions again. A set of three numbers, placed in parentheses, denote the plane ABCD as (hkl), or in this case, (210). Other plane examples are shown in Figure 2.11. Using Miller indices, we can also calculate the perpendicular distance between parallel planes, d_{hkl}, where one plane is closest to the origin and the other passes through the origin:

$$d_{hkl} = \frac{a}{\sqrt{h^2 + k^2 + l^2}} \tag{2.1}$$

Crystallographically equivalent planes also have a special notation represented as {hkl}. For example, the {111} family of planes includes (111), ($\bar{1}\bar{1}\bar{1}$), ($\bar{1}$11), (1$\bar{1}$1), (11$\bar{1}$), ($\bar{1}\bar{1}$1), ($\bar{1}$1$\bar{1}$), and (1$\bar{1}\bar{1}$). A summary of the common crystallographic notations used for planes and directions is provided in Table 2.3.

2.1.2.2 Single Crystal Silicon Fabrication

To obtain pure, single crystal silicon substrates for microfabrication, silicon must first be purified (Figure 2.12). Silicon never occurs alone in nature and is found as silica (impure SiO_2) or silicates (Si + O + other elements). Breakthroughs in purification and single crystal silicon growth in the early and mid-1950s were key developments that enabled the integrated circuit revolution.

First, silicon must be separated from its compounds, and then, this separated form, called metallurgical grade silicon (MGS), can be purified.

TABLE 2.3

Summary of Crystallographic Notations for Planes and Directions

Crystal Parameter	Notation
Plane	(hkl)
Equivalent planes	$\{hkl\}$
Direction	$[uvw]$
Equivalent directions	$\langle uvw \rangle$

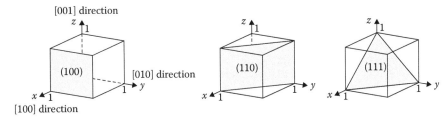

FIGURE 2.11
Three important planes in a cubic unit cell.

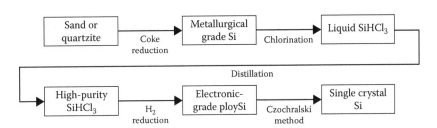

FIGURE 2.12
Purification process to obtain single crystal silicon from sand.

Quartzite, a relatively pure form of SiO_2, is loaded into a furnace in the presence of carbon (coal, coke, or wood chips) to produce silicon (solid) and carbon monoxide (gas); the overall reaction is [1]

$$SiO_2 + 2C \rightarrow Si + 2CO \qquad (2.2)$$

The silicon is now ~98% pure; it has to be chlorinated to change its solid state to a liquid state, because solids are difficult to purify. Changing solid silicon to the liquid form facilitates purification. To do this, the MGS is mechanically pulverized and is reacted with anhydrous hydrogen chloride to form trichlorosilane ($SiHCl_3$), which is liquid at room temperature

$$Si + 3HCl \rightarrow SiHCl_3 + H_2 \qquad (2.3)$$

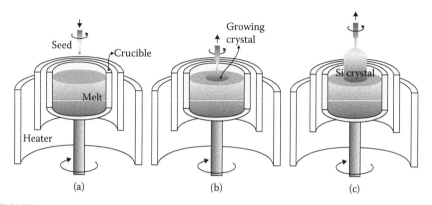

FIGURE 2.13
Czochralski method for single crystal silicon formation: (a) seed lowered down to melt, (b) seed dipped in melt and crystal growth begins, (c) partially grown crystal ingot.

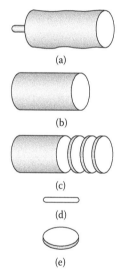

FIGURE 2.14
Mechanical processes to get wafer from ingot: (a) crystal ingot, (b) grind crystal ingot to precise diameter, (c) saw ingot into slices (not shown: flat grinding), (d) round edges of slices, (e) lap and polish the slice.

Trichlorosilane is then distilled to separate it from the impurities. The liquid is converted back into the solid state by a chemical reduction process in the presence of hydrogen, known as the Siemens process. This results in ultrapure electronic grade polycrystalline silicon (EGS).

To obtain single crystal silicon, a special technique called the "Czochralski" method is used (Figure 2.13). Single crystal silicon is obtained by solidifying the molten material at the liquid interface along a specified crystal direction. First, chunks of EGS are melted. A quartz crucible, capable of withstanding a melting point of 1421°C for silicon, holds the molten EGS. Quartz has a melting point of 1732°C. To start the single crystal formation, a small silicon seed crystal having the desired orientation is carefully aligned and clamped to a rod. The seed is dipped into the melt and is allowed to thermally equilibrate. Silicon begins to freeze onto the seed crystal, and the growing crystal is withdrawn at a controlled rate. The seed and the crucible are slowly rotated in opposite directions while pulling. This process proceeds slowly and results in a large cylindrical crystal known as an ingot. The uneven edges of the ingot are ground to a precise diameter, flats are ground along the ingot length to indicate the orientation and conductivity type, and individual wafers are sliced (Figure 2.14). Finally, the wafer edges are contoured to prevent chipping, the wafer surface is polished (on one side or both sides), and the entire substrate is cleaned. Now, the wafers are ready for device fabrication (Table 2.4).

TABLE 2.4

Typical Specifications for Polished Silicon Wafers

Parameters	100 mm Wafer	125 mm Wafer	150 mm Wafer
Diameter (mm)	100 ± 1	125 ± 1	150 ± 1
Thickness (mm)	0.5–0.55	0.6–0.65	0.65–0.7
Primary flat length (mm)	30–35	40–45	55–60
Secondary flat length (mm)	16–20	25–30	35–40
Bow (µm)	60	70	60
Total thickness variation (µm)	50	65	50
Surface orientation	$(100) \pm 1°$	Same	Same
	$(111) \pm 1°$	Same	Same

Source: Wolf, S., and R. N. Tauber. 1986. *Silicon Processing for the VLSI Era.* Sunset Beach, CA: Lattice Press.

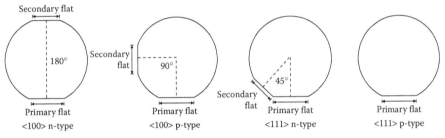

FIGURE 2.15
Primary and secondary flats of silicon wafers to indicate dopant type and orientation.

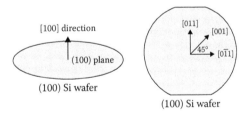

FIGURE 2.16
Crystal plane and direction orientations on a standard (100) silicon wafer.

2.1.2.2.1 *Wafer Orientations*

Silicon wafers can come in many different orientations. A standard system has been devised to easily determine wafer orientations by examining the size and orientation of flats, or cuts, that have been made on the wafer edges (Figures 2.15 and 2.16). Primary flats are longer and indicate the crystal orientation of the wafer. It can also be used as a registration key by automated wafer-handling equipment. Secondary flats are shorter and indicate the orientation and dopant type of the wafer.

2.1.2.3 Properties of Silicon

Silicon has been the leading substrate material of the microelectronics industry for decades. It is abundant, relatively inexpensive, and can be processed into high-quality crystals with exceptional purity and perfection. For the microelectronic industry, the key features of silicon include its tunable electrical conductivity, naturally occurring oxide insulation, and compatibility with microfabrication techniques.

2.1.2.3.1 Doping of Semiconductors

Semiconductors are versatile electronic materials because their conductive properties can be modified by adding impurities. This process of adding impurities to change the conductive properties of semiconductors is called doping. The foreign impurities are called dopants, and there are two types: (1) donors and (2) acceptors. Donors tend to give up electrons to the lattice and acceptors tend to receive electrons. There are two types of charge carriers in semiconductors: (1) electrons and (2) holes, which can be thought of as the absence of electrons and have a positive charge.

Silicon is a column IV element in the periodic table. Thus, a single silicon atom has four valence electrons and shares covalently bonded electrons with four adjacent silicon atoms. This unmodified silicon is referred to as an *intrinsic* semiconductor. By adding impurities, we create an *extrinsic* semiconductor (Figure 2.17). Common silicon dopants are listed in Table 2.5 and come from columns III and V of the periodic table.

Donors have five valence electrons, of which four participate in covalent bonds with neighboring atoms. A material modified with donors is referred to as n-type. Acceptors have three valence electrons, effectively creating a hole. A material modified with acceptors is referred to as p-type.

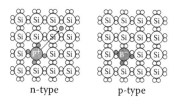

n-type   p-type

FIGURE 2.17
Visualization of donor (left) and acceptor (right) action to produce n- and p-type silicon.

2.1.2.3.2 Silicon as a Mechanical Material

From a mechanical point of view, silicon has many attractive features that make it a desirable material for sensors and actuators [2]. Its crystalline nature allows it to be sculpted using a number of different techniques to

TABLE 2.5

Common Silicon Dopants

Donors (Electron Increasing Dopants)		Acceptors (Hole Increasing Dopants)	
P		B	
As	Column V elements	Ga	Column III elements
Sb		In	

create either isotropic or anisotropic features such as wells, membranes, and beams. Although silicon is known to be a brittle material, these tiny microfabricated features can be quite strong; the Young's modulus of silicon is close to that of steel and above that of quartz, aluminum, and glass (Table 2.6). Also, silicon exhibits no plastic deformation or creep below 800°C and very little fatigue failure.

The important mechanical properties of silicon can be understood better if we start by examining the relationship between stress and strain in solids. For a normal stress ($\sigma = F/A$) imposed on a two-dimensional bar of silicon (Figure 2.18), there is a fractional change in length called strain, ε, which is given as follows:

$$\varepsilon = \frac{\Delta l}{l_0} \qquad (2.4)$$

where Δl is the change in length and l_0 is the initial starting length of the bar. The relationship between the stress and strain is given by Hooke's law (for small strains):

$$\sigma = \varepsilon E = \frac{\Delta l}{l_0} E \qquad (2.5)$$

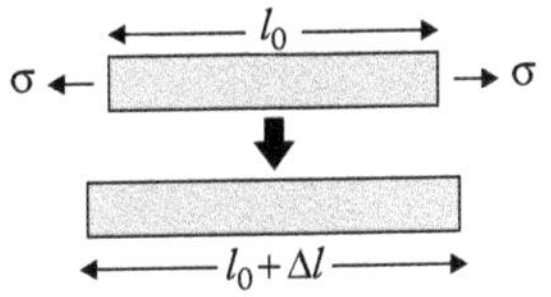

where E is Young's modulus.

So far we have considered only the axial strain. There must also be an associated transverse strain because cross-sectional area decreases under

FIGURE 2.18
Normal stress applied to the ends of a bar of a material.

TABLE 2.6
Properties of MEMS Materials

Material	Yield Strength (10^9 N/m²)	Young's Modulus (10^{11} N/m²)	Density (g/cm³)	Specific Heat (J/g · °C)	Thermal Conductivity (W/cm · °C)	Coefficient of Thermal Expansion (10^{-6}/°C)	Melting Point (°C)
Si	7.00	1.90	2.30	0.70	1.57	2.33	1400
SiC	21.00	7.00	3.20	0.67	3.50	3.30	2300
Si$_3$N$_4$	14.00	3.85	3.10	0.69	0.19	0.80	1930
SiO$_2$	8.40	0.73	2.27	1.00	0.014	0.50	1700
Al	0.17	0.70	2.70	0.942	2.36	25.0	660
Stainless Steel	2.10	2.00	7.90	0.47	0.329	17.30	1500
Cu	0.07	0.11	8.90	0.386	3.93	16.56	1080
GaAs	2.70	0.75	5.30	0.35	0.50	6.86	1238
Ge		1.03	5.32	0.31	0.60	5.80	937
Quartz	0.5–0.7	0.76–0.97	2.66	0.82–1.20	0.067–0.12	7.10	1710

Source: Madou, M. J. 1997. *Fundamentals of Microfabrication.* Boca Raton, FL: CRC Press.

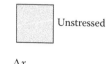

Unstressed

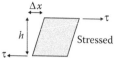

Stressed

FIGURE 2.19
Shear stress applied to a block.

tension. Poisson's ratio is a dimensionless ratio of the transverse (t) and axial (a) strains:

$$v = -\frac{\varepsilon_t}{\varepsilon_a} \tag{2.6}$$

Note that the negative sign contributed by the transverse strain component is due to a decrease in dimension. The maximum value of v is 0.5, with typical materials having values of 0.25–0.35.

Shear stress, τ, refers to stress applied to surfaces (Figure 2.19). In contrast to normal stress, shear stress creates shape change without any volume change. The shear modulus, G, is a ratio of the shear stress to the shear strain and is also related to Young's modulus and Poisson's ratio in the following manner:

$$G = \frac{\tau}{\frac{\Delta x}{h}} = \frac{E}{2(1+v)} \tag{2.7}$$

This is also an expression of Hooke's law and is true for small strains.

Because the silicon crystal is anisotropically arranged, its mechanical properties are orientation dependent. The relationship between stresses and strains is also more complex than the simple case considered thus far. Generic forms for these expressions in a matrix form are as follows:

$$\sigma_m = \sum_{n=1}^{6} E_{mn} \varepsilon_n$$
$$\varepsilon_m = \sum_{n=1}^{6} S_{mn} \sigma_n \tag{2.8}$$

where E is the stiffness coefficient matrix and S is the compliance coefficient matrix. Each 6×6 coefficient matrix contains only three independent constants due to a cubic symmetry. The matrices are given as follows:

$$E_{mn} = \begin{bmatrix} E_{11} & E_{12} & E_{12} & 0 & 0 & 0 \\ E_{12} & E_{11} & E_{12} & 0 & 0 & 0 \\ E_{12} & E_{12} & E_{11} & 0 & 0 & 0 \\ 0 & 0 & 0 & E_{44} & 0 & 0 \\ 0 & 0 & 0 & 0 & E_{44} & 0 \\ 0 & 0 & 0 & 0 & 0 & E_{44} \end{bmatrix}$$

$$S_{mn} = \begin{bmatrix} S_{11} & S_{12} & S_{12} & 0 & 0 & 0 \\ S_{12} & S_{11} & S_{12} & 0 & 0 & 0 \\ S_{12} & S_{12} & S_{11} & 0 & 0 & 0 \\ 0 & 0 & 0 & S_{44} & 0 & 0 \\ 0 & 0 & 0 & 0 & S_{44} & 0 \\ 0 & 0 & 0 & 0 & 0 & S_{44} \end{bmatrix}$$

(2.9)

We can show that $1/S_{11} = E$ (Young's modulus), $-S_{12}/S_{11} = \nu$ (Poisson's ratio), and $1/S_{44} = G$ (shear modulus). The values for these independent components are given in Table 2.7.

The values of Young's modulus and shear modulus for important crystal orientations are presented in Table 2.8.

2.1.2.3.3 Thermal Properties

The thermal expansion coefficient, α_T, is expressed as follows:

$$\alpha_T = \frac{d\varepsilon_x}{dT}$$

(2.10)

where ε_x denotes an unidimensional strain and T is the temperature. This quantity is the change in dimensions of solids due to temperature variations. Most solids expand in volume as temperature increases; the relationship between thermal expansion and temperature for silicon is depicted in Figure 2.20. When joining two materials together that will be subjected to thermal stress, the thermal expansion coefficient must be taken into account. For example,

TABLE 2.7

Stiffness and Compliance Coefficients for Silicon

Stiffness Coefficients	Compliance Coefficients
$E_{11} = 165.7 \times 10^9$ Pa	$S_{11} = 7.68 \times 10^{-12}$ Pa
$E_{12} = 63.9 \times 10^9$ Pa	$S_{12} = -2.14 \times 10^{-12}$ Pa
$E_{44} = 79.6 \times 10^9$ Pa	$S_{44} = 12.6 \times 10^{-12}$ Pa

Source: Brantley, W. A. 1973. *J Appl Phys* 44:534–5.

TABLE 2.8

Orientation Dependency of Silicon's Mechanical Properties

Orientation	Young's Modulus E (GPa)	Shear Modulus G (GPa)
[100]	129.5	79.0
[110]	168.0	61.7
[111]	186.5	57.5

Source: Greenwood, J. C. 1988. *J Phys E Sci Instrum* 21:1114–28.

there is a significant difference in the thermal expansion coefficients of silicon and silicon dioxide (Table 2.6). Thus, the thermal cycling of silicon-to-silicon dioxide interfaces may result in significant thermally induced interfacial strains. Therefore, special glass formulations have been made that closely match thermal expansion coefficients of silicon (Figure 2.21).

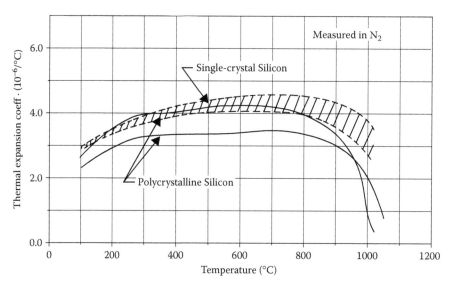

FIGURE 2.20
Thermal expansion coefficients of thin film polysilicon and single crystal silicon. (From Suzuki, T., A. Mimura, and T. Ogawa. 1977. *J Electrochem Soc* 124:1776–80. With permission.)

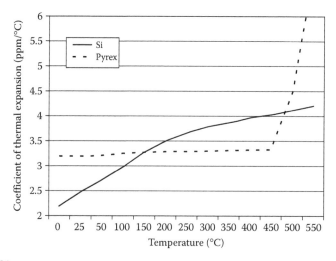

FIGURE 2.21
Thermal expansion coefficients of silicon and Pyrex 7740. (From Hsu, T.-R. 2004. *MEMS Packaging.* London: INSPEC, The Institution of Electrical Engineers. With permission.)

2.1.2.3.4 *Piezoresistivity*

In 1954 Charles Smith found that doped silicon exhibits piezoresistance, or a change in electrical resistance when subjected to a mechanical stress. This effect, present in both n- and p-type silicon, is particularly useful in piezoresistive sensors such as pressure and flow sensors.

The overall resistance change for a silicon piezoresistor subjected to a mechanical stress is

$$\frac{\Delta R}{R} = \frac{\Delta \rho}{\rho} + \frac{\Delta l}{l} - \frac{\Delta t}{t} - \frac{\Delta w}{w} \tag{2.11}$$

where ρ is the resistivity, l is the length, t is the thickness, and w is the width. The primary contribution comes from the change in resistivity, so we express Ohm's law in this case as follows

$$\{E\} = [r]\{j\} \tag{2.12}$$

where **E** is the electric field vector, **r** is the resistivity tensor, and **j** is the current density vector. So, for a piezoresistive crystal having a three-dimensional geometry, the change in resistivity due to a stress field is orientation dependent due to the anisotropy of the crystal structure (Figure 2.22).

The following equation shows this relationship:

$$[\Delta r] = [\pi][\sigma] \tag{2.13}$$

where $[\Delta r]$ is the resistivity change matrix, $[\pi]$ is the piezoresistive coefficient matrix, and $[\sigma]$ is the stress matrix containing six independent stress components for the cubic crystal structure of silicon. There are three normal stress components $(\sigma_{xx}, \sigma_{yy}, \sigma_{zz})$ and three shearing ones $(\sigma_{xy}, \sigma_{xz}, \sigma_{yz})$. By expanding Equation 2.13 and by inserting the piezoresistivity coefficient matrix we get

$$\begin{bmatrix} \Delta r_{xx} \\ \Delta r_{yy} \\ \Delta r_{zz} \\ \Delta r_{xy} \\ \Delta r_{xz} \\ \Delta r_{yz} \end{bmatrix} = \begin{bmatrix} \pi_{11} & \pi_{12} & \pi_{12} & 0 & 0 & 0 \\ \pi_{12} & \pi_{11} & \pi_{12} & 0 & 0 & 0 \\ \pi_{12} & \pi_{12} & \pi_{11} & 0 & 0 & 0 \\ 0 & 0 & 0 & \pi_{44} & 0 & 0 \\ 0 & 0 & 0 & 0 & \pi_{44} & 0 \\ 0 & 0 & 0 & 0 & 0 & \pi_{44} \end{bmatrix} \begin{bmatrix} \sigma_{xx} \\ \sigma_{yy} \\ \sigma_{zz} \\ \sigma_{xy} \\ \sigma_{xz} \\ \sigma_{yz} \end{bmatrix} \tag{2.14}$$

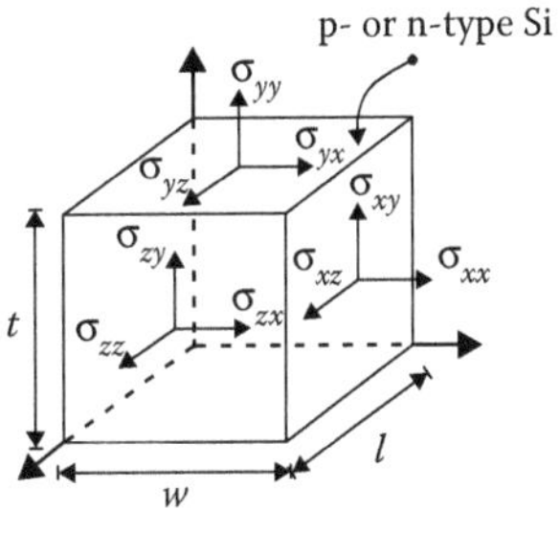

FIGURE 2.22
Silicon piezoresistor subjected to a stress field.

The final expression for the change of resistivity matrix is

$$
\begin{bmatrix}
\Delta r_{xx} \\
\Delta r_{yy} \\
\Delta r_{zz} \\
\Delta r_{xy} \\
\Delta r_{xz} \\
\Delta r_{yz}
\end{bmatrix}
=
\begin{bmatrix}
\pi_{11}\sigma_{xx} + \pi_{12}(\sigma_{yy} + \sigma_{zz}) \\
\pi_{11}\sigma_{yy} + \pi_{12}(\sigma_{xx} + \sigma_{zz}) \\
\pi_{11}\sigma_{zz} + \pi_{12}(\sigma_{xx} + \sigma_{yy}) \\
\pi_{44}\sigma_{xy} \\
\pi_{44}\sigma_{xz} \\
\pi_{44}\sigma_{yz}
\end{bmatrix}
\tag{2.15}
$$

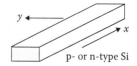

FIGURE 2.23
Silicon piezoresistor in which x and y correspond to the longitudinal and transverse directions, respectively.

The values of the piezoresistive coefficients depend on the orientation of the piezoresistor with respect to the silicon crystal, and values for <100> are given in Table 2.9.

From Equation 2.15, we can see that the change in resistance is dominated by the change in resistivity, which is in turn predominantly due to plane stresses. Thus, we can simplify the expression and say that

$$
\frac{\Delta R}{R} = \pi_l \sigma_l + \pi_t \sigma_t
\tag{2.16}
$$

The subscripts l and t correspond to longitudinal and transverse directions with respect to the resistor orientation, respectively, and are depicted in Figure 2.23. Some piezoresistive coefficients for the longitudinal and transverse directions for p-type silicon are listed in Table 2.10.

Piezoresistivity is also a strong function of temperature in silicon and deteriorates with increasing temperature. This dependency is illustrated in Table 2.11.

TABLE 2.9

Properties of Doped <100> Silicon at Room Temperature

Material	Resistivity ($\Omega \cdot$ cm)	π_{11} 10^{-1} (m²/N)	π_{12} 10^{-11} (m²/N)	π_{44} 10^{-11} (m²/N)
p-type Si	7.8	+6.6	−1.1	+138.1
n-type Si	11.7	−102.2	+53.4	−13.6

Source: Smith, C. S. 1954. *Phys Rev* 94:42–9.

TABLE 2.10

Piezoresistive Coefficients of p-Type Silicon Piezoresistors

Crystal Plane	x Orientation	y Orientation	π_l	π_t
(100)	<111>	<211>	$+0.66\pi_{44}$	$-0.33\pi_{44}$
(100)	<110>	<100>	$+0.5\pi_{44}$	0
(100)	<110>	<110>	$+0.5\pi_{44}$	$-0.5\pi_{44}$
(100)	<100>	<100>	$+0.02\pi_{44}$	$+0.02\pi_{44}$

Source: Hsu, T.-R. 2002. *MEMS and Microsystems: Design and Manufacture.* Boston: McGraw-Hill.

TABLE 2.11

Temperature Dependence of Resistivity and Piezoresistivity in Silicon

Doping Concentration (10^{16}/cm^3)	*p* Type		*n* Type	
	α_T (%/°C)	Temperature Coefficient of Piezoresistivity (%/°C)	α_T (%/°C)	Temperature Coefficient of Piezoresistivity (%/°C)
5	0.0	−0.27	0.01	−0.28
10	0.01	−0.27	0.05	−0.27
30	0.06	−0.18	0.09	−0.18
100	0.17	−0.16	0.19	−0.12

Source: French, P. J., and A. G. R. Evans. 1988. Piezoresistance in single crystal and polycrystalline Si. In *Properties of Silicon*, London: INSPEC, 94–103.

2.1.3 Properties of Thin Films

Silicon is the primary substrate material used in MEMS; other materials are deposited on the top as thin films. These films can take on properties that are quite different from their bulk counterparts. For instance, thin films have smaller grain size and result in surfaces with some degree of roughness. Density may also be lower due to porosity. Thin films are not crystalline but are instead either amorphous (e.g., oxide and nitride) or polycrystalline (metals). In addition, polycrystalline films can take on a preferred orientation or a fiber texture.

2.1.3.1 Adhesion

Excellent adhesion of films to the substrate is a necessity and, in part, a factor for processing these materials in a clean room environment to minimize contamination. In some cases, it is necessary to use surface treatments or intermediate adhesion promotion layers to obtain the required adhesive strength. Adhesion is quickly assessed using a simple tape test in which a pressure-sensitive adhesive is applied directly over the deposited film and is then peeled away. This qualitative test determines if adhesion is compromised. If any portion of the film is removed, a follow-up quantitative test can be performed. For instance, to extract the critical load, an abrasion or scratch test can be used to load a film until delamination occurs.

2.1.3.2 Stress

Thin films are all in a state of internal stress, either tensile or compressive, with typical values in the range of 10^8–5×10^{10} dynes/cm^2 [1]. A film in tensile stress tends to contract parallel to the substrate whereas one in compressive stress tends to expand. Thus, a tensile film causes concave bending in the substrate and a compressive film causes convex bending. These effects are illustrated in Figure 2.24, in which a deposited film has induced substrate bending due to stress.

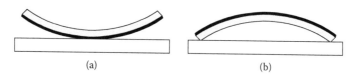

FIGURE 2.24
Substrate bending caused by deposition of stressed films for (a) tensile film, (b) compressive film.

Highly stressed films are undesirable and can lead to poor adhesion to the substrate; films are more susceptible to corrosion, cracks when under tensile stress, buckling of structures under compressive stress, and higher resistivity in the deposited film. Total stress in films is the sum of the external stress on the film, thermal stress, and intrinsic stress, which is expressed as follows:

$$\sigma = \sigma_{\text{external}} + \sigma_{\text{thermal}} + \sigma_{\text{intrinsic}} \qquad (2.17)$$

External stress may be caused by external temperature gradients or may be imposed by the packaging on a device.

Thermal stress is caused by the difference between the thermal expansion coefficients of the film (α_{film}) and the substrate $(\alpha_{\text{substrate}})$, which arise during the growth process:

$$\sigma_{\text{thermal}} = (\alpha_{\text{film}} - \alpha_{\text{substrate}})\Delta T E_{\text{film}} \qquad (2.18)$$

where ΔT is the difference between the process and measurement temperature and E_{film} is Young's modulus of the film. In the case of SiO_2 grown on silicon, the stress is negative or compressive because $\alpha_{\text{Si}} > \alpha_{\text{SiO}_2}$. Metal films typically exhibit tensile stress.

Intrinsic stress is not completely understood but is a result of many factors including film thickness, deposition rate, deposition temperature, ambient pressure, and type of the substrate.

The curvature of the substrate can be used to quantify the residual stress in films. Measurements are made before and after the film deposition. Then, the Stoney equation can be used to relate substrate curvature to the film stress:

$$\sigma_{\text{film}} = \frac{1}{6}\frac{E_{\text{substrate}}}{(1 - \nu_{\text{substrate}})}\frac{t_{\text{substrate}}^2}{t_{\text{film}}}\left(\frac{1}{R_{\text{before}}} - \frac{1}{R_{\text{after}}}\right) \qquad (2.19)$$

where E is Young's modulus, ν is Poisson's ratio, t is the thickness, and R is the radius of curvature. This will yield the average residual stress in the film. Other techniques can be used to obtain the variation in residual stress through the film thickness.

The tensile strength and Young's modulus can be determined from a film's stress–strain curves. The tensile strength is the maximum breaking force divided by the starting cross-sectional area and the modulus is measured from the slope of the linear region of the curve.

2.1.3.3 Resistivity

The resistivity of thin films can differ from that of bulk material. For example, thin film metals have a higher resistivity because they tend to contain more grain boundaries and more defects. The resistance, R, of a piece of film shown in Figure 2.25 is given as follows:

$$R = \rho \frac{l}{wt} = \rho \frac{l}{A} \qquad (2.20)$$

where ρ is the resistivity, l is the length, w is the width, and t is the thickness.

It would be tedious to calculate the resistance of a deposited or patterned film with this definition. Instead, it is more useful to use the concept of sheet resistance, R_s. In this case, we let $l = w$, so Equation 2.20 becomes

$$R = \frac{\rho}{t} = R_s \qquad (2.21)$$

The unit for sheet resistance is ohms per square ($\Omega/\square$) where the square refers to a square piece of the film of any size. If we break up a patterned resistor into squares of a known sheet resistance, then to calculate the total resistance we simply multiply the sheet resistance by the number of squares. In the resistor depicted in Figure 2.26, there are a total of six squares, so the resistance is $6R_s$.

Sheet resistance can be easily measured by several techniques. If we know the sheet resistance, then the film resistivity can be found using the following equation

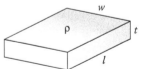

FIGURE 2.25
Piece of film having resistivity ρ.

$$\rho = R_s t \qquad (2.22)$$

The temperature coefficient of resistance (α) can be determined using

$$\alpha = \frac{R_2 - R_1}{R_1} \frac{1}{T_2 - T_1} = \frac{\Delta R}{R_1} \frac{1}{\Delta T} \qquad (2.23)$$

where $T_2 > T_1$.

FIGURE 2.26
Patterned resistor structure consisting of six numbered squares, each having a sheet resistance of R_s.

2.1.4 Silicon Compounds

2.1.4.1 Polycrystalline Silicon

Polycrystalline silicon is often referred to as polysilicon, poly-Si, or poly, and it has some properties similar to silicon. For example, polysilicon can be doped and oxidized. It also has similar mechanical properties and chemical resistance. In MEMS, polysilicon is often used as a structural material.

Polysilicon films are thermally grown or deposited over a wide temperature range and are made up of small single crystal silicon regions on the order of 1000 Å. The microstructure and orientation of these crystal grains is dependent on the process conditions such as the deposition temperature, dopant concentration, and thermal cycling [1]. For example, at temperatures below 580°C, deposited films are amorphous and at temperatures above 580°C, they are polycrystalline. The orientation of the grains is also set by the process temperature in which (110) grains are dominant for 600–650°C and (100) is preferred for 650–700°C [11].

2.1.4.2 Silicon Dioxide

Silicon's widespread popularity as a semiconductor and a MEMS material is partly due to its propensity to form a stable oxide that can be used as an insulator (Table 2.12). In fact, a thin silicon dioxide (SiO_2) layer (<20 Å) is formed on virtually all silicon surfaces exposed to an oxidizing ambient (i.e., presence of O_2 or H_2O). Thicker oxide layers are thermally grown and are amorphous. This amorphous form is referred to as fused silica, whereas the crystalline form is called quartz. Silicon dioxide finds an additional use in

TABLE 2.12

Selected Properties of Thermal Silicon Dioxide

DC resistivity at 25°C ($\Omega \cdot$ cm)	10^{14}–10^{16}
Density (g/cm³)	2.27
Dielectric constant	3.8–3.9
Dielectric strength (V/cm)	5–10×10^6
Energy gap (eV)	~8
Linear expansion coefficient (cm/cm · °C)	5.0×10^{-7}
Etch rate in buffered HF (Å/min)	1000
Melting point (°C)	~1700
Molecular weight	60.08
Molecules/cm³	2.3×10^{22}
Refractive index	1.46
Specific heat (J/g°C)	1.0
Thermal conductivity (W/cm · °C)	0.014
Stress in film on Si (dyne/cm²)	2–4×10^9, compression

Source: Wolf, S., and R. N. Tauber. 1986. *Silicon Processing for the VLSI Era.* Sunset Beach, CA: Lattice Press.

MEMS as a sacrificial material because it is readily dissolved in hydrofluoric acid (HF), which does not etch polysilicon or nitride.

Silicon dioxide films are typically obtained either by thermal oxidation or low-pressure chemical vapor deposition (LPCVD). Thermal oxidation requires a silicon surface to form oxide (44% of the oxide thickness is due to consumption of silicon). Through this self-limiting process, films up to 2 μm thick can be obtained. Oxides can also be deposited by LPCVD to form even thicker films. These low-temperature oxides (LTOs) exhibit a higher etch rate in HF and are also used as insulators. By adding phosphorous to the gas composition, phosphosilicate glass (PSG) is formed. These films make excellent sacrificial layers because they exhibit a higher etch rate than LTOs. PSG consists of P_2O_5 and SiO_2, and this is a binary glass or binary silicate. It can also be flowed at high temperatures (1000–1100°C) to smooth the surface and achieve better step coverage.

Silicon dioxide films can be formed by spin coating. Such films are useful for planarization in addition to insulation. Spin glass (SOG) starts off as a liquid formulation containing a solvent and SiO_2. Curing is required to obtain the final oxide film. Quartz is the crystalline form of silicon dioxide and is often used as a substrate or structural material. It possesses optical transparency, insulating properties, and piezoelectric properties. In addition, its thermal stability is greater than that of silicon.

Other popular silicon dioxide substrates include soda lime glass and Pyrex 7740. Soda lime glass is a common and inexpensive glass that is composed mainly of silica, sodium carbonate (Na_2CO_3), and calcium carbonate ($CaCO_3$) or dolomite ($MgCO_3$). Soda lime glass is not as resistant to high temperatures as quartz. Pyrex 7740 is a low-expansion borosilicate glass that is engineered to have a thermal expansion coefficient close to that of silicon. It is more resistant to high temperatures and heat shock than soda lime glass. In addition to silica, borosilicate glass contains boric oxide (B_2O_3) and small amounts of alkalis. The addition of boron also increases the resistance to chemical corrosion.

2.1.4.3 Silicon Nitride

Silicon nitride is an amorphous film often used as an insulator, for surface passivation, as a mask for selective oxidation of silicon, or as an etch mask. In MEMS, it has also been used as a structural material. It is an excellent diffusion barrier against moisture and sodium. The most common deposition methods are LPCVD and plasma enhanced chemical vapor deposition (PECVD). LPCVD is more cost efficient and operates at higher temperatures (700–800°C). This process yields films that are uniform, chemically resistant, excellent insulators, and exhibit good step coverage. However, these films can have high tensile stress. On the other hand, PECVD occurs at lower temperatures (200–400°C) and can result in almost stress-free films. The trade-off is that films are usually nonstoichiometric and can have a significant hydrogen incorporation (up to 18–22%) [1]; these films are sometimes referred to as Si_xN_y:H. The incorporation of hydrogen impacts the wet and dry etch rates of the PECVD film; for this, a high etch rate in HF is obtained while LPCVD nitrides etch slowly in HF.

2.1.5 Piezoelectric Crystals

The piezoelectric effect was discovered in the 1880s by brothers Pierre and Paul-Jacques Curie. Certain crystals would exhibit a surface charge approximately proportional to an externally applied mechanical stress. Likewise, these materials also respond to an applied electrical stress by exhibiting a mechanical deformation in the crystal. Piezoelectric crystals include lead zirconate titanate (PZT), zinc oxide (ZnO), quartz, and barium titanate (BaTiO$_3$; Table 2.13). These crystals are noncentrosymmetric or possess no center of symmetry. This enables positive and negative charges to form when the material is subjected to a mechanical load. For symmetric crystals, applied loads do not result in a net polarization.

When stress is applied to a slab of piezoelectric material as in Figure 2.27, the strain, S, is

$$S = s\sigma \qquad\qquad (2.24)$$

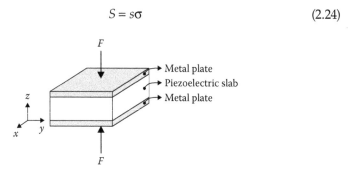

FIGURE 2.27
Piezoelectric slab sandwiched between two metal plates.

TABLE 2.13

Comparison of Piezoelectric Materials

Material	Piezoelectric Constant (pC/N)			Dielectric Constant	Curie Temperature (°C)	Maximum Coupling Coefficient
Quartz	$d_{11} = 2.31$	$d_{14} = 0.73$	-	$\varepsilon_1 = 4.52$ $\varepsilon_3 = 4.63$	550	0.1
PZT	$d_{33} = 80$ to 593	$d_{31} = -94$ to −274	$d_{15} = 494$ to 784	$\varepsilon_3 = 425$ to 1900	193–490	0.69–0.75
Sol-gel PZT	$d_{31} = 23$	$d_{32} = 4$	$d_{33} = -35$	4	>150	0.2
PVDF	$d_{15} = -12$	$d_{33} = 12$	$d_{31} = -4.7$	$\varepsilon_3 = 8.2$	-	-
ZnO	$d_{33} = 220$	$d_{31} = -88.7$	-	$\varepsilon = 1300$	-	0.49

Source: Madou, M. J. 1997. *Fundamentals of Microfabrication.* Boca Raton, FL: CRC Press.

where s is the compliance and σ is the applied stress. Equation 2.24 is a representation of Hooke's law, where $1/s$ is Young's modulus. If we apply a potential across the slab, then the resulting displacement vector, D, is

$$D = \varepsilon E = \varepsilon_0 E + P \tag{2.25}$$

where ε is the dielectric constant, E is the electric field, ε_0 is the permittivity of vacuum, and P is the polarization vector.

Let us consider a simple one-dimensional case in which the field, stress, strain, and polarization are in the same direction as the piezoelectric material. At a low frequency, and taking the principle of energy conservation into account, we can obtain the following equations:

$$
\begin{aligned}
D &= d\sigma + \varepsilon_\sigma E \\
S &= s_E \sigma + d'E
\end{aligned}
\tag{2.26}
$$

where ε_σ is the permittivity under constant stress and s_E is the compliance for constant electric field. When the surface area is constant under applied stress, then $d = d'$ (not true for polymers), where d is the piezoelectric charge coefficient or piezoelectric constant. Rearranging Equation 2.26, we can solve for the electric field and stress:

$$
\begin{aligned}
E &= \frac{D}{\varepsilon_\sigma} - \frac{d\sigma}{\varepsilon_\sigma} = \frac{D}{\varepsilon_\sigma} - g\sigma \\
\sigma &= -\frac{d}{s_E}E + \frac{1}{s_E}S = \frac{1}{s_E}S - eE
\end{aligned}
\tag{2.27}
$$

where $g = d/\varepsilon_\sigma$ is the piezoelectric voltage coefficient and $e = d/s_E$ is the piezoelectric stress coefficient.

To compare piezoelectric materials, it is more useful to use the electromechanical coupling coefficient, which is the square root of the ratio of the energy available at the output to the stored energy available at the input. This constant, k, is the geometric mean of the piezoelectric voltage coefficient and the piezoelectric stress coefficient:

$$k = \sqrt{ge} = \sqrt{\frac{d^2}{\varepsilon_\sigma s_E}} \tag{2.28}$$

2.1.6 Metals

Metals provide a range of functions in microelectronics and MEMS, from serving as etch masks to interconnects to structural elements. Several methods are used to deposit metals, including evaporation, sputtering, chemical vapor deposition (CVD), and electroplating. Aluminum and gold are commonly used. Chromium and titanium are often used in conjunction with gold as adhesion layers. Other popular metals include nichrome (NiCr) for thermal sources, shape memory alloys (e.g., TiNi), and permalloy (NiFe) for magnetic devices. For implantable devices, titanium and platinum are preferred. In the case of electronic neural prostheses, iridium and platinum are used for their charge delivery capabilities.

2.2 Polymeric Materials for MEMS

Polymers are of particular importance in bioMEMS as many are categorized as biomaterials (along with metals, ceramics, and other materials). Their inert and biocompatible properties make them attractive candidates in biological and chemical applications. Recently, many polymers have been explored as MEMS materials—either as substrates, coatings, sacrificial layers, or as structural layers. When compared to silicon and other traditional microelectronic or MEMS materials, polymers offer increased fracture strength, a low Young's modulus, and high elongation; polymers also have a relatively low cost.

Polymers are long-chain molecules consisting of repeating units called "mers." They are formed from monomers through a polymerization process. The resulting materials have a wide variety of properties that are influenced by their molecular properties and morphology. In general, polymers can be divided into biological and synthetic polymers. Biological polymers are materials collected from animals and can be reconstituted to form another usable polymer. Examples include deoxyribonucleic acid, cellulose, starch, and proteins. Synthetic polymers are entirely manmade and are usually formed from petroleum-based products. In bioMEMS, it is possible that both biological and synthetic polymers can be used. We discuss some common synthetic polymers used in MEMS in this section. Biological polymers and other biological components for MEMS will be discussed in Chapters 6 and 7.

Polymers can be classified by structure, physical state, thermal behavior, chemical structure, and end use. Unfortunately, a uniform classification system does not exist. Based on their chain structure, polymers can be classified as linear or branched and two or three dimensional. The physical state among polymers varies greatly with temperature, molecular weight, and chemical structure. Possible physical states include semicrystalline or amorphous and glassy or rubbery.

Thermal classification of polymers is widely used and is based on a polymer's response to heat exposure. The two distinct thermal classes are *thermoplastics* and *thermosets*. Thermoplastics typically soften and flow under heat or pressure above their glass transition temperature (T_g). When they are cooled, they will become solid. However, this process is reversible and reapplication of heat will again soften these polymers. Polycarbonate and acrylic are examples of thermoplastic polymers.

The glass transition temperature is best understood at a molecular level. At this temperature, enough thermal energy is imparted to the amorphous polymer chains such that they can slide past one another at a noticeable rate. This is distinct from the melting temperature (T_m) associated with semicrystalline polymers. At this temperature, ordered regions are broken up, or melted, and become disordered. Note that completely amorphous polymers do not have an associated melting temperature but all polymers exhibit T_g.

Thermosets, on the other hand, contain cross-linked chains and thus are not responsive to heat or pressure. Instead of softening or flowing, thermosets tend to decompose or break. Thermoset polymers generally cure (or set) to a final shape through the application of heat. This curing process is an irreversible chemical reaction that results in cross-linking. These structural features confer improved mechanical properties, thermal stability, and chemical resistance to thermosets compared to thermoplastics. An example of a thermoset is silicone rubber.

Chemical classifications might consider the polymer's elemental composition, chemical groups that are present, or synthesis method. Examples of end-use classifications include plastics, rubbers, fibers, adhesives, and coatings.

2.2.1 Polydimethylsiloxane (Silicone Rubber)

Polydimethylsiloxane (PDMS) is more popularly known as silicone rubber. In microelectronics, PDMS is well-known as a potting or encapsulation material. It has proven extremely useful as a rapid prototyping material for producing simple microstructures inexpensively. PDMS has a low glass transition temperature, high elasticity, low stiffness, and high resistance to oxidation and chemical attack. It is also optically transparent.

PDMS is a thermoset consisting of a repeating backbone of silicon and oxygen with two organic functional groups attached to the silicon (Figure 2.28). Although available in bulk, the most commonly used PDMS involves a two-part system of an elastomer and a curing agent that are mixed in a 10:1 ratio to form a prepolymer. The prepolymer is casted and then cured (addition of heat is optional). Submicron features can be patterned, and devices entirely contained within a slab (few millimeters thick) of PDMS are possible.

2.2.2 Polyimide

Polyimides have a long history of use as insulators and packaging materials in microelectronics.

These flexible polymers can have a linear or aromatic structure (Figure 2.29), and the cured material can have either a thermoset or thermoplastic behavior. Polyimides first became popular in MEMS as flexible

FIGURE 2.28
Chemical structure of polydimethylsiloxane.

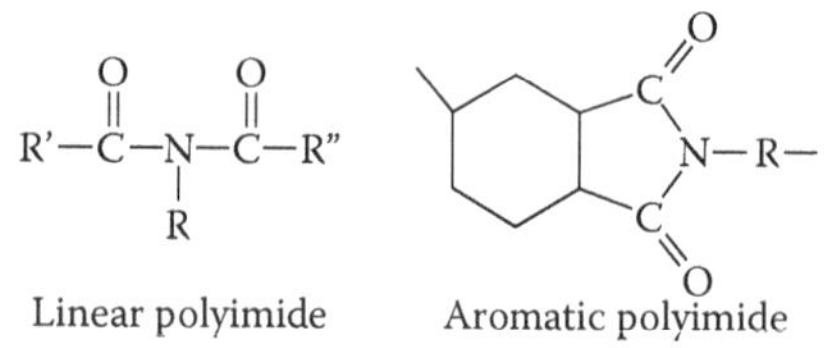

FIGURE 2.29
Chemical structure of polyimide.

substrates for sensors and multielectrode arrays for their potential biocompatibility and biostability. Its key properties include a high glass transition temperature, high thermal stability, high mechanical strength, low moisture absorption, chemical stability, and solvent resistance. In addition, its flexibility, inertness, and low cytotoxicity make it an attractive material for biological and medical applications.

Polyimides are available in many forms. Bulk polyimide is supplied as films or tapes backed with pressure-sensitive adhesives. It is also available in liquid form in both photodefinable and nonphotodefinable versions. Liquid formulations consist of a polyamic acid precursor that undergoes a thermal treatment called *imidization* to form the cured polyimide material.

2.2.3 SU-8

SU-8 was developed by IBM in 1989 and is a photodefineable epoxy-type resist (Figure 2.30). This highly cross-linked thermoset is particularly useful for inexpensively creating thick structures with high aspect ratios. Its use in MEMS became popular in the late 1990s as an alternative to expensive X-ray lithography techniques to produce molds for replicating metal or plastic parts. More recently, it has gained popularity as a versatile MEMS material. SU-8 can be used as an insulator, a structural material, mold, adhesive bonding layer, and etch mask. Some argue that SU-8 has biocompatible properties.

SU-8 is available in multiple viscosities for creating structures up to 2 mm high. The liquid formulation contains the SU-8 resin, a solvent, and a photosensitizer. The final polymer results after processing, which involves baking and exposure to ultraviolet (UV) light. The high degree of cross-linking confers excellent chemical compatibility. SU-8 is also relatively transparent. The electrical, magnetic, optical, and mechanical properties of SU-8 are tunable through the addition of functional materials to the prepolymer liquid.

2.2.4 Parylene

Parylene (poly (*p*-xylylene)) was well known to both microelectronics and medical industries long before it became a popular MEMS material (Figure 2.31). In earlier applications, Parylene was used primarily as

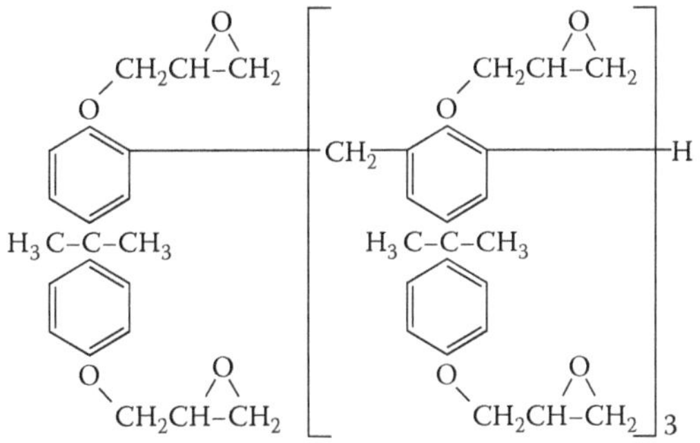

FIGURE 2.30
Chemical structure of SU-8 resin (modified from [12]).

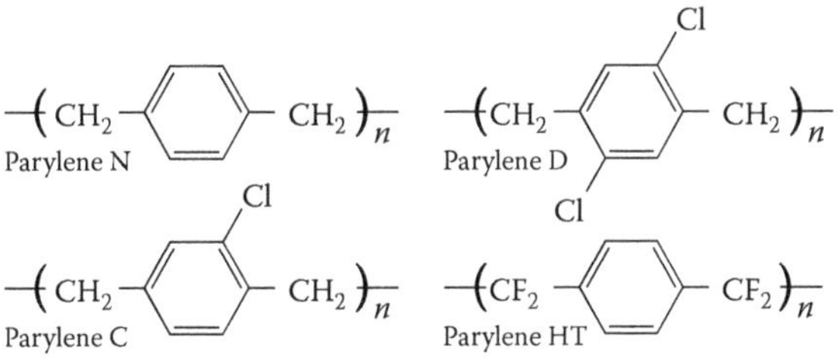

FIGURE 2.31
Chemical structure of Parylenes.

a protective coating. For example, Parylene C in particular is a coating for stents, cardiac assist devices, catheters, and surgical tools. This particular type of Parylene is noted for its USP (United States Pharmacopeia) Class VI designation, which is the highest biocompatibility level possible for a polymer. Parylene C is documented as having low cytotoxicity, biostability, and chemical inertness. These are thermoplastic polymers.

Unlike many of other polymers used in MEMS, Parylene is deposited as a thin film using a unique CVD process and dedicated equipment. This results in conformal and pinhole-free coatings over all exposed surfaces regardless of their orientation or local topography. Transparent films of thicknesses up to tens of micrometers are routinely possible.

2.2.5 Conductive Polymers

Although polymers are usually considered insulators, we can modify their electrical properties to create *conductive polymers* (Figure 2.32). The conductive properties of certain polymers have been recognized since the 1800s; however, a full understanding of conductive polymers was attained much later. A high conductivity iodine-doped polyacetylene was discovered in 1974 by Alan Heeger, Alan MacDiarmid, and Hideki Shirakawa, who were later awarded the 2000 Nobel Prize in chemistry for their thorough

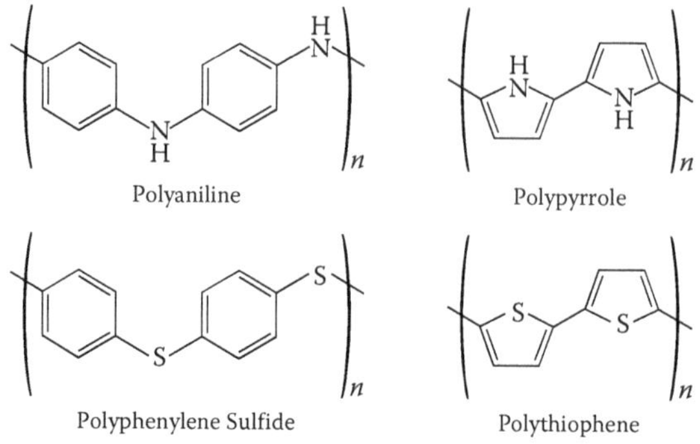

FIGURE 2.32
Chemical structure of some conductive polymers.

characterization of the nature behind the conduction. Conductive polymers possess conjugated double bonds and are doped to achieve high electrical conductivity, even approaching that of metals. These polymers become conductive when they are oxidized or reduced by donor or acceptor electrons. Other popular monomers are aniline, pyrrole, phenylene sulfide, and thiophene, which result in polyaniline, polypyrrole, polyphenylene sulfide, and polythiophene, respectively.

2.3 Biomaterials

Certainly, new materials are required for bioMEMS that differ drastically from those typically encountered in microelectronics or more historic MEMS applications. A bioMEMS engineer must demonstrate not only the efficacy of a device but also its biocompatibility, which ultimately means safety. These factors are often determined by the interaction between the material or device and the biological system.

For the purpose of this discussion, we will limit the term "biomaterials" to refer to materials used in medical implants, extracorporeal devices, and disposable systems (others may use the term to refer to materials that are biological in origin or *biological materials*). For example, the National Institutes of Health uses the following definition [13]: "Any substance (other than a drug) or combination of substances, synthetic or natural in origin, which can be used for any period of time, as a whole or as a part of a system which treats, augments or replaces any tissue, organ, or function of the body." Regardless of what definition we use, biomaterials must meet biocompatibility requirements for their intended application. The concept of biocompatibility is covered in depth in Section 2.3.3.

There are three major classes of biomaterials: (1) metallic, (2) ceramic and glass, and (3) polymeric [14]. While bulk metals are commonly used in load-bearing biomedical applications (e.g., hip prostheses and fracture fixation screws), in bioMEMS, we typically use thin metal films. Inside the human body (37°C and pH of 7.3) or other corrosive environments, metals may degrade via corrosion and release by-products that induce adverse biological responses. Examples of commonly used metals in biomedical implants include titanium and its alloys, cobalt-chromium alloys, and stainless steel.

Ceramics and glasses may be bioinert (do not induce immunologic host reactions), bioactive (form bonds with tissue), or biodegradable (degrade by hydrolytic breakdown in the body and metabolically absorbed) but suffer from low tensile strength and brittleness. Examples of common biomedical ceramics include alumina (Al_2O_3), zirconia (ZrO_2), bioglass ($Na_2OCaOP_2O_3$-SiO), and hydroxyapatite [$Ca_{10}(PO_4)_6(OH)_2$]. Common applications include coatings on dental implants and filler for bone defects in the middle ear or nasal septal bone.

Polymeric biomaterials have diverse properties that can be tuned by altering the composition, structure, and atomic arrangement. Therefore, polymers are the most widely used biomaterials. Their applications include cardiovascular and orthopedic implants, tissue engineering scaffolds, contact lenses, sutures, and coatings for pharmaceutical tablets or capsules. Polymers are also susceptible to degradation, which involves disruption of structure or alteration of bonds or chains. Unintentional degradation through leaching of low molecular weight species can be a concern. Thus, careful selection of the appropriate sterilization method is necessary when using polymers. In contrast, biodegradable polymers undergo a controlled degradation, which is useful in sutures and controlled drug delivery. Polylactic acid (PLA) and polyglycolic acid (PGA) are common biodegradable polymers. Other common biomedical polymers include polymethylmethacrylate (PMMA), polypropylene, polytetrafluoroethylene (PTFE), and silicone rubber.

Regardless of their type, all biomaterials possess properties that are governed by their microstructure. In addition, the atomic arrangement of constituent atoms at the surface of a biomaterial has different properties than that of the bulk. This arises due to the lack of near neighbor atoms and has major implications in tissue-material interactions in which surface phenomena dominate.

2.3.1 Material Selection for Biological and Medical Applications

Material selection in biomedical applications is difficult due to the lack of complete biocompatibility criteria and data; in fact, appropriate selection has benefited from serendipity (as in the case of the contact lens). Rigorous selection of appropriate materials should include the following criteria:

- Biocompatibility
- Physicochemical properties
- Durability and length of use (and any other relevant mechanical requirements)

- Desired function of device
- Nature of physiological environment
- Sterilization method
- Adverse effect in the event of failure
- Cost and production

Examples of medical applications and the types of biomaterials used are presented in Table 2.14.

The biocompatibility of devices and the materials from which they are constructed is a key requirement and must be demonstrated and approved before marketing and clinical use. Prior to delving deeper into the concept of biocompatibility, it is useful to understand medical device regulation.

2.3.2 Food and Drug Administration and Medical Device Regulation

In the United States, medical devices are regulated by the Food and Drug Administration (FDA), which is part of the U.S. Department of Health and Human Services. The approval of new medical devices falls under the purview of the FDA Center for Devices and Radiological Health (CDRH). This regulatory body along with several enacted laws has played critical roles in ensuring the high level of safety of medical devices marketed in the United States.

Historically, there are two important laws that outline what is considered a medical device in the United States: (1) the Federal Food, Drug, and Cosmetic Act (FFDCA; 1962) and (2) the Medical Device Amendments to the FFDCA (1976; Public Law 94-295). These laws define a medical device as

TABLE 2.14

Applications of Various Biomaterials in Medical Devices

Application	Materials
Joint replacements (hip and knee)	Ti, Ti-Al-V alloy, stainless steel, polyethylene
Fracture fixation plate	Stainless steel, cobalt-chromium alloy
Bone defect repair	Hydroxyapatite
Artificial tendons and ligaments	Teflon, Dacron
Heart valve	Stainless steel, carbon
Catheters	Silicone rubber, Teflon, polyurethane
Drug release	Polylactide-co-glycolide
Intraocular lens	Polymethylmethacrylate, silicone rubber, hydrogel
Contact lens	Silicone acrylate, hydrogel
Tissue engineering	Polylactic acid, polyglycolic acid, polylactide-co-glycolide

Sources: Dee, K. C., D. A. Puleo, and R. Bizios. 2002. *An Introduction to Tissue-Biomaterial Interactions.* Hoboken, NJ: Wiley-Liss; and Ratner, B. D. 1996. *Biomaterials Science: An Introduction to Materials in Medicine.* San Diego, CA: Academic Press.

An instrument, apparatus, implement, machine, contrivance, implant, in vitro reagent, or other similar or related article, including a component, part or accessory, which is recognized in the official National Formulary, or the United States Pharmacopeia (USP), or any supplement to them, intended for use in the diagnosis of disease or other conditions, or in the cure, mitigation, treatment, or prevention of disease, in man or other animals, or intended to affect the structure or any function of the body of man or other animals, and which does not achieve any of its primary intended purposes through chemical action within or on the body of man or other animals and which is not dependent upon being metabolized for the achievement of any of its primary intended purposes.

These laws resulted in regulations set in place by the FDA and in medical device categorization as Class I, II, or III. Increasing class number corresponds to increasing regulatory controls to assure reasonable safety and effectiveness of the products on humans. In other words, devices that potentially pose greater risk are subject to stricter regulation.

- Class I—general controls
- Class II—special controls
- Class III—premarket approval

Generic device types, of which there are over 1700, are grouped by the FDA into 16 medical specialty panels and also are assigned a regulatory class. For example, a specialty panel may be cardiovascular or neurology.

Class I devices are subject to least regulatory control as they pose minimal risk to users. These "general controls" also apply to Class II and III devices. General controls include manufacturer registration, listing devices to be marketed with the FDA, good manufacturing practices (GMP), proper labeling, submission of premarket notification, and reporting. Class I devices include bandages, examination gloves, and hand-held surgical instruments

Class II devices are subject to general controls and special controls. These special controls may include additional labeling requirements, demonstration of performance standards, and postmarket surveillance. Powered wheelchairs, infusion pumps, and surgical drapes are examples of noninvasive devices in Class II.

Class III devices are subject to the regulations of Classes I and II as well as premarket approval. These stricter regulations pertain to devices that are involved in supporting or sustaining human life and that may be important for preventing impairment of human health. Companies must obtain premarket approval for these devices that entails a thorough scientific review to ensure safety and effectiveness. Class III devices include heart valve replacements, silicone breast implants, and implanted brain stimulators.

New technologies are often placed into Class III. Regulations exist to address primarily safety concerns. In particular, the FDA is concerned with

(1) misuse or unintended use, (2) device failure, and (3) adverse biological effects resulting from device materials. Thus, regardless of the potential benefits of a new technology, the device approval process can be long or even indefinite if unanswered questions or concerns exist. These issues, in particular, impact the potential translation of bioMEMS devices into practical medical products.

A second form of classification was also issued by the FDA, in which seven device categories were established: preamendment, postamendment, substantially equivalent, implant, custom, investigational, and transitional. Each of these categories is described in Table 2.15.

Other important medical device legislation includes the following:

- Safe Medical Devices Act (1990; Public Law 101-629)
- Quality System Regulation (2002)
- Medical Device Amendments of 1992 to FFDCA
- Food and Drug Administration Modernization Act (1997; Public Law 105-115)

Other countries also have legislation and regulations that must be satisfied prior to bringing a device to the market.

2.3.3 Biocompatibility

The biocompatibility of an implanted device is in large part attributed to the constituent biomaterials although other factors can play a significant role (e.g., surface condition, surface treatment, device shape). To understand biomaterials and better evaluate their suitability for a particular application, we should establish a definition of biocompatibility. Many definitions have been proposed, such as

> Acceptance of an artificial implant by the surrounding tissues and by the body as a whole. The biomaterial must not be degraded by the body environment, and its presence must not harm tissues, organs, or systems. If the biomaterial is designed to be degraded, then the products of the degradation should not harm the tissue and organs. [17]

Although this definition is more descriptive, a simpler and more inclusive definition we can use is "Biocompatibility is the ability of a material to perform with an appropriate host response in a specific application." [18]

A critical metric for regulatory agencies is the establishment of biocompatibility of the device and the materials from which it is constructed. To do this, a battery of tests must be conducted to determine performance of materials under appropriate conditions including duration of contact (with body fluids, tissues, and organs). The test battery includes in vitro, ex vivo, animal model, and clinical evaluations of biocompatibility. In vitro tests

TABLE 2.15
FDA Medical Device Categories

Category and Description	Classification Rules	Examples
Preamendment devices (or older devices)		
Devices on the market before May 28, 1976, when the Medical Device Amendments were enacted	Devices are assigned to one of three classes. A presumption exists that preamendment devices should be placed in Class I unless their safety and effectiveness cannot be ensured without the greater regulation afforded by Classes II and III. A manufacturer may petition the FDA for reclassification.	Analog electrocardiography machine; electrohydraulic lithotripter; contraceptive intrauterine device and accessories; infant radiant warmer; contraceptive tubal-occlusion device; automated heparin analyzer; automated differential cell counter; automated blood-cell separator; transabdominal amnioscope
Postamendment devices (or new devices)		
Devices put on the market after May 28, 1976	Unless shown to be substantially equivalent to a device that was on the market before the amendments took effect, these devices are automatically placed in Class III. A manufacturer may petition the FDA for reclassification.	Magnetic resonance imager; extracorporeal shock-wave lithotripter; absorbable sponge; YAG laser; AIDS-antibody test kit; hydrophilic contact lenses; percutaneous catheter for transluminal coronary angioplasty; implantable defibrillator; bone-growth stimulator; alpha fetoprotein RIA kit; hepatitis B-antibody detection kit
Substantially equivalent devices		
Postamendment devices that are substantially equivalent to preamendment devices	Devices are assigned to the same class as their preamendment counterparts and subject to the same requirements. If and when the FDA requires testing and approval of preamendment devices, their substantially equivalent counterparts will also be subject to testing and approval.	Digital electrocardiography machines; YAG lasers for certain uses; tampons; ELISA diagnostic kits; devices used to test for drug abuse

(Continued)

TABLE 2.15 (*Continued*)

FDA Medical Device Categories

Category and Description	Classification Rules	Examples
Implanted devices		
Devices that are inserted into a surgically formed or natural body cavity and intended to remain there for ≥30 days	Devices are assumed to requirement placement in Class III unless a less-regulated class will ensure safety and effectiveness.	Phrenic-nerve stimulator; pacemaker pulse generator; intracardiac patch; vena cava clamp
Custom devices		
Devices not generally available to other licensed practitioners and not available in finished form. Product must be specifically designed for a particular patient and may not be offered for general commercial distribution.	Devices are exempt from premarketing testing and performance standards but are subject to general controls.	Dentures; orthopedic shoes
Investigational devices		
Unapproved devices undergoing clinical investigation under the authority of an Investigational Device Exemption.	Devices are exempt if an Investigational Device Exemption has been granted.	Artificial heart; ultrasonic hyperthermia equipment; DNA probes; laser angioplasty devices; positron emission tomography machines
Transitional devices		
Devices that were regulated as drugs before enactment of the statute but are now defined as medical devices.	Devices are automatically assigned to Class III but may be reclassified in Class I or II.	Antibiotic susceptibility disks; bone heterografts; gonorrhea diagnostic products; injectable silicone; intraocular lenses; surgical sutures; soft contact lenses

Source: Kessler, D. A., S. M. Pape, and D. N. Sundwall. 1987. *N Engl J Med* 317:357–66.

allow simple, rapid, and low cost screening of materials and devices. Many useful in vitro models exist due to advances in cell culture techniques but provide only a limited perspective of the whole body. Thus, care must be taken when extrapolating the results. Animal models may provide a more realistic insight into the many events that occur at the tissue-biomaterial interface and allow the determination of in vivo compatibility of materials or devices. However, like in vitro models, extrapolation may be unreliable and even dangerous due to species differences; positive results in animals do not translate directly to human biocompatibility. These studies are expensive and more complex than in vitro studies but are a necessary and useful step prior to human clinical trials. Animal research works are governed by the Animal Welfare Act (1985) and related amendments. These laws regulate the use of laboratory animals and animal research in the United States. Animal research protocols are required prior to initiation of any study and is subject to review and approval by local Institutional Animal Care and Use Committees (IACUC).

In the United States, a test selection table (10993-1) is used that is defined by the American National Standards Institute, Inc. (ANSI), Association for the Advancement of Medical Instrumentation (AAMI), and the International Standards Organization (ISO). The series of standards is further modified by the FDA and are listed in Table 2.16. Devices are grouped by body contact type into three categories: (1) surface devices (e.g., skin and mucosal membranes), (2) external communicating devices (e.g., indirect path to blood), and (3) implant devices (e.g., bone or tissue or blood contact). Tests are performed in concordance with the FDA good laboratory practice (GLP) regulations.

Finally, clinical trials are necessary to evaluate the performance of these devices in humans prior to their commercial availability to the general public. Clinical trials are subject to approval by the FDA and involve a sequence of three phases designated as I, II, or III. In each phase, a small number of human subjects receive the device and are monitored. The outcome of these trials will be used by the FDA to determine whether the device is approved for marketing.

2.3.4 Sterilization

An additional requirement of biomedical devices is that sterilization be performed prior to use to avoid infection and its associated side effects. Sterilization processes are intended to remove microorganisms (e.g., bacteria, yeast, molds, and viruses). A variety of methods exist, and it is imperative to determine the compatibility of the sterilization method chosen with the materials contained in a particular device. Common sterilization methods are summarized in Table 2.17.

Biomedical Microsystems

TABLE 2.16

FDA Initial Evaluation Tests for Biocompatibility

Device Categories			Initial Evaluation								Supplemental Evaluation	
	Body Contact	Contact Duration	Cyto-toxicity	Sensi-tization	Irritation or Intracutaneous Reactivity	Systemic Toxicity (Acute)	Subchronic Toxicity (Subacute Toxicity)	Geno-toxicity	Implant-ation	Hemocom-patibility	Chronic Toxicity	Carcino-genicity
Surface devices	Skin	A	●	●	●							
		B	●	●	●							
		C	●	●	●							
	Mucosal membrane	A	●	●	●							
		B	●	●	●	○	○		○			
		C	●	●	●	○	●	●	○		○	
	Breached or compromised surfaces	A	●	●	●	○						
		B	●	●	●	○	○		○			
		C	●	●	●	○	●	●	○		○	
External communicating devices	Blood path indirect	A	●	●	●	●						
		B	●	●	●	●	○			●		
		C	●	●	○	●	●	●		●	●	●
	Tissue, bone dentin communicating	A	●	●	○	○			○			
		B	●	●	○	○	○	●	●			
		C	●	●	○	○	○	●	●		○	●

Circulating	A	•	•	•	•	•	○	•	•	•		
blood	B	•	•	•	•	•	•	•	•	•	○	•
	C	•	•	•	•	•	•	•	○	•	•	•
Implant	A	•	•	○	○	•	•	•				
devices — Bone or	B	•	○	○	•	○	•	•				
tissue	C	•	○	○	•	○	•	•				
	A	•	•	•	•	•	•	•	○	•	•	•
Blood	B	•	•	•	○	•	○	•	○	•	○	•
	C	•	•	•	•	•	•	•	•	•	•	•

Source: Anand, V. P. 2000. *Med Device Diagn Ind* 22:206–19.

A = limited exposure (≤24 hours); B = prolonged exposure (24 hours–30 days); C = permanent contact (>30 days); • = FDA and ISO evaluation tests; ○ = additional tests required by FDA.

TABLE 2.17

Common Sterilization Methods (adapted from [15])

Sterilization Method	Method Details
High energy radiation	Isotope, electron beam, or X-ray radiation doses
Autoclaving	Saturated steam at high pressures and elevated temperatures (~130–140°C)
Dry heat	High temperatures (>140°C)
Ethylene oxide [(CH$_2$)$_2$O] gas	Exposure to flammable toxic gas at or near ambient temperatures; EtO is a suspected carcinogen

2.4 Problems

1. Print the 3D single crystal silicon model found at http://robotics.eecs.berkeley.edu/~pister/crystal.pdf. Assemble the model.

2. Draw the (100), (110), and (111) planes for silicon. In your drawings, include the atomic arrangement within each plane and indicate the silicon crystal lattice constant. Label any other key features needed to fully describe the plane. Using the drawings, calculate the surface atomic density of silicon in these planes. Which plane has the highest atomic density? [Hint: consider atoms contained in the "center" of the unit cell planes to be entirely within the plane. Likewise, atoms at the boundaries are shared.]

3. (a) Draw the following atomic planes and label all important identifying features: (011), $(\bar{1}11)$, (021), (112), and $(\bar{1}01)$.

   (b) Draw the following directions and label all important identifying features: $[\bar{1}11]$, $[021]$, $[112]$, $[\bar{1}01]$, and $[131]$.

4. Many bioMEMS devices use anodically bonded silicon-to-glass substrates to create microchannels. What are the thermal expansion coefficients of single crystal silicon (bulk) and standard soda lime glass? Does a thermal mismatch exist? What alternative glass(es) can be used for bonding to silicon (provide thermal expansion coefficients as an argument)? Cite your reference(s). [Hint: use journal papers and other references to answer this question.]

5. What is the difference between a thermoset polymer and a thermoplastic polymer? What is the difference in their chemical structures?

6. What is the glass transition temperature? How should this material property be used in MEMS design from a thermal point of view? Recall that the Space Shuttle Challenger disaster was caused by using a rubber o-ring below its glass transition temperature.

7. Obtain and read Kurt Petersen's paper "Silicon as a mechanical material" by using library resources (Reference [2] below). What is

Young's modulus for silicon and how can this parameter be used in mechanical design? How does silicon compare with steel in terms of mechanical properties? Describe some applications of mechanical beams and diaphragms. How could a biomedical engineer utilize these elements?

8. Conduct additional research on the common sterilization methods presented in Table 2.17. Compare and contrast the respective advantages and disadvantages for each. Also, suggest appropriate methods for sterilization of each class of biomaterial.

9. There are three classes of medical devices designated by the FDA. What are they? Briefly explain the classification system, especially in terms of regulatory control. List examples for each class. CardioMEMS, Inc. makes the EndoSure wireless pressure measurement system for monitoring abdominal aortic aneurysms. What is the FDA classification for this device? [Hint: do your own research for this question.]

References

1. Wolf, S., and R. N. Tauber. 1986. *Silicon Processing for the VLSI Era*. Sunset Beach, CA: Lattice Press.
2. Petersen, K. E. 1982. Silicon as a mechanical material. *Proc IEEE* 70:420–57.
3. Madou, M. J. 1997. *Fundamentals of Microfabrication*. Boca Raton, FL: CRC Press.
4. Brantley, W. A. 1973. Calculated elastic-constants for stress problems associated with semiconductor devices. *J Appl Phys* 44:534–5.
5. Greenwood, J. C. 1988. Silicon in mechanical sensors. *J Phys E Sci Instrum* 21:1114–28.
6. Suzuki, T., A. Mimura, and T. Ogawa. 1977. The deformation of polycrystalline-silicon deposited on oxide-covered single crystal silicon substrates. *J Electrochem Soc* 124:1776–80.
7. Hsu, T.-R. 2004. *MEMS Packaging*. London: INSPEC, The Institution of Electrical Engineers.
8. Smith, C. S. 1954. Piezoresistance effect in germanium and silicon. *Phys Rev* 94:42–9.
9. Hsu, T.-R. 2002. *MEMS and Microsystems: Design and Manufacture*. Boston: McGraw-Hill.
10. French, P. J., and A. G. R. Evans. 1988. Piezoresistance in single crystal and polycrystalline Si. In *Properties of Silicon* London: INSPEC, Institution of Electrical Engineers, 94–103.
11. Bhushan, B. 2004. *Springer Handbook of Nanotechnology*. Berlin: Springer.
12. LaBianca, N. C., J. D. Gelorme, and K. Y. Lee, et al. 1995. High aspect ratio optical resist chemistry for MEMS applications. In *Proceedings of the Fourth International Symposium on Magnetic Materials, Processes, and Devices*. The Electrochemical Society, 386–396.

13. Boretos, J. W., M. Eden, and National Institutes of Health (US) 1984. *Contemporary Biomaterials: Material and Host Response, Clinical Applications, New Technology, and Legal Aspects*. Park Ridge, NJ: Noyes Publications.
14. Dee, K. C., D. A. Puleo, and R. Bizios. 2002. *An Introduction to Tissue-Biomaterial Interactions*. Hoboken, NJ: Wiley-Liss.
15. Ratner, B. D. 1996. *Biomaterials Science: An Introduction to Materials in Medicine*. San Diego, CA: Academic Press.
16. Kessler, D. A., S. M. Pape, and D. N. Sundwall. 1987. The federal regulation of medical devices .*N Engl J Med* 317:357–66.
17. Park, J. B., and R. S. Lakes. 2007. *Biomaterials: An Introduction*. New York: Springer.
18. Williams, D. F., and European Society for Biomaterials. 1987. *Definitions in Biomaterials: Proceedings of a Consensus Conference of the European Society for Biomaterials, Chester, England, March 3-5, 1986*. Amsterdam: Elsevier.
19. Anand, V. P. 2000. Biocompatibility safety assessment of medical devices: FDA/ ISO and Japanese guidelines. *Med Device Diagn Ind* 22:206–19.

3

Microfabrication Methods and Processes for BioMEMS

3.1 Introduction

Gordon Moore made a revolutionary prediction in 1965 that integrated circuit complexity would double every 24 months with respect to the minimum component cost. This prediction, later known as Moore's law, has driven an exponential reduction of transistor feature size and resulted in the portable, yet computationally powerful, electronic tools that we enjoy today. These incredible feats of miniaturization are in large part due to advances in the microfabrication technology used to construct individual transistors starting from a plain silicon (Si) wafer.

To understand the underlying processes used to create biomedical micro-electromechanical systems (bioMEMS), we will study only a subset of the processes that originated from the integrated circuit industry as well as processes that were specifically developed for the fabrication of microelectromechanical systems (MEMS). The interested reader can refer to the many texts available on the various aspects of microfabrication for further study.

Fundamental *microfabrication* processes are simply categorized as follows:

- Lithography
- Etching
- Deposition
- Cleaning
- Other

Etching may more generally be referred to as a *subtractive* process and deposition as an *additive* process. These microfabrication (or *micromachining*) processes lead to the creation of microstructures that, when combined in a meaningful design, result in MEMS (Figure 3.1).

Compared to the traditional integrated circuit microfabrication, there are several unique aspects of MEMS fabrication that can be attributed to the

FIGURE 3.1
Generalized process by which MEMS are created.

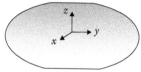

(100) Silicon wafer

FIGURE 3.2
Reference coordinate system
for a (100) silicon substrate.

integration across different modalities including electrical, mechanical, chemical, and biological ones. Some of these unique features are listed below:

- Sacrificial etching
- Thick films
- Deep etching and etching into the bulk of the substrate
- Double-sided lithography (processing on both sides of a substrate)
- Three-dimensional assembly
- Wafer bonding (or devices constructed of multiple stacked dies)
- Molding

MEMS devices typically start with a concept or an idea, which is then refined through careful design and often computer-aided simulation. Once a complete device design is obtained, the next step is to begin micro-fabrication. Microfabrication starts with the selection of an appropriate substrate. Substrate materials were reviewed in Chapter 2. Single crystal silicon and glass are perhaps the most widely used substrates in bioMEMS. A typical single crystal silicon wafer is shown in Figure 3.2 with a coordinate system that will facilitate our discussion on microfabrication. The *x-y* plane is arranged such that the *z*-axis origin is located on the substrate surface.

Starting with a substrate, a number of processing steps are performed using combinations of microfabrication processes. The precise order of steps used to obtain a final device is called a *process sequence* or *process flow*. These steps contain information detailing modifications to the substrate in the *z*-direction such as the thicknesses of deposited layers and depths of etch steps. Information about features in the *x*- and *y*-directions is contained in a *mask set*. Patterns contained in a mask are transferred to the substrate using a technique called *lithography*. For example, a simple process flow might include the deposition of a material onto a substrate, patterning of the material via lithography, and finally its selective etching. A typical process sequence will include a number of these steps that are concatenated in a particular order such that the desired structure is obtained.

3.2 Microlithography

Lithography is a process in which relief patterns are transferred to a substrate. In ancient times, stone was used as the substrate. In MEMS, lithography is typically performed using a light-sensitive material, or a *photoresist*, as the "writing tool" and a silicon wafer or other suitable substrate as the "stone." This process is more accurately referred to as photolithography due to the optical method of transferring the written information (down to the nanometer scale) to the substrate. The precise arrangement of the features to be written is contained in the engineering design. Due to the scale of features used in bioMEMS, sometimes this process is also referred to as *microlithography* or even *nanolithography* (for features in the range of nanometers).

3.2.1 Photolithography Process

Standard photolithography steps are summarized as follows. The process sequence is also illustrated using cross-section drawings taken through part of the substrate (Figure 3.3):

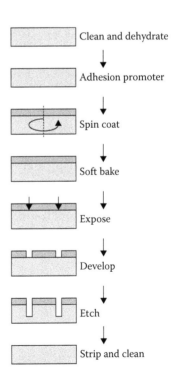

1. Substrate cleaning and dehydration
2. Adhesion promotion step
3. Spin on photoresist to apply it to the substrate
4. Soft bake to drive off the solvents
5. Photomask alignment
6. Optical exposure through photomask
7. Develop to reveal transferred pattern
8. Descum
9. Hard bake (or post bake)
10. Process using the photoresist film
11. Photoresist stripping
12. Cleaning

These steps capture the *z*-direction information. The *x*- and *y*-direction information is contained in *photomasks*, or *masks*. These are produced from a two-dimensional layout in a computer-aided design (CAD) file. They contain artwork with clear and opaque features to be transferred to the photoresist layer. A more detailed discussion on photomasks is provided in Section 3.2.4.

Clean and dehydrate

Adhesion promoter

Spin coat

Soft bake

Expose

Develop

Etch

Strip and clean

FIGURE 3.3
Example of standard steps for photolithographic processing.

3.2.1.1 Surface Preparation

Surface preparation often precedes microfabrication steps. For lithography, substrate cleanliness is paramount. Contaminated surfaces commonly result in poor adhesion or defects. Thus, photolithography and other microfabrication processes are carried out in a cleanroom environment (see Section 3.2.5).

Adhesion between the photoresist and the substrate affects the pattern transfer fidelity and the robustness of the photoresist as a masking layer in subsequent processes. In general, photoresists adhere weakly to hydrophobic surfaces covered by absorbed water. Poor adhesion may cause loss of the pattern, loss of resolution, or undercutting of the pattern during etching, especially in wet etching processes. Dehydration baking prior to the coating process is used to promote the adhesion by removing moisture adsorbed to surfaces. Following dehydration, silane promotion agents, such as hexamethyldisilazane (HMDS) and trimethylsilylchloride (TMSC), are applied by dipping, spraying, spin coating, or vapor-phase priming to reduce the surface tension of the substrate and further facilitate the adhesion. HMDS reacts to silylate silanol groups and thus link hydroxylated silicon dioxide (SiO_2) and the photoresist. The reaction is as follows (Figure 3.4):

$$2SiOH + \left[(CH)_3 Si \right]_2 NH \rightarrow 2SiOSi(CH_3)_3 + NH_3 \tag{3.1}$$

TMSC, on the other hand, reacts with the surface water to form a polysiloxane polymer film on the surface.

3.2.1.2 Spin Coating

Photoresist is deposited evenly over a flat substrate by a process known as spin coating (Figure 3.5). The liquid photoresist is dispensed onto a static substrate and is spread at low spin speeds, or alternatively, onto a substrate spinning at low speeds. The substrate spin speed is ramped quickly to high speeds (~krpm) and is spun at a constant speed. This allows centrifugal forces to spread the solution to the final thickness and simultaneously dry the film as solvent evaporates. The resulting thickness of the coating is a function of the

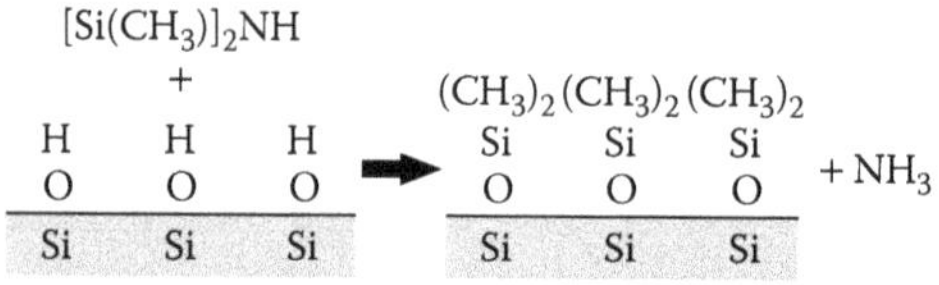

FIGURE 3.4
Hexamethyldisilazane adhesion promotion reaction on an oxidized silicon wafer.

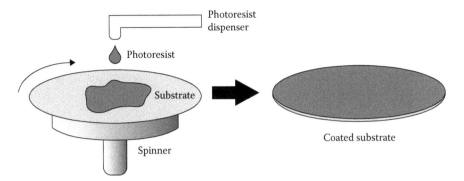

FIGURE 3.5
Spin coating process.

solution concentration, viscosity, spin speed, and spin time. Film thickness, t, can be predicted with the following empirical expression [1]:

$$t = \frac{kC^\beta \eta^\gamma}{\omega^a} \tag{3.2}$$

where k is a calibration constant, C is the concentration of solids, η is the intrinsic viscosity, and ω is the spin speed in rotations per minute (rpm). The exponential factors β, γ, and a are empirically determined from slopes obtained by least-squares fitting log-log plots of t versus C, η, and ω. This equation is true for speeds less than 6000 rpm in which case the thickness is independent of the wafer diameter.

A typical spin speed versus thickness curve is shown in Figure 3.6.

Defects and other undesirable features may arise during the spin coating process. Common issues associated with the spin coating and their potential causes are given in Table 3.1.

3.2.1.3 Soft Baking

Residual solvent is expelled from the photoresist through a process known as soft baking or prebaking (75–100°C). This process reduces the solvent content in the film from ~20–30% to ~4–7%. This process also improves adhesion and anneals latent stresses accumulated during the spinning process. The thickness of the resist is further reduced in this step due to the solvent loss.

3.2.1.4 Exposure

The photoresist films are then exposed to light of an appropriate wavelength that is preset to a particular intensity and exposure time. An appropriate

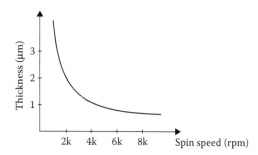

FIGURE 3.6
Typical spin speed versus thickness curve.

TABLE 3.1

Common Spin Coating Problems and Potential Causes (After [2])

Problem	Potential Causes
Comets	Spun-off dirt
Craters	Silicones in resist, splashback
Cobwebs	High molecular weight, air flow
Cloudy films	Moisture in air
Backside coating	Excess liquid, noncentering
Edge dewetting	Poor priming
Pinholes	Contaminants
Lip buildup	Air flow

exposure dose is necessary to completely expose the entire thickness of the photoresist film. Dose, D, expressed in J/cm^2 is calculated as follows:

$$D = It \qquad (3.3)$$

where I is the incident light intensity (W/cm^2) and t is the exposure time in seconds.

For positive resists, the exposure induces a chemical reaction that increases the solubility of the resist in the exposed areas. For negative resists, the exposure renders the photoresist insoluble. Thus, in this case, a latent image is transferred from the mask to the photoresist.

3.2.1.5 Development

Soluble portions of the photoresist are removed either by immersion or by spray development to reveal the transferred image. During this process, it is critical that fresh developer should be supplied to the wafer surface for uniform results. For example, the solution should be agitated for thorough development by immersion processes. Following the development, optical inspection of the patterns transferred is critical to catch mistakes, defects, or unacceptable process variations.

3.2.1.6 Descum

Thin residual photoresist films, or *scum*, may be left after the development. This unwanted material can act as an unintentional masking layer during the subsequent etching processes. To remove these layers, a short oxygen plasma etch, known as *descum*, is used. Plasma etching processes are described in Section 3.4.1.2.

3.2.1.7 Hard Baking and Post Baking

An additional baking step can be used to further improve the hardness of the photoresist film, improve the adhesion, and drive out any remaining moisture from solvents or swelling during the development. This step is performed at higher temperatures (120°C), which can result in reflow due to the softening

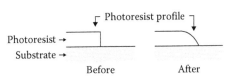

FIGURE 3.7
Photoresist profile modification by reflow during hard bake.

and flattening of the photoresist film for a tapered profile (Figure 3.7). Reflow of the photoresist film occurs close to its glass transition temperature.

3.2.2 Photoresists

Liquid photoresist formulations typically consist of three components: (1) a polymer resin, (2) a sensitizer (or inhibitor), and (3) a carrier solvent. The sensitizer gives the resist its radiation absorption and developer resistance properties. There are two tones of photoresist: (1) positive and (2) negative photoresists. As mentioned in Section 3.2.1.4, the exposed regions in positive photoresists become more soluble. This photochemically induced solubility occurs due to the weakening of the polymer chains via rupture or scission in the exposed areas. The opposite occurs in negative resists in which polymer chains are strengthened by photochemically induced cross linking. This difference is illustrated in Figure 3.8. Another key difference is that positive photoresists are developed in alkaline solutions and are subsequently cleaned by a water rinse, whereas negative photoresists are usually developed in organic solvents and are therefore rinsed clean in other organic solvents.

   Traditionally, there are two ways in which lithography is used to achieve patterning: etch-back and lift-off (Figure 3.9). In the etch-back process, photoresist is applied on top of the layer to be patterned. The unwanted or

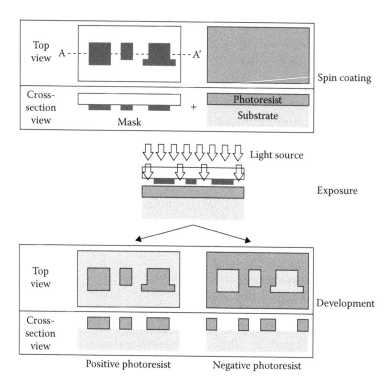

FIGURE 3.8
Lithography process steps highlighting the difference between positive and negative photoresists. The cross-section is taken through A–A'.

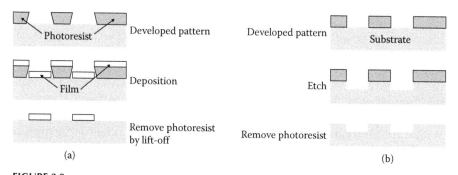

FIGURE 3.9
(a) Lift-off and (b) etch-back techniques.

unmasked material is then selectively removed. In the lift-off process, the layer to be patterned is deposited on the patterned photoresist. The deposition process must be line-of-sight such that there are discontinuities in the layer. These intentional defects allow the unwanted material on top of the photoresist to be removed when the photoresist is removed. The difference between these two techniques lies in the photoresist sidewall profile. The possible profiles and their corresponding uses are depicted in Figure 3.10.

We will also explore how photoresists can also be used as sacrificial layers in a MEMS process known as surface micromachining.

3.2.3 Photolithography Tools and Resolution

Commonly used photolithography tools are contact, proximity, and projection. Graphical representations of their optical layouts are shown in Figure 3.11.

3.2.3.1 Contact and Proximity Printings

Contact printing is the oldest method of pattern transfer in microfabrication. The mask is clamped directly to the photoresist-coated substrate during

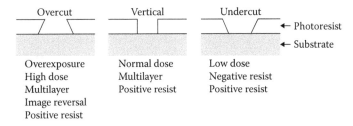

FIGURE 3.10
Possible photoresist profiles and methods by which to produce them (after [2]).

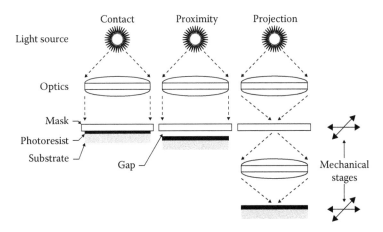

FIGURE 3.11
Diagrams of lithography tools (after [3]).

exposure. Although this method enables faithful image transfer and potential for infinite resolution, contact between the mask and the substrate results in wear and possible transfer of contaminants. These defects are reproduced on the next substrate exposed with this mask.

In *proximity printing*, contact between the mask and the substrate is eliminated by introducing a small gap. Although damage to the mask is reduced, the trade-off is that the resolution is decreased because the gap induces diffraction effects on the transferred image.

3.2.3.2 Projection

Additional optical elements enable the separation of the mask and the substrate by large distances. The mask image is focused on the substrate surface at the same scale (1X) or at a reduced scale (5X or 10X). Some systems also enable step-and-repeat printing in which the image is projected to only a portion of the substrate (a single die) and then mechanically positioned to the next desired location. This process is repeated to scan the image across the substrate (Figure 3.12).

3.2.3.3 Resolution

Theoretically, infinite resolution is achievable by contact printing, but, in practice, resolutions of ~0.1 μm are possible with high-quality masks. In proximity printing, the theoretical resolution is limited by diffraction occurring at pattern edges, which results in pattern distortion. This technique also

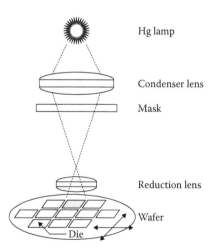

FIGURE 3.12
Step-and-repeat projection printing (after [3]).

suffers from a small depth of focus (DOF). In both methods, closely spaced features may be distorted by a constructive interference. A practically achievable resolution for proximity printing is ~1–2 μm for a gap of 5–10 μm [2].

An approximation of the resolution limit for proximity printing, R, is given by

$$R \sim \sqrt{0.7\lambda z} \qquad (3.4)$$

where z is the gap between the mask and the substrate and λ is the wavelength [2]. Higher resolutions can be achieved with projection methods.

The minimum feature size, R, in a projection system is determined by the Rayleigh equation

$$R = \frac{k_1 \lambda}{\text{NA}} \qquad (3.5)$$

where k_1 is the resolution factor, λ is the wavelength, and NA is the numerical aperture of the imaging lens. k_1 is an empirical observation that depends on the resist and process parameters and typically has values between 0.5 and 1.0; a typical value of 0.81 is assumed in production environments [1]. NA is the geometric ratio between the focal length and the aperture of the lens and quantifies the ability of the lens to collect diffracted light from the photomask and transmit it to the photoresist. It is expressed as follows:

$$\text{NA} = n \sin \alpha \qquad (3.6)$$

where n is the index of refraction and α is the half-angle of angular acceptance of the lens. Our definition of resolution can further be refined by considering diffraction. A small ideal point source appears as a blurry disk, known as the *Airy disk*, as a result of diffraction associated with the lens. So, if spaced closely enough, two Airy disks will appear as a single blurred disk (Figure 3.13). Thus, it is useful to apply the Rayleigh criterion in which two images are defined as being resolved when the intensity between the Airy disks decreases to 80% of the image intensity, or mathematically speaking

$$R = \frac{0.6\lambda}{\text{NA}} \qquad (3.7)$$

where R, in this case, is the separation between the two images.

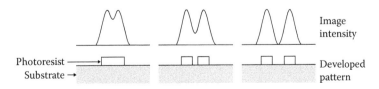

FIGURE 3.13
Pattern transfer of two objects near the diffraction limit (after [4]).

DOF and NA are also related. The Rayleigh DOF criterion states that

$$\text{DOF} = \pm\frac{k_2\lambda}{(\text{NA})^2} = \pm\frac{k_2 R^2}{k_1^2 \lambda} \tag{3.8}$$

where k_2 is an empirical process dependent factor with a typical value of 0.5. DOF provides a quantitative measure of the range in which an image is in focus and is a critical factor in achieving the desired line width. Note that the DOF decreases as NA increases (Table 3.2). Because substrates, especially patterned ones, are not perfectly flat, an image may fall outside the focal plane. Therefore, a compromise is necessary between high-resolution and large DOF.

Projection resolution is calculated as 0.5 (λ/NA) and can be 1 µm or less. The associated DOF is a few micrometers.

The resolution can be improved by decreasing the wavelength and resolution factor or by increasing the numerical aperture. The different wavelengths used in lithography and their achievable resolution assuming an NA of 0.7, k_1 of 0.5, and k_2 of 1.0 are calculated using Equations 3.5 and 3.8, which are displayed in Table 3.3. Practically speaking, reducing the resolution factor is difficult and dependent on a number of photoresist and process conditions.

TABLE 3.2

Numerical Aperture and Depth of Focus for k_2 of 0.5

	Depth of Focus (µm)	
NA	400 nm	200 nm
0.2	5.0	2.5
0.3	2.0	1.0
0.4	1.5	0.8
0.5	0.8	0.4

Source: Moreau, W. M. 1988. *Semiconductor Lithography: Principles, Practices, and Materials.* New York: Plenum Press.

TABLE 3.3

Wavelengths for Optical Lithography and the Associated Minimum Feature Size and Depth of Focus

Wavelength	R_{min} (nm)	DOF (nm)
436 nm (g-line)	311	850
405 nm (h-line)	290	790
365 nm (i-line)	260	730
248 nm (KrF Excimer)	175	500
193 nm (ArF Excimer)	140	400
157 nm (F2)	112	320
126 nm (Ar2)	90	257

Source: Brunner, T. 1997. *IEEE City 9.*

The physical limit of NA is 1, but technology limits achievable NA to ~0.9. Further improvement in numerical aperture is achieved by replacing the medium between the optics and the substrate with water (NA ~ 1.4) instead of air (as in *immersion lithography*).

3.2.3.4 Double-Sided Lithography

In MEMS, it is often necessary to pattern features on both sides of the substrate. Registration or alignment marks are used to ensure a precise relationship between the front- and back-side patterns. For silicon, infrared imaging can be used to facilitate the process (silicon is transparent at infrared wavelengths). Specialized tools have been developed for double-sided lithography and alignment.

3.2.3.5 Multilayer Resists

Multiple resist layers are often used in MEMS either to increase the thickness of the resulting resist structure or to planarize the surfaces having rough topography. Large topography changes are common in MEMS devices in which thick films are deposited or deep etching is performed. Subsequent patterning is affected by focusing difficulties and potential failed transfer of features on covering large step height differences (Figure 3.14).

For example, two types of resist can be used to planarize a surface using the multilayer resist process (Figure 3.15). The first layer is used to smooth

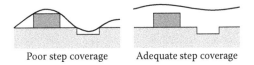

Poor step coverage   Adequate step coverage

FIGURE 3.14
Single-layer resist coating over deep topography in which the left figure exhibits inadequate coverage of deep topography features and the right figure shows photoresist planarization of topography.

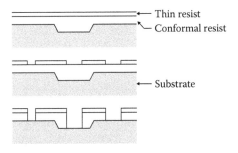

Thin resist
Conformal resist

Substrate

FIGURE 3.15
Multilayer resist process for surface planarization in which two layers of resist are applied to planarize the wafer topography and two exposure steps are used to transfer the image through the two resist layers.

out large height deviations and the second resist, having different composition and spectral response, is spin coated on top of the first. This second layer is processed independently of the first, and the desired pattern is easily transferred down to the first layer.

3.2.3.6 Other Lithography Methods

3.2.3.6.1 Direct Write Lithography

It is possible to bypass the need for a photomask by directly exposing the photoresist in a spot-by-spot fashion. Excimer lasers (1–2 µm), electron beams (0.1–0.2 µm), and focused ion beams (FIBs; 0.05–0.1 µm) can create submicron feature sizes. Patterns are transferred directly from a computer to a substrate through a beam steered to positions on the substrate surface. The pattern is written into the resist by raster or vector scanning. Raster scanning moves the beam across the entire substrate as shown in Figure 3.16. Vector scanning selectively exposes portions of the resist. A mask is not required; however, the serial nature of the process and the small spot size necessitate much longer process times. Nonoptical lithography techniques will be discussed in Section 3.4.5.

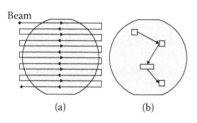

FIGURE 3.16
(a) Raster and (b) vector scanning methods of direct write lithography.

3.2.4 Photomasks

3.2.4.1 Mask Layout

To create a photomask for lithography, first an engineer lays out the desired design using mask design software. Each mask layer in a layout contains two-dimensional design information that corresponds to a particular process step. When the masks, containing x- and y-direction information, are combined with the appropriate sequence of process steps, each containing z-direction information, the desired device results.

3.2.4.2 Producing a Photomask

The mask artwork contained in the software needs to be transferred to a physical photomask that will contain the master patterns transferred during the exposure step in lithography. The simplest and least expensive form of this is a printed transparency. These polymer films are suitable for research and low production applications. More durable masks are formed on glass substrates; typical materials used in the order of least to most expensive include photographic emulsion on soda lime glass, iron oxide (Fe_2O_3) on soda lime glass, chrome on soda lime glass, and chrome on quartz.

3.2.4.3 Mask Alignment

For processes with multiple masks, accurate registration between each mask and the substrate must be achieved. There are three degrees of freedom between the mask and the substrate: x, y, and θ. Alignment marks on the mask and the substrate enable an accurate registration of patterns prior to exposure (Figures 3.17 and 3.18). Although we can perform this task manually, modern production systems use automatic pattern recognition and alignment systems to reduce the time this process takes (~1–5 seconds by machines versus 30–45 seconds by human operators). Common types of misregistration between layers in projection systems are shown in Figure 3.19.

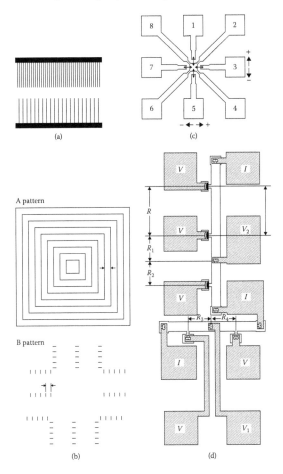

FIGURE 3.17
Alignment features for level-to-level registration : (a) a set of verniers, (b) a pattern set requiring individual line-width measurement, (c) and (d) allow determination of x, y, and θ misregistration by measuring the electrical resistance of the patterned resistors. (From Runyan, W. R., and K. E. Bean. 1990. *Semiconductor Integrated Circuit Processing Technology*. Reading, MA: Addison-Wesley. With permission.)

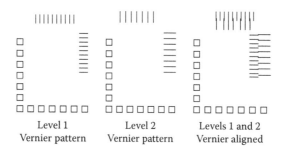

FIGURE 3.18
Demonstration of good level-to-level registration using alignment marks. (From Moreau, W. M. 1988. *Semiconductor Lithography: Principles, Practices, and Materials.* New York: Plenum Press. With permission.)

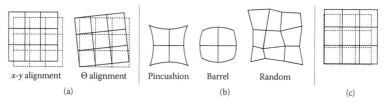

FIGURE 3.19
Types of misregistration: (a) feature overlay error, (b) optical distortion, (c) magnification error. (From Runyan, W. R., and K. E. Bean. 1990. *Semiconductor Integrated Circuit Processing Technology.* Reading, MA: Addison-Wesley. With permission.)

3.2.5 Cleanroom Processing

Microfabrication, and lithography in particular, is preferably performed in a cleanroom environment. This controlled environment is necessary to minimize particulate and other forms of contamination that could adversely affect microfabrication. For example, contaminants such as dust and hair measure ~1–5 μm and 100 μm (in diameter), respectively. Their accidental presence on a substrate can cover and destroy tiny but critical device features. Major sources of contamination include process equipment, process chemicals, process environment, and operating personnel. To control the contamination, processing is performed in a cleanroom, tools are made of designated materials, chemicals of high purity are used, and operating personnel are covered from head to toe with particle-containing garments.

Cleanrooms have associated ratings (e.g., class 100) that provide a quantitative measure of the number of particles of a given size (0.5 μm) in a standard volume of air. Table 3.4 lists Federal Standard 209E, which designates the different clean-room classes. This classification has since been replaced with the ISO 14644-1 classification system in Table 3.5. Cleanrooms can protect against solid contaminants such as hair, lint, and other particulate matter; however, liquid and gaseous contaminants must also be considered.

TABLE 3.4

Federal Standard 209E Cleanliness Class Designations

Cleanroom Class	Number of Particles of Corresponding Diameter per Cubic Foot				
	0.1 μm	0.2 μm	0.3 μm	0.5 μm	5.0 μm
100,000	–	–	–	100,000	700
10,000	–	–	–	10,000	70
1,000	–	–	–	1,000	7
100	–	750	300	100	–
10	350	75	30	10	–
1	35	7.5	3	1	–

TABLE 3.5

ISO 14644-1 Cleanliness Class Designations and Comparison to Federal Standard 209E

ISO Class	FS 209E Class	Number of Particles of Corresponding Diameter per Cubic Meter					
		0.1 μm	0.2 μm	0.3 μm	0.5 μm	1.0 μm	5.0 μm
9	–	–	–	–	35,200,000	8,320,000	293,000
8	100,000	–	–	–	3,520,000	832,000	29,300
7	10,000	–	–	–	352,000	83,200	2,930
6	1,000	1,000,000	237,000	102,000	35,200	832	293
5	100	100,000	23,700	10,200	3,520	83	29
4	10	10,000	2,370	1,020	352	8	–
3	1	1,000	237	102	35	–	–
2	–	100	24	10	4	–	–
1	–	10	2	–	–	–	–

3.3 Doping

As discussed in Chapter 2, doping is a process through which the electrical conductivity of semiconductors is modified through the addition of foreign impurities. Doping to achieve electrical conduction may be used in either integrated circuits or MEMS. However, in MEMS, doping is also used to create etch-resistant regions known as *etch stops*. Doping is commonly accomplished either by *diffusion* or by *ion implantation*. More importantly, these techniques may be used to selectively dope specific material regions.

3.3.1 Diffusion

Diffusion, in the case of semiconductors, may take place in a number of different forms depending on the availability of dopants in gas, liquid, or solid form. Most commonly, diffusion proceeds with a gaseous source in which impurities are constantly supplied at a particular concentration to a surface such as silicon. These impurities initially are present in a larger concentration in the gas phase and a lower concentration in the solid substrate; thus, the impurities move from high to low concentration regions. This process takes place at elevated temperatures, which enhances the diffusive transport of impurities into silicon. In contrast to gases and liquids, in which significant diffusion can occur at room temperature, diffusion in solids requires high temperatures for motion of impurities (typically ~1000°C). There is a limit on the number of impurities that can be introduced into a solid without precipitation, called the *solid solubility*; these limits are determined experimentally and vary somewhat with temperature. In practice, diffusion is usually limited to select regions. Other areas are masked with materials such as silicon dioxide in the case of silicon.

The diffusion process is mathematically captured by Fick's second law:

$$\frac{\partial C(x,t)}{\partial t} = D \frac{\partial^2 C(x,t)}{\partial x^2} \tag{3.9}$$

where C is the concentration, D is the diffusivity (also known as the diffusion coefficient or diffusion constant; in cm²/s), x is the distance, and t is the time of diffusion. This expression is obtained by combining Fick's first law with the continuity equation. We also assume that the diffusion constant is independent of the concentration. The solution under the initial and boundary conditions of

$$C(x,0) = 0$$
$$C(0,t) = C_s \tag{3.10}$$
$$C(\infty,t) = 0$$

is given as follows:

$$C(x,t) = C_s \text{erfc}\left[\frac{x}{2\sqrt{Dt}}\right] \tag{3.11}$$

where C_s is the surface concentration (atoms/cm³) and erfc is the complementary error function. The expression $\sqrt{Dt}$ is also known as the diffusion length. It is possible to define junction depth, x_j, as the depth in silicon at which the concentration is equal to that of the substrate background doping concentration (C_{sub})

$$C(x_j,t) = C_{sub}$$

$$x_j = 2\sqrt{Dt} \; \text{erfc}^{-1}\left[\frac{C_{sub}}{C_s}\right] \tag{3.12}$$

This particular case is referred to as *predeposition* and is usually carried out at lower temperatures (800–900°C).

In an alternate method, diffusion proceeds with a thin layer of dopant present at the surface (total quantity fixed at Q_0 atoms/cm^2). This process is also called drive-in diffusion and typically follows a predeposition process to position the impurities at the desired junction depth. Here, the initial and boundary conditions are modified

$$C(x,0)=0$$

$$\int_0^\infty C(x,t)\,dx = Q_0 \tag{3.13}$$

$$C(x,\infty)=0$$

The solution for this case takes on a Gaussian profile

$$C(x,t)=\frac{Q_0}{\sqrt{\pi Dt}}\exp\left[-\frac{x^2}{4Dt}\right] \tag{3.14}$$

The surface concentration (setting $x = 0$) is given by

$$C_s = C(0,t)=\frac{Q_0}{\sqrt{\pi Dt}} \tag{3.15}$$

As in the first case, a junction depth can be defined as

$$x_j = \left(4Dt\ln\left[\frac{Q_0}{C_{sub}\sqrt{\pi Dt}}\right]\right)^{1/2} \tag{3.16}$$

The diffusivity is also a function of temperature

$$D = D_0\exp\left[-\frac{E_a}{kT}\right] \tag{3.17}$$

where D_0 is the frequency factor (in cm^2/s), E_a is the activation energy (in eV), k is the Boltzmann constant, and T is temperature (in K).

3.3.2 Ion Implantation

Doping may also be achieved by bombarding a substrate with a beam of ions at sufficient acceleration energy (1 keV–1 MeV) to implant them into the substrate (100 Å–10 µm). However, specialized equipment is required for *ion implantation*. First, ions are accelerated as a high-energy beam into a substrate. Some ions penetrate, undergo collisions, and eventually come to rest within the substrate. The total length of the path that the ions follow is referred to as the range (R)

and is quite broad given that ions follow a zigzag path due to collisions. The total range is not of interest; instead, we focus on the final distance below the surface at which a dopant ion comes to rest, or the projected range (R_p). The total number of ions implanted per unit area, or *implantation dose* (φ in ions implanted/cm^2), can be controlled. The resulting distribution of implanted ions is Gaussian and the ion concentration, n, as a function of depth is

$$n(x) = \frac{\phi}{\sqrt{2\pi}\Delta R_p} \exp\left[-\frac{(x - R_p)^2}{2\Delta R_p{}^2} \right] \tag{3.18}$$

where x is the distance and ΔR_p is the projected straggle (standard deviation of the Gaussian distribution).

Ion implantation has many advantages over diffusion and is now the primary doping method. The process occurs near room temperature, and so minimal thermal stress and strain are imposed on the substrate. Many profiles can be obtained, and a variety of masking materials can be used because the process does not require elevated temperatures. Compared to diffusion, ion implantation allows precise control of the number of dopant atoms introduced and the boundary of the doped region (less lateral distribution exists). Also, ion implantation through thin surface layers, such as silicon dioxide, is possible.

However, dopant distribution is not as uniform as in diffusion, so a diffusion step is required to drive the impurities to specified depths. Bombardment of the substrate also produces defects. To remove these and to electrically activate the dopants, a heating step is performed (>900°C for silicon).

3.4 Micromachining

3.4.1 Subtractive Processes

Subtractive processes include etching techniques and other methods to selectively remove material. When choosing a subtractive process, the following factors should be considered:

- Desired geometry and ability of the process to achieve it
- Feature depth and uniformity
- Surface roughness resulting from the process
- Compatibility of the process with material exposed to the active agents
- Safety, cost, availability, and environmental impact of the active agents

Etching is the most frequently used subtractive process, and can be classified by the type of mechanism used to remove the material. Etching in liquid chemicals is known as *wet etching. Dry etching* is also possible. This latter mode is dominated by physical etching processes and is typically performed in a vacuum. Many forms of plasma etching involve physical processes but may also possess a chemical etching component.

Etching processes occur at surfaces, and the rate at which etching reactions occur can be limited by either a reaction or diffusion. First, reactants are transported to the surface to be etched (adsorption). A surface reaction takes place in which removal of material occurs. Then, these products are transported away from the etched surface (desorption). *Reaction-limited* etching is typically a function of process temperature. *Diffusion-limited* etching depends on a number of factors influencing the distribution and transport of the reactants and products including the device layout and geometry. These factors are typically referred to as "loading" effects. The slowest step necessarily dominates the overall etching rate.

3.4.1.1 Wet Etching

Wet chemical etching is a relatively simple process in which selectively patterned substrates are immersed into a liquid etching bath for a specified time and then are rinsed clean. In some cases, it may be necessary to halt the etching process with an additional chemical rinse step. Masked areas are protected during etching, and only the areas exposed to the etchant are removed. The chemical etching rate is a function of temperature and concentration of reactants and products. In addition, there are a number of other factors to consider, including anisotropy, selectivity, etch rate, uniformity across the substrate, surface quality, and reproducibility. Some of these factors are explained next, and also apply to dry etching processes.

The resulting profile of the etched sidewalls can be categorized as either *isotropic* or *anisotropic*. Most chemical etches are isotropic, that is, etching proceeds at equivalent rates in all directions. However, in an anisotropic etch, a significant difference exists between the etch rates in different directions. Typical cross sections for isotropically and anisotropically etched trenches are shown in Figure 3.20. For wet etching, anisotropy refers to the etch rate variations along the different crystallographic directions. For dry etching, anisotropic etching corresponds to the case where there is little lateral or horizontal etching, and etching is predominantly in the direction normal to the substrate.

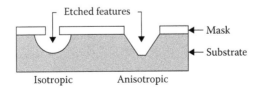

FIGURE 3.20
Examples of isotropic and anisotropic trenches in silicon produced by wet etching.

The degree of anisotropy, A, can be expressed as follows [6]:

$$A = 1 - \frac{R_h}{R_v} \qquad (3.19)$$

where the subscripts h and v indicate horizontal and vertical etch rates, respectively. In isotropic etching, the horizontal and vertical etch rates are equal ($R_h = R_v$), so $A = 0$. For an ideal anisotropic etching, the horizontal etch rate is zero ($R_h = 0$), so $A = 1$.

Isotropic etching results in undercutting of the mask as a result of horizontal etching. This effect is depicted in Figure 3.21. When etching thin films, undercutting sets a limit to the closest spacing between features and the width of the etched trench. For example, the final width of an etched line is smaller than its original dimensions on the mask. This difference between the etched structure and the original mask image is called the *bias*. In Figure 3.22, we define the width of the original mask feature as d_m. The etched structure has a final feature width of d_f, and the thickness of the film to be etched is h. The difference between the mask and final feature dimensions is the bias, B, or

$$|B| = |d_m - d_f| \qquad (3.20)$$

For a pure isotropic etching, the bias will be twice the film thickness when the etch reaches the end of the film.

From the parameters used to define the bias, we can also define the aspect ratio, which is a comparison of the height to the width of etched structures

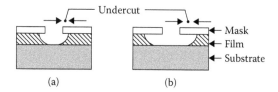

(a)   (b)

FIGURE 3.21
Isotropic etching of a film as a function of etch duration: (a) barely etched through film, (b) overetching of film.

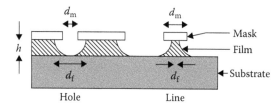

Hole   Line

FIGURE 3.22
Definition of bias in etched features.

(i.e., aspect ratio may be expressed as $h{:}w$). In general, structures having a high aspect ratio are desirable but often difficult to obtain. The isotropic nature of most wet chemical etching prohibits the creation of high aspect ratio structures. In Figure 3.23, there are two types of structures on the substrate. The taller structures clearly have a large aspect ratio because $h_1 > w_1$, but the shorter structures have low aspect ratio because $w_2 > h_2$.

For etching applications, a masking material is used to protect designated areas of the substrate from etching. For example, photoresist is a common masking material. It is not common to etch the entire surface area of a substrate except in select processes. Thus, it is important that the masking material chosen is resistant to the etchant or at least has a lower etch rate than that of the material to be etched. The suitability of a masking material is quantified using a parameter known as selectivity, S, which is mathematically defined as

$$S = \frac{R_{\text{substrate}}}{R_{\text{mask}}} \tag{3.21}$$

where R is the etch rate. The subscript "substrate" refers to the material to be removed and "mask" refers to the intended masking material. Selectivity can also be expressed as a ratio $R_{\text{substrate}} : R_{\text{mask}}$. High selectivity is desirable and is characteristic of wet etching due to the selective nature of chemical reactions. Selectivity can also impact the dimensions of the final structure because the mask is eroded during etching. This is illustrated in Figure 3.24.

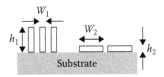

FIGURE 3.23
Structures having high (left) and low (right) aspect ratio.

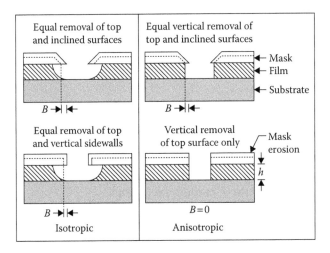

FIGURE 3.24
Mask erosion and dimensional stability.

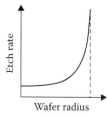

Etch rate (y-axis)

Wafer radius (x-axis)

FIGURE 3.25
Bull's eye effect.

The uniformity of the etch depth across a substrate impacts the selectivity, thickness of the masking material required to complete the etch, and the total etch duration. The etch duration, t, is related to the desired etching depth, d, as follows:

$$t = \frac{d}{R} \quad (3.22)$$

This represents the ideal case, in which the etch rate and etching reaction are assumed to be uniform over the entire substrate. Practically speaking, this is usually not the case, and overetching may be required. A commonly encountered example of nonuniform etching is the "bull's eye" effect, in which the etch rate varies as a function of the substrate radius (Figure 3.25).

Overall, wet chemical etching is simple to implement and requires minimal equipment (etch chemicals, etch resistant containers, timer, hot plate [optional], and a fume hood are required). Wet etching proceeds quickly; however, speed is sometimes achieved at the expense of surface quality. Specific examples of isotropic and anisotropic etchings will be discussed in Section 3.4.3.1.

3.4.1.2 Dry Etching

The inability of wet etching to produce an accurate pattern transfer with straight sidewalls and minimal undercut necessitates the exploration of other techniques. Dry etching, in which the chemically active species are in a gaseous state, is an alternative method to remove materials. In contrast to wet etching, dry etching may use chemical, physical, or a combination of both these mechanisms. These options enable anisotropic etching of high aspect-ratio structures, which is not possible with wet etching. Overall, dry etching processes are typically slower than wet ones. *Plasma* and *gas-phase etching* are two common types of dry etching.

3.4.1.2.1 Plasma Etching

Plasma etching is a general term used to refer to a number of mechanisms used to remove material. There are four fundamental plasma processes: (1) sputtering, (2) pure chemical etching, (3) ion driven etching, and (4) ion inhibitor etching. These processes are illustrated in Figure 3.26. To understand the operational principles behind these processes, it is necessary to introduce the fourth state of matter, the plasma.

A relatively inert gas is the starting ingredient for generating a plasma. A gas or combination of gases is selected on the basis of chemical reactivity of its constituent species with the material to be etched. By subjecting the gas feed to a low pressure and either a direct current (DC) or a high-frequency (radio frequency [RF] range) voltage, a plasma is formed. This plasma consists of a

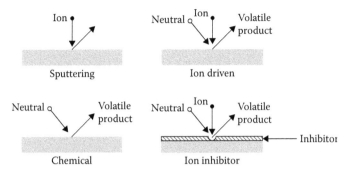

FIGURE 3.26
Four fundamental plasma processes (after [6]).

partially ionized gas composed of ions, electrons, and neutral species. The ionized gas is neutral and has equal numbers of free electrons and positively charged ions.

Six processes occur during chemical plasma etching: (1) generation of reactive etchant species, (2) diffusion to the etch surface, (3) adsorption by the surface, (4) chemical reaction, (5) desorption of the by-products, and (6) diffusion of the by-products into the bulk gas. While all these steps are necessary for plasma etching, the desorption step is a critical requirement. Because the mass loss is evaporative, the by-products must be volatile to prevent redeposition. The volatile by-products are easily removed in a low-pressure system. Chemical etching is a result of reactions with radicals at the surface and is a very selective process. However, it also can be isotropic or nondirectional.

Plasmas also consist of ionic species, which can be used to further enhance the etching. The charged ions can be used to bombard the surface to physically remove materials. Because the ion flux is perpendicular to the surface to be etched, plasma processes that use these ionic species are directional and can produce anisotropic structures. This method is called *sputtering*.

Energetic ions can also be used to damage the surface to produce reactive species that are chemically removed. This ion-driven etching mode increases the etch rate beyond that of sputtering alone. Ion inhibitor etching involves etchants and inhibitor species. The etchants produce isotropic chemical etching by themselves. When combined with the inhibitors that form thin films on sidewalls of the etched structures, vertical sidewalls are formed because the sidewalls have little or no ion bombardment. Limiting the attack of the sidewalls greatly improves the anisotropy.

A typical parallel-plate plasma etcher is configured as shown in Figure 3.27. Wafers are placed between the two parallel electrodes, and a plasma is produced between the electrodes by an RF generator. Plasma formation occurs over a specific pressure range (few torrs to fractions of a millitorr), and the chemical species generated depend on the process gases selected.

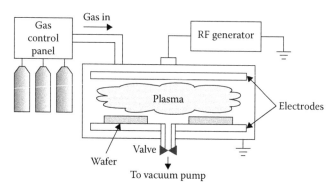

FIGURE 3.27
Standard configuration for a plasma etching tool.

3.4.1.2.2 *Sputter Etching Mode*

Plasma etching by sputtering is a purely physical process in which surface material is ejected by energetic ions through momentum transfer. This process depends on the bonding forces and the surface structure of the substrate. Thus, this is not a selective process, and it removes all exposed material. However, the etch rate is low, and damage to the substrate can result. Also, low pressure is necessary to ensure a long mean free path to prevent reflection and redeposition of sputtered material. Sputtering is often used for surface cleanup.

3.4.1.2.3 *Plasma and Reactive Ion Etching Modes*

The two common modes used in microelectronics are *plasma etching* and *reactive ion etching* (RIE) modes. Electrode areas are symmetric in plasma etching mode, and the etching is predominantly chemical. In RIE mode, a negative self-bias DC voltage occurs between the plasma and the wafer electrode to accelerate ions to the wafer. This adds a physical etching component in addition to the chemical one. This type of voltage occurs for asymmetric electrode areas. In plasma etching mode, there exists an identical and relatively small DC voltage between the plasma and either electrode. However, for RIE, a large negative bias occurs between the plasma and the smaller electrode. *Reactive ion beam etching* (RIBE) is another, less common mode. Ions that are extracted from the plasma and then accelerated at the wafer are the primary etching mechanism here.

Plasma etching is isotropic, whereas RIE and RIBE are more anisotropic but suffer from poor selectivity due to physical etching. Variables that affect etching include operating pressure, wafer temperature, electrode voltage, gas composition, gas flow rate, and loading. Loading effects are encountered when the area of the surface to be etched varies from run to run. For example, because more wafers are placed in the plasma etching tool, the area to be etched increases and the etch rate consequently decreases compared to the cases where fewer wafers are loaded. This is also true for runs in which single wafers are loaded, but each wafer has a different masked, or exposed, areas. Physically, loading is related to the depletion of the reactive species during etching.

3.4.1.2.4 Deep Reactive Ion Etching

High aspect ratio structures (>5:1) with vertical sidewalls are difficult to achieve by typical dry etching processes. *Deep reactive ion etching* (DRIE) combines a high density plasma with a switched-chemistry process to overcome these limitations. DRIE uses the patented Bosch process in which deposition of a protective film on the sidewalls is alternated with etching of silicon (Figure 3.28). Because the reactive species predominantly attack the bottom of the etched silicon feature, near vertical sidewalls and a high degree of anisotropy are possible. The cyclical nature of the etch process leaves an artifact known as scallop on the etched sidewalls. Each scallop corresponds to one full etch and deposit cycle.

3.4.1.2.5 Vapor Phase Etching

Gases or vapors react directly with certain materials without requiring a plasma. Hydrofluoric acid (HF) vapor etching of silicon dioxide at room temperature is the simplest form of this type of etching. Silicon also reacts with gas-phase halogen fluorides, including chlorine trifluoride (ClF_3), bromine trifluoride (BrF_3), bromine pentafluoride (BrF_5), iodine pentafluoride (IF_5), and xenon difluoride (XeF_2). Of these, XeF_2 and BrF_3 vapor etchings have been found to be simple high-rate etching methods to release free-standing MEMS structures without expensive tools for plasma etching and the *stiction* issues associated with wet etching.

XeF_2 is a solid at room temperature with a vapor pressure of 4 torr, whereas BrF_3 is a liquid at room temperature with a vapor pressure of 7 torr. Both

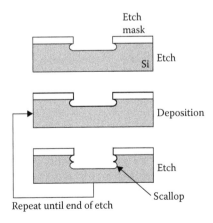

FIGURE 3.28
Cyclical nature of deep reactive ion etching process.

chemicals pose potential hazards although XeF_2 is arguably safer due to the lack of Br but still forms HF in the presence of water. Etching reactions are carried out in a vacuum chamber between 1 and 7 torr, and the overall etching reactions for these fluorine-based etchants are hypothesized to be

$$2XeF_2 + Si \rightarrow 2Xe + SiF_4 \qquad (3.23)$$

$$4BrF_3 + 3Si \rightarrow 2Br_2 + 3SiF_4 \qquad (3.24)$$

Both chemicals exhibit good selectivity over common masking materials such as silicon dioxide, silicon nitride (Si_3N_4), and photoresists.

3.4.1.2.6 Ion Beam Milling

Ion beam milling techniques are seldom used, but may prove useful in specialized applications. Here, energetic ions are focused into a large-diameter beam and accelerated toward a substrate. In its simplest form, ion milling resembles sputtering. However, three electrodes, instead of two, are used. Electrons are emitted by a filament (cathode) and accelerated toward an anode in one chamber to initiate gas ionization. A perforated electrode separates this chamber from a sample chamber. The perforated electrode extracts ions from the first chamber, collects them into a wide beam, and accelerates the beam toward the sample. Inert ions (usually argon [Ar]) are used to increase sputtering yield and prevent unwanted chemical reactions.

High etch rates are achieved when a denser ion plasma is present. By integrating a magnetic field into the ionization chamber (magnetically enhanced ion beam etching), electrons are forced to take an indirect helical path between the filament and the anode. This increases the ionization efficiency which in turn produces a denser ion plasma. By replacing argon ions with a reactive species (e.g., fluorine [F]), chemical removal processes are included in otherwise purely physical surface interactions (called RIBE).

Ion beams are usually several centimeters in diameter. It is possible to focus the beam into a very narrow (~50 nm) stream of ions that are then used to selectively machine a surface. In this FIB technique, the ion source is a liquid metal (e.g., gallium [Ga]) instead of argon.

3.4.1.2.7 Laser Ablation

Light energy emitted by a laser is a simple method to selectively remove material without using lithography. A laser beam is focused onto a substrate mounted on a computer-controlled motorized translation stage (x, y, z, θ). Control of the z-direction allows the creation of three-dimensional structures, which can greatly simplify what would otherwise be a multistep fabrication process. Thus, laser ablation is useful for a rapid prototyping of virtually any material even though the material removal process is serial due to the scanning of the small cutting laser spot. The overall resolution is limited by the laser spot size; advanced techniques can produce submicron features. Common high-power output lasers used for laser ablation include carbon dioxide (CO_2) and neodymium-doped yttrium aluminum garnet (Nd:YAG). One drawback is the generation of debris during ablation. This may obscure features, contaminate laser optics, or cause other undesirable events. By adding an appropriate gas feed, volatile etch products are produced that are easily removed (laser photochemical etching). For example, silicon can be etched by adding chlorine gas to produce silicon tetrachloride ($SiCl_4$) gas.

3.4.1.2.8 Electrodischarge Machining

Electrodischarge machining (EDM), like laser ablation, is an electrothermal process. Material is removed following the input of thermal energy. The electrothermal cutting tool is either machined from a durable conductive material in the form of the desired shape or is a wire that is manipulated to cut the desired shape (wire EDM [WEDM]). In WEDM, new wire is constantly supplied. The workpiece must also be conductive. The tool and workpiece are the cathode and anode, respectively, and are immersed together in a dielectric bath. Rapid, high-voltage pulses (~100 Hz–100 kHz) are applied to the closely spaced tool and workpiece (~25 μm), which induces intense sparking that melts and vaporizes small amounts of material (~1 μm surface finish). Debris is removed in the fluid. This process is not selective and also wears down the tool, which is typically made of a material with high wear resistance (e.g., carbon, zinc, and brass). Simple EDM and WEDM setups are shown in Figure 3.29. A comparison between wet and dry etching techniques based on different parameters is presented in Table 3.6 and may serve as a convenient guide for choosing one process type of over the other.

3.4.2 Additive Processes

Additive processes involve the buildup of layers on a substrate using techniques such as oxidation, chemical vapor deposition (CVD), physical vapor

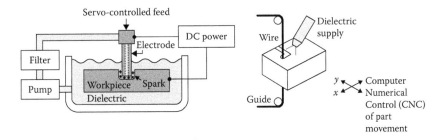

FIGURE 3.29
Schematic showing major components for electrodischarge machining (left) and wire electrodischarge machining (right).

TABLE 3.6

Comparison of Wet and Dry Etching Techniques

Parameter	Dry Etching	Wet Etching
Directionality	Good for most materials	Good only for single crystal materials
Automation	Good	Poor
Environmental impact	Low	High
Mask film adhesion	Not as critical	Very critical
Selectivity	Poor	Very good
Process compatible materials	Only certain materials	Nearly all materials
Scalability of process	Difficult	Easy
Cleanliness	Conditionally clean	Good to very good
Critical dimension control	Very good (<0.1 µm)	Poor
Equipment cost	Expensive	Relatively inexpensive
Typical etch rate	Slow (~0.1 µm/min) to fast (~6.0 µm/min)	Fast (~1.0 µm/min and above)
Operational parameters	Many	Few
Etch rate control	Good for slow etch	Difficult

Source: Madou, M. J. 1997. *Fundamentals of Microfabrication.* Boca Raton, FL: CRC Press.

deposition (PVD), spin coating (introduced in Section 3.2.1.2), and epitaxy. Of these, oxidation consumes the silicon substrate for the growth of silicon dioxide. The other techniques directly add thin or thick film materials to the substrate, which is not limited to silicon. Factors such as adhesion, film stress, conformality, and pinholes affect the films deposited by additive processes.

3.4.2.1 Oxidation

Silicon dioxide, which is traditionally used as an insulator in microelectronics, is also a versatile MEMS material that is used as an etch mask and a sacrificial layer. Although a number of different methods are available to produce silicon dioxide films, *thermal oxidation* is the least expensive. Even at room temperature, a thin ~20-nm *native oxide* forms on all exposed silicon surfaces by oxidation. Thermal oxidation is classified by the manner of introduction of oxygen to heated silicon substrates; *dry* oxidation uses oxygen gas and *wet* oxidation uses water vapor as the oxygen carrier. The overall reactions for these two processes are

$$\text{Dry: } Si_{(solid)} + O_{2(gas)} \rightarrow SiO_{2(solid)} \tag{3.25}$$

$$\text{Wet: } Si_{(solid)} + 2H_2O_{(steam)} \rightarrow SiO_{2(solid)} + 2H_{2(gas)} \tag{3.26}$$

Reactions are performed in quartz tube furnaces at temperatures of 900–1200°C, which are similar to the setup shown in Figure 3.30.

In both of these reactions, silicon is required for the formation of oxide and is consumed during the process. Oxygen diffuses through existing oxide to the silicon–silicon dioxide interface, where it reacts to form new oxide and consumes silicon in the process. Then, this moves the interface further into the silicon substrate. Silicon consumption is determined by the density and relative molecular weights of silicon and silicon dioxide; 44% of the final oxide thickness was originally silicon [3] and is expressed mathematically as

$$x_{silicon} = 0.44 x_{oxide} \tag{3.27}$$

and is shown in Figure 3.31. Overall, wet oxidation yields thicker oxide layers, and 1–2 µm-thick conventionally grown oxides are practical. Oxide growth can be predicted by the Deal-Grove model, which is accurate for oxide thickness of ~300–20,000 Å deposited at temperatures of 700–1300°C [8]. For sufficiently long oxidation times, the oxide thickness and time follow a parabolic behavior:

$$x = \sqrt{Bt} \tag{3.28}$$

where B is the parabolic rate constant and t is the time. B can be expressed as (in $\mu m^2/h$)

$$B = \frac{2DC}{N} \tag{3.29}$$

where D is the oxidant diffusivity (in cm^2/s), C is the equilibrium oxidant concentration in the oxide (in cm^{-3}), and N is the number of molecules of oxidant per unit volume of oxide (in cm^{-3}). In this parabolic regime, oxidation is

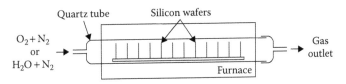

FIGURE 3.30
Typical furnace setup for thermal oxidation of silicon.

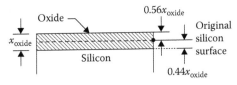

FIGURE 3.31
Consumption of the silicon surface by the thermal oxidation process.

diffusion-limited. For short oxidation times, the oxide thickness and process time are linearly related

$$x = \frac{B}{A}(t + \tau)$$ (3.30)

where B/A is the linear rate constant (in $\mu m/h$). A is given by (in μm)

$$A = 2D\left(\frac{1}{k_s} + \frac{1}{h}\right)$$ (3.31)

where k_s is the silicon oxidation rate constant and h is the gas-phase mass transport coefficient. In Equation 3.30, τ (in hours) is

$$\tau = \frac{X_i^2 + AX_i}{B}$$ (3.32)

where X_i is the initial oxide thickness; this is the time corresponding to the initial oxide thickness. In this linear regime, oxidation is reaction limited. The rate constants are temperature dependent and are higher for wet oxidation, leading to higher oxidation rates [3].

Silicon oxide forms on all exposed silicon surfaces, so it is possible to selectively grow the oxide by using patterned thin films that are impervious to oxygen diffusion. Silicon nitride has this property and is used in a technique called *local oxidation of silicon* (LOCOS) for selective oxidation (Figure 3.32). This is primarily used for forming insulating regions in microelectronics.

3.4.2.2 Chemical Vapor Deposition

Silicon dioxide can also be deposited by CVD. CVD is a versatile deposition method that can be used to deposit a variety of substrates, both organic and inorganic. These materials include silicon nitride, polysilicon, metals (e.g., aluminum [Al], silver [Ag], gold [Au], titanium [Ti], tungsten [W], copper [Cu], and platinum [Pt]), shape memory alloys (e.g., nickel titanium [NiTi]), and piezoelectrics (e.g., zinc oxide [ZnO]). In fact, some films can only be deposited by CVD. The deposition method uses somewhat

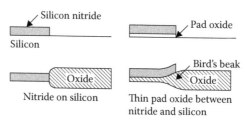

FIGURE 3.32
Selective masking of thermal oxidation by nitride masks (after [4]).

complex chemical reactions at surfaces involving mass and heat transfer. When depositing thin films, it is desirable to produce films with a reasonable thickness uniformity, high purity, high density, controlled composition, high structural perfection, excellent adhesion, and good step coverage. Overall, CVD processes have higher growth rates and quality compared to PVD methods. CVD produces films having a variety of chemical compositions with good process control.

In a CVD reactor, a given composition of gaseous reactants and diluent inert gases is flowed over a hot substrate. Reactants adsorb to the substrate surface, react to form the desired film, and the desorbed by-products are removed. Only heterogeneous reactions are desired, in which the film-forming chemical reaction takes place only at the substrate surface. Reactions occurring in the gas phase, or homogeneous reactions, usually result in undesirable particulate-based films with a low density or pinhole defects. These lesser-quality films also have poor adhesion.

The energy to drive the chemical reactions in CVD can be generated thermally or can be imparted by energetic electrons (e.g., from a plasma). The different types of CVD reactors may use one or more of these energy sources to promote the film deposition. The rate at which a film is deposited depends on a number of parameters, including process temperature, pressure of the carrier gas, and gas flow velocity. The deposition rate is dominated by the slowest step in the process, which is either dependent on the rate of reactants arriving at the surface (mass transport–limited) or the rate of the surface reaction (reaction-limited). Generally speaking, CVD processes are mass transport–limited at higher temperatures and reaction-limited at lower temperatures.

In CVD, silicon dioxide is deposited, for example, by reacting silane (SiH_4) and oxygen gas at 400–500°C as follows:

$$SiH_4 + O_2 \rightarrow SiO_2 + 2H_2 \qquad (3.33)$$

This reaction is typical of atmospheric pressure CVD (APCVD), which is a mass transport–limited process. In APCVD equipment, gas flow is typically introduced parallel to the substrate (Figure 3.33). Silicon dioxide can also be deposited by low pressure CVD (LPCVD) or plasma-enhanced CVD (PECVD) processes. These are all specialized cases of the basic CVD process.

In APCVD systems, films have poor uniformity and step coverage and are also prone to particulate contamination. By changing the process pressure, it is possible to achieve a higher film quality or even increase the growth rate. LPCVD is carried out at ~1 torr, which minimizes diffusion effects, and hence is a surface reaction–limited process. The change in pressure gives rise to a higher growth rate by increasing diffusivity. The resulting films are less dense but are more uniform. In contrast to APCVD systems, in LPCVD gas flow is introduced perpendicular to the substrate (Figure 3.34).

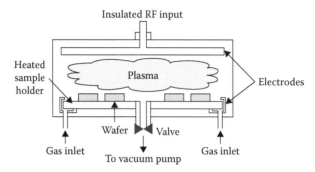

FIGURE 3.33
Typical parallel plate chemical vapor deposition reactor.

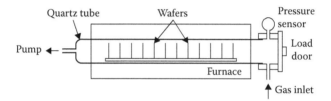

FIGURE 3.34
Typical tube style chemical vapor deposition reactor.

An example of an LPCVD oxide reaction is

$$SiH_2Cl_2 + 2N_2O \rightarrow SiO_2 + 2N_2 + 2HCl \qquad (3.34)$$

in which dichlorosilane (SiH_2Cl_2) and nitrous oxide are reacted near 900°C at but at lower pressures than in APCVD. LPCVD silicon nitride is deposited using ammonia as a carrier gas. A common reaction uses dichlorosilane (650–750°C)

$$3SiH_2Cl_2 + 4NH_3 \rightarrow Si_3N_4 + 6HCl + 6H_2 \qquad (3.35)$$

Polysilicon is deposited at 600–650°C by a pyrolysis process

$$SiH_4 \rightarrow Si + 2H_2 \qquad (3.36)$$

Finally, PECVD, as the name would imply, uses plasma to generate ions or radicals used in the chemical reaction. The purpose of the plasma is to further decrease the process temperature by energetic electrons that facilitate reactions that would otherwise only occur at higher temperatures. While this may seem an unnecessary complication of an otherwise simple process, the merit of reducing overall process temperature becomes important when evaluating an entire process flow. For example, when depositing new materials on top of a substrate with existing materials, the temperature of the deposition process must be considered so as not to degrade the underlying materials. With PECVD, it is possible to deposit films on top of materials with low thermal stability.

The plasma contains high energy electrons that facilitate the dissociation of the process gas molecules. These species eventually adsorb to the substrate surface and recombine to form the desired film. Gas flow may be introduced either parallel or perpendicular to the substrate. The lower temperatures of this process compared to other CVD methods result in amorphous films. Due to the presence of defects from extraneous reactions, the quality of these films is also lower. However, the films possess good adhesion, low pinhole density, and good step coverage.

Silicon dioxide films are formed by PECVD at lower temperatures (200–400°C) but may be nonstoichiometric due to hydrogen incorporation

$$SiH_4 + 2N_2O \rightarrow SiO_2 + 2H_2 + 2N_2 \tag{3.37}$$

PECVD nitride also suffers from hydrogen incorporation (200 – 400°C)

$$SiH_4 + NH_3 \rightarrow Si_xN_yH_z + H_2 \tag{3.38}$$

A comparison of the three major CVD processes is provided in Table 3.7. Regardless of the method of deposition, surface coverage is an important factor to consider and varies with each technique. This is particularly important when trying to deposit films uniformly over step features. Typically, it is desirable to obtain a completely *conformal* coating that has uniform thickness on all exposed features regardless of the surface topology (Figure 3.35).

TABLE 3.7

Comparison of Three CVD Processes)

CVD Process	Pressure	Temperature (°C)	Materials	Considerations
APCVD	100–760 torr	350–400	Polysilicon, low temperature oxide	Poor step coverage, low temperature, mass transport limited
LPCVD	0.25–2 torr	500–900	Polysilicon, nitride, oxide, tungsten, PSG	Low deposition rate, high temperature, high purity and uniformity, conformal step coverage, reaction-limited
PECVD	0.1–5 torr	200–400	α-silicon, nitride, oxide	Low temperature, fast deposition, good step coverage, contamination, reaction-limited

Sources: Wolf, S., and R. N. Tauber. 1986. *Silicon Processing for the VLSI Era*. Sunset Beach, CA: Lattice Press; Madou, M. J. 1997. *Fundamentals of Microfabrication*. Boca Raton, FL: CRC Press; and Sze, S. M. 1994. *Semiconductor Sensors*. New York: Wiley.

CVD = chemical vapor deposition; APCVD = atmospheric pressure CVD; LPCVD = low pressure CVD; PSG = phosphosilicate glass; PECVD = plasma enhanced CVD.

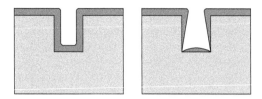

FIGURE 3.35
Conformal coating (left) and nonconformal coating (right).

The degree to which a coating is conformal depends on the physics of the deposition process in relation to the arrival angle of the material to be deposited. Conformal coatings usually result when surface migration of adsorbed reactants occurs prior to formation of a permanent bond. It follows that if molecules have adequate energy to reach the surface but undergo no surface migration, the resulting uniformity of coverage of textured surfaces, for example, the bottom of a trench, is poor. In other words, the *mean free path* is short, and only partial surface coverage is achieved. This is typical of PVD metals, whereas CVD polysilicon and nitride are quite conformal. Mean free path is explained in detail in Section 3.4.2.3.1.

3.4.2.3 Physical Vapor Deposition

PVD relies on the direct impingement of particles on surfaces to produce thin films. The two major PVD processes are *evaporation* and *sputtering*. Both have similar deposition methods. First, the material to be deposited must be converted into a gaseous or vapor phase, typically from a solid source. The gaseous phase is then transported to a substrate where condensation, nucleation, and growth occur. The deposition of compounds is sometimes referred to as *reactive* evaporation or sputtering (e.g., silicides, nitrides, or carbides), and typically involves a gas feed.

3.4.2.3.1 Evaporation

The oldest and simplest PVD method is *evaporation*. The desired material is heated to a high temperature at a high vacuum (<10^{-5} torr), and the vapor is allowed to condense on a surface held at a lower temperature than that of the vapor. Metals, for example, readily evaporate at high temperatures and are commonly deposited by evaporation. The evaporation rate (R) is dependent on the equilibrium vapor pressure (P_e) of the source material ($R \propto P_e$). Reasonable deposition rates are achieved for vapor pressures >1.5 Pa (>10 mtorr) [3], and a wide range of temperatures satisfy this condition for various metals (e.g., aluminum at 1200 K and tungsten at 3230 K).

The high vacuum at which evaporation is carried out also serves a purpose. Extraneous gas molecules and vapor species are removed to allow the evaporated species to reach the substrate without colliding. By evacuating the chamber in which evaporation is performed, the mean free path is increased such that it is greater than the distance between the substrate and the source and scattering is minimized. A vacuum is any pressure below atmospheric pressure and is categorized in several ranges listed in Table 3.8. Conversions between commonly used pressure units are provided in Table 3.9.

The mean free path, λ, is the average distance a molecule travels between collisions and is calculated from the following equation

$$\lambda = \frac{1}{\sqrt{2}\pi d^2 n} \tag{3.39}$$

where d is the molecular diameter and n is the gas concentration. It is more convenient to express the mean free path in terms of pressure. For air at 300°K, the following expression provides λ in millimeters:

$$\lambda = \frac{6.6}{P(\text{in Pa})} = \frac{0.05}{P(\text{in torr})} \tag{3.40}$$

The mean free path for commonly used pressure ranges are tabulated in Table 3.10.

The simplest evaporation system uses a resistively heated tungsten wire to support small pieces of metal to be evaporated. A high current is passed through the wire to heat the source metal; the wire is also heated in the process, so refractory metals which have high melting points are used. The substrate is mounted some distance away from the source and as a result, is near room temperature. A typical resistive evaporation system is shown in Figure 3.36.

TABLE 3.8

Vacuum Ranges and Corresponding Pressures (After [4])

Range	Pressure Range
Low	$10^5 - 3.3 \times 10^3$ Pa (750 – 25 torr)
Medium	$3.3 \times 10^3 - 10^{-1}$ Pa (25 – 7.5 10^{-4} torr)
High	$10^{-1} - 10^{-4}$ Pa (7.5 × 10^{-4} – 7.5 × 10^{-7} torr)
Very high	$10^{-4} - 10^{-7}$ Pa (7.5 × 10^{-7} – 7.5 × 10^{-10} torr)
Ultra high	$10^{-7} - 10^{-10}$ Pa (7.5 × 10^{-10} – 7.5 × 10^{-13} torr)

TABLE 3.9

Common Pressure Units and Their Conversions

Unit	Conver sion
Pascal (Pa)	1 Pa = 1 N/m² = 7.5 × 10^{-3} torr = 7.5 μmHg
Torr	1 torr = 133.3 Pa = 1 mmHg
Bar	1 bar = 1 × 10^5 Pa = 750 torr
Atmosphere (atm)	1.013 × 10^5 Pa = 760 torr

TABLE 3.10

Mean Free Path for Various Vacuum Ranges
Calculated Using Equation 3.40

Pressure Range	Mean Free Path (mm)
Low	$6.6 \times 10^{-5} - 2 \times 10^{-3}$
Medium	$2 \times 10^{-3} - 6.6 \times 10^{1}$
High	$6.6 \times 10^{1} - 6.6 \times 10^{4}$
Very high	$6.6 \times 10^{4} - 6.6 \times 10^{7}$
Ultra high	$6.6 \times 10^{7} - 6.6 \times 10^{10}$

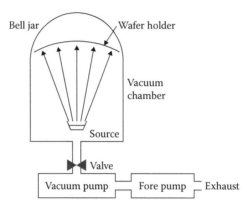

FIGURE 3.36
Thermal evaporation system.

For depositing refractory metals that require higher temperatures, electron beam evaporation is used. A high power stream of accelerated electrons (5–30 keV) melts and evaporates the source material. While deposition rates are much higher, X-ray damage and the ejection of molten droplets may occur due to such high energies. Inductive (radio frequency) heating offers a compromise; large deposition rates are possible but with minimal radiation damage.

Evaporation is primarily limited to metals. Multicomponent materials are difficult to evaporate due to differing vapor pressures. Instead, coevaporation from two or more independently controlled evaporation sources is performed. Overall, evaporation proceeds with high deposition rates and produces pure films with a purity approaching that of the source. Little residual gas incorporation and surface damage occurs, unlike sputtering. However, due to the line-of-sight style deposition, step coverage is poor.

3.4.2.3.2 *Sputtering*

Sputtering is far more versatile when one considers that it can be used on nearly all materials (metals, alloys, semiconductors, and insulators).

Examples of materials that can be deposited by sputtering include thin metal films (aluminum, platinum, gold, titanium, tungsten, titanium/tungsten, aluminum alloys), molybdenum (Mo), silicon, silicon dioxide, and metal silicides. Compared to evaporation, sputtering provides good step coverage and greater control of alloy composition. Also, it is possible to clean substrates by sputtering prior to deposition.

First, ions are generated and carried by argon gas to a target. The accelerated ions bombard the target and sputter the atoms from the surface. These ejected atoms are directed toward the substrate, where some atoms condense to form a film. The bombarding ions must possess adequate energy to sputter the atoms from a surface. Energies lower than the binding energy result in reflection or physisorption. Just above the binding energy, surface reactions such as surface migration and surface damage can occur without sputtering. At much higher energies, bombarding ions are embedded into the target. The correct energy setting results in sputtering, or the ejection of material from the target.

Ions are generated from a plasma, so the process pressure is set by the plasma requirement. High pressures (0.1–1 Pa) result in the incorporation of a small amount of gas that is present in the bombarding species. These pressures also set the mean free path, which is typically less than the distance between the target and the substrate. Thus, backscattering or other modes of energy loss are possible. Ion bombardment may also produce some damage that can be annealed out at elevated temperatures.

A typical sputtering system is shown in Figure 3.37. The target is negatively biased, and the sputtered ions are positive. This bias is produced by either DC or radio frequency excitation. A magnetic field can enhance the numbers of ions and increase the rate of sputtering. This is referred to as magnetron sputtering. The substrate is near room temperature, and the deposition is independent of temperature. However, this limits the deposition rate. By increasing the temperature of the substrate (~200°C), films with

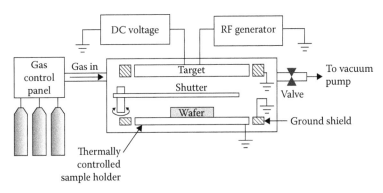

FIGURE 3.37
Major parts of a sputtering system.

better adhesion and less cracking are produced. In reactive sputtering, it is possible to obtain two-component materials from a target consisting of just one component and an appropriately selected sputter gas. For example, silicon nitride may be deposited using a silicon target in the presence of nitrogen gas.

3.4.2.4 Epitaxy

Epitaxy involves the growth of a thin crystalline layer on a crystalline substrate in which the substrate can be thought of as a seed crystal. However, this process occurs below the melting point, unlike the Czochralski process. It is possible to deposit the same material on a substrate (homoepitaxy, e.g., silicon on silicon) or a different crystalline material (heteroepitaxy, e.g., silicon on sapphire [Al_2O_3]). Epitaxial growth of compounds is also possible (e.g., gallium arsenide [GaAs]).

A number of epitaxy techniques exist that resemble CVD (as in vapor-phase epitaxy [VPE]) or physical evaporation (as in molecular beam epitaxy [MBE]). Processes may originate from the vapor, liquid (liquid phase epitaxy [LPE]), or solid phase (solid phase epitaxy [SPE]). Of these three processes, VPE is the most popular and is widely used. In VPE, silicon is deposited from vapors diluted in hydrogen carrier gas in a hydrogen reduction process (800–1150°C and 10–100 torr)

$$SiCl_4 + 2H_2 \rightarrow Si + 4HCl \qquad (3.41)$$

Although other sources can be used, silicon tetrachloride is the most popular. In all epitaxial processes, the goal is to produce defect-free thin single crystalline films. Thus, surface nucleation and surface reaction kinetics are important. A number of reactions that are not captured in the simple reaction in Equation 3.39 occur at the surface.

To obtain the crystal growth, nucleation must occur; this happens more readily on crystalline substrates (e.g., silicon) than amorphous ones (e.g., silicon dioxide). This difference in nucleation can be exploited for selective epitaxial growth in which patterned oxide or nitride serve as masks against the epitaxial growth of silicon on silicon.

As the name would suggest, MBE produces thin films directing e-beam evaporated atoms at a high velocity to a substrate. The supersaturated state of the atoms readily results in nucleation on the substrate surface and subsequent film growth. This process is conducted in an ultrahigh vacuum (10^{-11} torr) and at high temperatures (600–900°C). MBE is also used for silicon epitaxy, but not as frequently as VPE.

Gallium arsenide films are produced by either metal organic chemical vapor deposition (MOCVD) or LPE. In MOCVD, an organometallic compound supplies gallium. These materials are readily vaporized in a bubbler and are transported by a carrier gas for reacting with arsine to form the film.

The LPE produces layers from melts. The material to be deposited is added to the molten metal, applied to a substrate, and allowed to cool so that the material freezes onto the surface. The LPE methods also exist for silicon epitaxy.

3.4.2.5 Electrochemical Deposition

Electroplating of metal films and structures is done by an electrochemical deposition process in which a voltage applied between an anode and a cathode immersed in an electrolyte solution results in metal deposition at the cathode (Figure 3.38). In this configuration, electrochemical reactions occur at each electrode. For example, nickel is commonly deposited by electroplating, and the following reactions occur (in nickel chloride [$NiCl_2$] and potassium chloride [KCl] solution)

$$Ni^{2+} + 2e^- \rightarrow Ni\,(\text{cathode reaction})$$
$$2Cl^- \rightarrow Cl_2 + 2e^-\,(\text{anode reaction}) \tag{3.42}$$

Nickel deposits on the cathode (hydrogen evolution also occurs) and the overall reaction is

$$NiCl_2 \rightarrow Ni + Cl_2 \tag{3.43}$$

Whole wafers can be electroplated. First, a conductive seed layer is deposited; this seed layer is attached to the cathode. For selective plating, the seed layer is patterned with an insulating mold (e.g., photoresist) that can remain in the final structure or be removed after plating. Metal will plate only on the exposed metal areas delineated by the mold. Depending on the desired thickness, the plated metal may extend above the mold. Special polishing processes such as chemical mechanical polishing (CMP) may be performed following plating to remove extraneous metal above the mold and smooth thickness variations of the plated metal across the substrate. Electroplating depends on pH, temperature, and current density; proper process control reduces the incorporation of voids due to gas bubbles. Metals that are

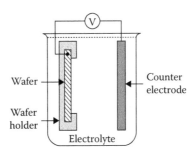

FIGURE 3.38
Basic electroplating setup.

commonly deposited by electroplating include nickel, copper, gold, and per-malloy (NiFe).

Electroless plating, a form of electrochemical deposition, does not require voltage or current to reduce a metal. In electroless plating, a single electrode is used, and both anodic and cathodic reactions occur at this electrode. In the absence of voltage, a chemical reductant is required to initiate the deposition, and the deposition in turn catalyzes the reduction reaction. Thus, electroless plating is an autocatalytic process. Usually, electroless plating is performed at higher temperatures than electroplating. Electroless nickel, for example, is usually deposited using a hypophosphite reductant. Both conductors and insulators can be deposited by electroless plating, and metal alloys also can be deposited through codeposition.

3.4.3 Bulk and Surface Micromachining

The core micromachining processes for creating MEMS are bulk and surface micromachining. Although newer techniques such as LIGA and specialized lithographic processes have been developed, the majority of MEMS devices are still fabricated using one or a combination of these micromachining processes.

3.4.3.1 Bulk Micromachining

Bulk micromachining is a subtractive process specifically used for the fabrication of MEMS devices. Features are sculpted into the substrate by either wet or dry as well as by chemical or physical techniques. Some of these techniques date back to the 1950s or 1960s and were originally used in microelectronics. Silicon is the most popular substrate for bulk micromachining, although glass, quartz, and other materials have been investigated. The following discussion will be limited to isotropic and anisotropic etching of silicon.

3.4.3.1.1 Isotropic Etching

Wet isotropic etching of silicon is usually a diffusion-limited process and is performed using acidic solutions. An example of silicon profiles resulting from isotropic wet etching with a silicon dioxide mask is shown in Figure 3.39. Agitation is required to achieve the ideal isotropic shape. The dry isotropic etching processes, which were described in Section 3.4.1.2, produce

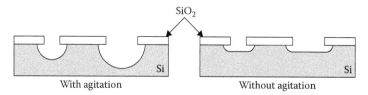

FIGURE 3.39
Isotropic wet etching of silicon and the importance of solution agitation.

similar etch profiles. For example, plasma etching is purely isotropic due to the chemical nature of the etching process. However, reactive ion etching incorporates both physical and chemical processes. The lateral etch rate, and thus the undercut of the mask, is less compared to plasma etching, and the profiles can be anisotropic.

The most common system for wet isotropic etching is called HNA and is so named for the three chemicals in the etch mixture: hydrofluoric acid (HF), nitric acid (HNO_3), and either acetic acid (CH_3COOH) or water. Hydrofluoric and nitric acids are active etchants, whereas acetic acid and water mainly serve as diluents. Acetic acid is typically preferred over water because it preserves the oxidizing power of nitric acid by preventing its dissociation. The overall reaction for HNA etching is as follows:

$$Si + HNO_3 + 6HF \rightarrow H_2SiF_6 + HNO_3 + H_2 + H_2O \qquad (3.44)$$

Nitric acid oxidizes silicon to form a layer of silicon dioxide. Then, the oxidized products are dissolved away by hydrofluoric acid. There are a number of mixture compositions; the rate-limiting factor is typically the lowest concentration chemical. Isoetch curves for the HNA system are shown in Figure 3.40. The isotropic nature of this etching system has made it popular for polishing applications.

3.4.3.1.2 Anisotropic Etching

In general, anisotropic wet etching is carried out at elevated temperatures and is slower than isotropic etching. Anisotropic etching processes are

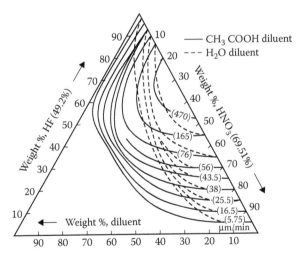

FIGURE 3.40
Isoetch curves for silicon using the HNA system. (From Wolf, S., and R. N. Tauber. 1986. *Silicon Processing for the VLSI Era*. Sunset Beach, CA: Lattice Press. With permission.)

usually reaction limited. The most common system used is a mixture of potassium hydroxide (KOH), isopropyl alcohol, and water (with KOH at ~50% by weight). This is an orientation-dependent etch and will remove some planes faster than others. For example, the (111) plane is more densely packed than the (100) plane, so it logically follows that the (100) plane will etch faster than the (111) plane. However, the etch rates do not always follow this logic, and the relationship may be reversed with respect to the packing density for certain etch chemistries.

If a (100) wafer is used, the resulting shape of an etched trench will have (111) plane sidewalls at an angle of 54.74° from the (100) surface. A V-shaped trench results for short etch durations or small mask openings. A flat-bottom trench with angled sidewalls will result for longer etch durations and sufficiently large mask openings. The resulting etch geometries are shown in Figure 3.41. In both cases, (111) planes, also the slowest etching planes, form the sidewalls. Silicon dioxide and silicon nitride are common masks used for etching silicon using KOH. The typical etching rates for (100) silicon and these masking materials are displayed in Table 3.11.

In some cases, other planes can also appear with intermediate etch rates. This is especially true when etching mesas or structures of convex corners (e.g., {331} planes may be exposed). The corner faceting can be reduced by using corner compensation structures to retard etching at convex corners.

If a (110) wafer is used, the resulting shape of an etched trench will have vertical sidewalls aligned with (111) planes; (110) and (111) planes are perpendicular (Figure 3.41). In all cases, careful alignment of the mask to the appropriate crystal plane is necessary to achieve a precise dimensional control. Other commonly used wet anisotropic etching systems include

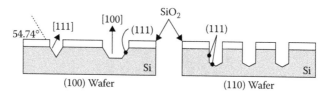

FIGURE 3.41
Anisotropic wet etching of (100) and (110) silicon wafers showing orientation.

TABLE 3.11

Typical Etch Rates for Silicon and Common Masking Materials in KOH (1 KOH: 2 H_2O by Weight, 80°C) and Calculated Selectivity Ratio to Silicon

Material	Etch Rate (Å/min)	Selectivity
<100> Silicon	14,000	1:1
Silicon dioxide	41–77	~102:1
Silicon nitride	0	>104:1

Source: Williams, K. R., and R. S. Muller. 1996. *J Microelectromech Syst* 5:256.

ethylenediamine $(NH_2(CH_2)_2NH_2)$/pyrocatechol $(C_6H_4(OH)_2)$ + water (EDP) and tetramethylammonium hydroxide (TMAH). EDP offers high etch selectivities for masking films such as SiO_2 (which etches faster in KOH), silicon nitride, and some metals. Also, the etch rate for boron-doped regions in EDP is extremely slow, allowing these regions to serve as "etch stops." However, EDP is toxic and is typically not used unless absolutely necessary. Compared to EDP, KOH offers greater selectivity of {110} over {111} planes but etches silicon dioxide quickly. Silicon nitride is a better masking material for long etches. Also, the etch stop performance of KOH is inferior. TMAH is used when aluminum is present but can leave hillocks behind on the etched surfaces. Both KOH and EDP attack aluminum.

In anisotropic etching systems, the etch rate, R, is temperature dependent and obeys the Arrhenius law

$$R = R_0 \exp(-E_a/kT) \qquad (3.45)$$

where R_0 is the pre-exponential factor, E_a is the activation energy, k is the Boltzmann constant $(86.1 \times 10^{-6}$ eV/K), and T is the temperature.

Note that in bulk micromachined structures created using anisotropic wet etching, the dimensions of a structure on the opposite side of the substrate are enlarged by a distance of $\sqrt{2}t$ where t is the thickness of the substrate. For example, if features are patterned onto a 400-µm-thick silicon substrate, the final feature size is over 560 µm larger than the surface feature.

Anisotropic dry etching occurs only when the lateral etch rates are small or inhibited by the presence of a passivation layer. Sputtering, RIE, RIBE, and DRIE may be used to achieve vertical sidewalls.

3.4.3.1.3 Resistivity-Dependent Etching

We can use the etch rate variation resulting from doping silicon to automatically stop an anisotropic wet etch process. Doped silicon, either p- or n-type, generally etches slowly in alkaline etchants such as EDP and KOH. This etch behavior is useful in creating very thin membranes of precise thickness, and it is an attractive alternative to the careful endpoint detection otherwise used in creating thin silicon membranes in undoped substrates.

A variation in etch rate is also encountered for acidic etchants such as HNA: doped silicon etches much faster than pure silicon. For example, if a 1:3:8 volume ratio of $HF:HNO_3:CH_3COOH$ is used at room temperature, the etch rate of heavily doped regions is about 50–200 µm/h $(>5 \times 10^{18}$ cm$^{-3})$. A selectivity of about 150 is possible over lightly doped regions $(<10^{17}$ cm$^{-3})$ [12].

3.4.3.1.4 Electrochemical Etch Stops

We can also create etch stops using electrochemical techniques and can avoid heavily doped regions, which may result in stress in the silicon membrane. Either side of a p-n junction can be etched; the side to be etched is connected to an anode. A 0.5–0.6 V bias is applied between the junction and reference cathode (usually a platinum electrode) immersed in a strong alkaline solution (e.g., EDP or KOH). Etching proceeds until the exposed side is etched away. Etching is thought to stop due to the SiO_2 passivation layer formed at the junction by anodic oxidation. The etching setup is shown in Figure 3.42.

3.4.3.2 Surface Micromachining

In contrast to bulk micromachining, *surface micromachining* relies on the selective addition and subtraction of layers to build structures on top of a substrate. Additive layers are called *structural layers*, whereas temporary layers that support these structural layers are removed to form the final structure are called *sacrificial layers*. An example of a common structural layer is polysilicon. Low temperature oxide (LTO) and phosphosilicate glass (PSG) are often used as sacrificial layers. Photoresist and polyimide may also be used as sacrificial layers because they are easily removed by oxygen plasma; however, these polymers limit the temperature of subsequent processes.

Although the term was coined more recently, the first example of surface micromachining dates back to 1952 with the invention of a method for creating electrostatic shutters [13]. This was followed by the resonant gate transistor invented by Nathanson in 1965 [14]. The first commercial surface micromachined device did not appear on the market until 1991, when Analog Devices introduced the ADXL-50, a 50-gravity accelerometer for airbag deployment, to the automobile industry.

A classic example of a surface micromachining process involves polysilicon as a structural layer, LPCVD PSG as a sacrificial layer, and silicon as a starting substrate. The simple process flow depicted in Figure 3.43 describes the fabrication of simple mechanical structures, such as a cantilever or a bridge, using this technique. For example, oxide is first deposited, which serves as a sacrificial layer. It is selectively patterned and then a layer of polysilicon is

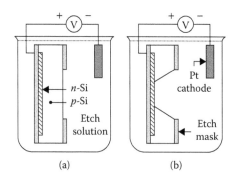

(a)   (b)

FIGURE 3.42
Electrochemical etch-stop setup in which n-type silicon serves as (a) the etch stop and (b) the membrane formed after etching. (after [9]).

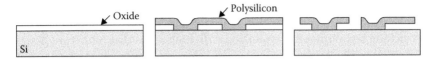

FIGURE 3.43
Example of a surface micromachining process in which oxide is the sacrificial material and polysilicon is the structural material.

deposited. By removing the oxide, free-standing polysilicon structures are formed. Many materials have been added to the MEMS toolbox; thus, today many choices of structural and sacrificial layers exist.

Due to the thin film nature of additive processes, surface micromachined structures are smaller than bulk micromachined devices that are fabricated by etching much thicker substrates. However, the reduced size affords increased feature density. Furthermore, surface micromachining is not dependent on crystallography, which places size restrictions on features fabricated by wet bulk micromachining. A variety of deposition methods exist for layer addition; for example, CVD techniques yield 2- to 5-μm-thick films. These methods also enable the deposition of layers with a range of compositions and consequently, an assortment of mechanical properties. However, thin film structures may be fragile and require specialized packaging following fabrication to protect devices from environmental factors.

The layer-by-layer process allows complex multilayer structures to be fabricated with relative ease but at the cost of multiple processing steps. However, this may simplify the overall process or impart other advantages. For example, sealed cavities can be formed that are much smaller than those in bulk micromachining, which usually requires the joining of two substrates. Hermetically sealed cavities are required in devices such as pressure sensors. A cavity formed following removal of sacrificial layers is simply sealed by plugging the small opening that is used for the release process with another layer of material. Examples of processes used to create sealed cavities are shown in Figure 3.44. Methods include (1) growth of silicon dioxide on the interior of a polysilicon cavity, (2) deposition of a conformal seal over the cavity opening, and (3) plugging of the cavity opening using a line-of-sight deposition.

The selection of sacrificial layers involves careful consideration of many factors. Typically, wet etching is used to remove these layers

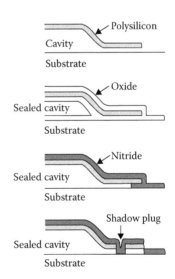

FIGURE 3.44
Examples of methods for sealed cavity or channel formation by surface micromachining.

although dry etching processes may also be used. The etch chemistry and technique are selected with a preference for high selectivity and high etch rate. Wet etching techniques are associated with high etch rates, which are desirable because openings to remove the sacrificial layers can be quite small resulting in long etch times. However, a phenomenon called *stiction* may prevent the use of wet etching in some applications; this is discussed in detail in Section 3.6.1.1. Selectivity must take into account all materials present so that the etch process does not accidentally attack or remove important features.

The mechanical properties of the constituent layers play an important role in the overall success of a surface micromachining process and the resultant device. Unfortunately, the mechanical properties of thin films may differ from those of the bulk material; yet, the thin film mechanical properties are often inadequately characterized. The properties of these films depend in large part on the method by which they are deposited. Adhesion, interfacial stresses, residual stresses, and the mechanical strength of layers are examples of properties that are in some part determined by the deposition process.

3.4.4 LIGA

Due to the limited aspect ratios of structures produced using traditional silicon micromachining (a function of substrate thickness and thin films), a method to create high aspect ratio metal or polymer structures was devised. LIGA is a German acronym for Lithographie Galvanoformung Abformung, which literally translates as lithography, electroforming, and molding. Although the process was originally developed in Germany at the Karlsruhe Nuclear Research Center for creating microfluidic nozzles for uranium enrichment, LIGA is now synonymous with processes for creating high aspect ratio structures through a combination of lithography and molding. Traditional LIGA is performed with X-ray photoresist, which is patternable in thicknesses from micrometers to centimeters. If this process is combined with metallization steps, thick metal structures can be created that are otherwise not possible with surface or bulk micromachining techniques.

LIGA involves three major steps: (1) formation of a pattern, typically with deep X-ray lithography; (2) electroplating of a metal on the patterns to create a master mold; and (3) creating a plastic molding from the metal master (Figure 3.45). We can skip the final step and end with the metal structure as the final part. Another alternative is to use the molded plastic parts as the final product or as templates for more metal or ceramic replicas.

High aspect ratios of more than 100:1 are possible due to the higher penetration of the shorter X-ray wavelengths into photoresist. This is achieved with a significant start-up expense; a synchrotron radiation source, which

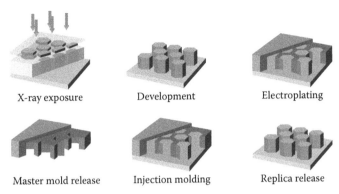

X-ray exposure   Development   Electroplating

Master mold release   Injection molding   Replica release

FIGURE 3.45
Example of a typical LIGA process.

costs approximately $30 million [7], is required. Due to the expense of these systems, only a few facilities exist in the world. The photoresist is also specially formulated (e.g., based on polymethylmethacrylate [PMMA]). X-ray photomasks, as one would expect, require regions that are transparent and opaque to X-rays. Masks may be constructed of exotic materials such as thin membrane beryllium or titanium with X-ray absorber regions of gold, tungsten, tantalum, or alloys. In the electroplating step, nickel is most commonly used; copper, gold, and titanium are also possible candidates. Overall, the traditional LIGA process is very expensive due to these special material requirements.

Some applications may require a free standing plastic part. In traditional LIGA, the plastic part is not obtained until the final step. To shorten the process, a sacrificial layer can be introduced between the PMMA resist and the substrate. Following X-ray lithography, the PMMA structure is released by simply removing the sacrificial layer. This modified process is called sacrificial LIGA (SLIGA).

With the availability of advanced photoresist formulations, it is possible to perform a variation of LIGA using conventional UV lithography equipment. In this so-called poor man's LIGA, the PMMA resist is replaced with conventional UV-sensitive thick polyimide or SU-8 resists. While high aspect ratios can be formed with this method, in general, the structures produced have lower resolution and less precision.

3.4.5 Micromolding and Imprint Lithography

Most microfabrication methods discussed so far are geared toward traditional MEMS materials such as silicon and oxide. Micromolding and imprint lithography are a family of fabrication techniques primarily focused on polymers.

3.4.5.1 Casting and Soft Lithography

Molding of polymers is achieved by a number of different processes. Both casting and soft lithography rely on molding to produce the final part. While casting may be performed with nonpolymer materials, soft lithography is limited to polymer microfabrication. These techniques are well-known for rapid prototyping of microfluidic systems.

The most frequently used material is silicone rubber or polydimethylsiloxane (PDMS). Casting has also been performed with polystyrene and other materials. These materials are distinct from those used to produce the master mold. The master contains a relief structure, typically with features of identical height, onto which a liquid prepolymer mixture is poured. Long-lasting masters are fabricated from etched silicon surfaces; however, masters may be rapidly produced using photoresists or other photosensitive polymers. Speed is gained at the cost of mold longevity; these polymer molds may be reused, but fewer castings are possible compared to silicon masters. The liquid prepolymer becomes a negative replica of the master and is separated from the master. A typical process for producing a microchannel structure by casting is summarized in the following steps (Figure 3.46):

1. Mix PDMS elastomer and a curing agent (10:1).
2. Pour the prepolymer mixture over the mold. The mold may be pretreated with a release agent to facilitate the removal of the replicated part.
3. Degas the prepolymer in a low vacuum to remove air bubbles trapped during the mixing process.
4. Cure to cross-link the polymer.
5. Peel the cured PDMS off of the mold.
6. Punch the liquid-access holes.
7. Bond the molded PDMS slab to a substrate to cap the microchannels.

This process is easily repeated to produce complex multilayer structures.

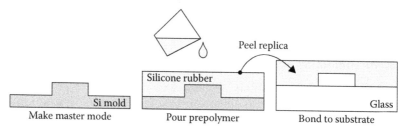

FIGURE 3.46
Process for creating a simple microchannel by replica molding with PDMS.

The negative replica produced contains only part of the desired microchannel, necessitating an additional step to close the microchannel. This is accomplished by simply bonding the molded PDMS slab to another surface through either an irreversible or reversible process. One method for the irreversible bonding of PDMS to surfaces is achieved by oxidizing the PDMS surface to create silanol groups. These groups react to form permanent bonds with the mating material. Methods to oxidize the PDMS include oxygen plasma, RIE, and ozone plasma; regardless of the method, special tools are required. Irreversible seals are formed with a variety of materials including PDMS, glass, silicon, silicon nitride, and polyethylene.

Irreversible bonding between two PDMS surfaces is achieved using an alternate method that does not require any additional equipment. This strategy adjusts the elastomer to curing agent ratio (usually 10:1) in mating PDMS pieces such that one layer has an excess of curing agent (e.g., 3:1) and the other elastomer (e.g., 30:1). Thus, when two layers are joined and cured a second time, excess components react at the interface to result in cross-linking and formation of a homogenous slab of PDMS. More modest changes in the elastomer-to-curing-agent ratio can also be used to modify PDMS mechanical stiffness. For example, if more curing agent is present (e.g., 9:1), the PDMS will be stiffer. Even without surface treatments or changing the mix ratio, PDMS adheres well to other smooth surfaces and forms reversible bonds. The strength of these bonds is sufficient to support pressurized channel structures (up to 1 bar) [15].

The discussion thus far has focused only on casting. Soft lithography [16] includes casting (also known as replica molding) but also a number of other fabrication technologies. Molded replicas may in turn be used to stamp structures in a process known as microcontact printing (µCP). Micromolding in capillaries (MIMIC) uses capillary action to wick fluids into micromolded capillaries placed on a substrate. Once these fluids are cured, the mold is removed. Many other variations of these processes exist and fall under soft lithography.

3.4.5.2 Injection Molding

Injection molding is a common conventional machining technique used to rapidly replicate plastic parts. For example, CDs and DVDs are formed using injection molding. Many electronic packages are made using this technique. Unlike casting and soft lithography, which are essentially room-temperature processes, injection molding requires heat to bring the raw plastic material above its glass transition temperature. The molten plastic is then injected into the mold form and hardens when cooled (Figure 3.47). This cooling process is rapid and may be faster than the curing step used in casting. The completed part is then removed. Many polymers, including both thermosets and thermoplastics, can be molded in this manner. For example, it is common to use electroplated nickel molds created from silicon masters for the injection molding of acrylic parts. Polystyrene and polycarbonate are also frequently used injection molding polymers. Like casting, rapid reproduction of parts is achieved once a mold is obtained. Molds created through the LIGA process or by DRIE of silicon can be used to injection mold tiny parts.

3.4.5.3 Compression Molding

Compression molding is also referred to as *relief printing* or *hot embossing*. As the names suggest, a polymer plate and mold are brought into contact under compression at elevated temperatures to transfer relief structures on the master mold into the polymer (Figure 3.48). This thermal imprinting process is usually performed under vacuum to remove any air bubbles and water vapor. For example, a PMMA substrate may be imprinted with a silicon master under a 4000-lb force. Nickel molds and polycarbonate substrates are also common.

3.4.6 Other Micromachining Techniques

3.4.6.1 Microstereolithography

Conventional stereolithography is often used for rapid prototyping. This process involves spot-by-spot polymerization of a liquid resin by a laser into a three-dimensional shape that is defined in a CAD software package. The CAD file dissects the three-dimensional shape into a stack of two-dimensional

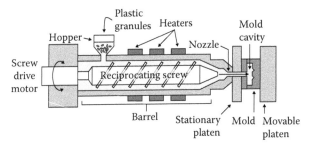

FIGURE 3.47
Injection molding system.

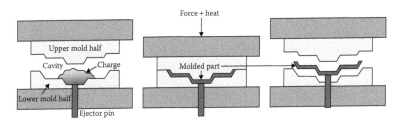

FIGURE 3.48
Compression molding.

cross sections. Each cross section is fed to the computer-controlled laser such that the structure is constructed in a layer-by-layer manner. Since the laser can be focused into a μm-scale spot, very small structures can be produced by microstereolithography. The overall process is summarized in Figure 3.49.

3.4.6.2 Probe-Based Lithographic Techniques

If one had a tiny pen or stylus, it would be possible to directly modify a surface without being limited by wavelength. Many lithographic techniques are based on this simple concept and rely on the widespread availability of atomic force microscopes (AFMs) and the probes they use to interact with surfaces (Figure 3.50). At the atomic level, probes can be used to manipulate and position atoms directly. However, more practical methods of probe-based lithography include engraving and direct writing of photoresist or other materials using the probe as a pen (as in dip-pen lithography). The dip-pen lithography technique allows many materials, including biological material, to be printed onto substrates as single or even multilayer structures. A probe is dipped into "ink" containing the molecules (e.g., conductive polymers, gold, DNA, antibodies, or alkanethiols) of interest and dragged across a surface. Single probes obviously necessitate long processing times

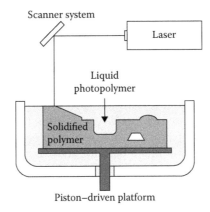

FIGURE 3.49
Stereolithography process and apparatus.

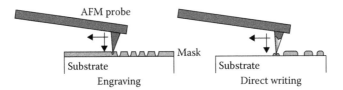

FIGURE 3.50
Lithographic techniques enabled by atomic force microscope probes.

due to the serial nature of the lithographic technique. Arrayed probes promise much higher throughput.

3.4.6.3 Chemical Mechanical Polishing

The buildup of multiple layers of processing often results in a highly nonplanar substrate surface. The surface texture may negatively impact subsequent lithography or other process steps, and thus planarization is required. A typical planarization technique might include deposition of a thick resist over the surface topology followed by etch-back. During the etch-back process, the resist and the underlying material are etched at equal rates.

Planarization may also occur through CMP. Many materials may be planarized with this technique. The substrate is securely mounted and brought into contact with a polishing pad; these opposing surfaces are counter-rotated. A slurry, consisting of an abrasive and chemical components, is introduced between the wafer and pad. The combination of these elements results in removal of the wafer surface by both chemical and physical forces.

3.5 Wafer Bonding, Assembly, and Packaging

Most devices are not ready to use immediately after fabrication and may require several additional steps, including assembly and packaging. In fact, packaging accounts for the majority of the total cost of a consumer MEMS device (anywhere between 20 and 95% of the total cost [17]). Packaging may also significantly increase the overall device size.

Packaging, in particular, presents many interesting challenges for MEMS devices. In integrated circuits, packaging serves the primary purpose of protecting chips from the environment and also provides mechanical support, electrical connections, and a path for heat dissipation. In MEMS, packaging serves many of the same purposes but must also allow the interaction between the chip and the environment (e.g., fluidic interconnections) or integration of modalities (e.g., optical, mechanical, thermal, or chemical ones) beyond the electrical ones present in integrated circuits. However, despite the decades-old history of MEMS, MEMS packaging has not matured and industry standards are still largely absent. MEMS packaging is largely application-specific, and many devices require custom solutions. All these reasons contribute to a significant cost attributed to the packaging.

The two most popular encapsulation materials to protect microelectronic devices are ceramics and plastics. Ceramics offer great performance benefits (hermetic sealing, high resistance to environmental factors, and durability) but are expensive. Plastics are inexpensive and used in most applications where some vulnerability to moisture is tolerated. The temperature at which plastic packages are used is also important because high temperatures induce degradation. Many common failure modes of microelectronic packaging (i.e., cracking, delamination, and warping) stem from adverse thermomechanical forces due to a mismatch of the coefficients of thermal expansion (CTE), fatigue fracture from thermal cycling, intrinsic stresses and strains stemming from fabrication processes, and degradation from environmental exposure [17]. These same issues exist in MEMS packaging.

Strategies for packaging may take place at the die- or wafer-level. Individual devices, or *dies*, are obtained through a process called *dicing*. Die-level packaging requires handling and assembly of individual dies into packages. This method is preferred for devices containing delicate components. Packaging may also be performed simultaneously over an entire wafer in a batch process.

3.5.1 Dicing

One of the key advantages of microfabrication is the ability to batch fabricate devices. Thus, an entire substrate is seldom used to create a single device. Instead, many individual devices known as dies exist on the substrate and need to be separated. *Dicing* is performed by mounting the wafer on a pressure-sensitive adhesive film in a specialized saw. The thin saw blade, usually diamond/resin or diamond/nickel, is passed through the substrate but does not cut all the way through to maintain the integrity of the wafer. The score line left by the saw allows each die to be easily separated from the wafer after dicing. High cutting speeds are used (~30–40 kprm) and the blade follows a predetermined cutting path pattern (Figure 3.51). For sufficiently small (<.5 mm on a side) or nonrectangular dies, dicing is impractical for separating individual parts. Instead, DRIE may be used if the substrate is silicon. Conventional machining methods may be used for other substrates.

3.5.2 Wafer Bonding Methods

Bonding processes are frequently used for assembly and packaging of devices. For instance, a complete device may require multiple pieces that are

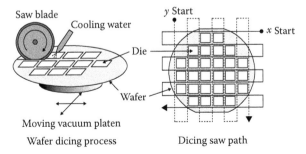

FIGURE 3.51
Dicing process to separate individual dies.

joined together at the wafer level in a batch process or at the die level. First, nonadhesive–based technologies for joining are discussed followed by a brief discussion of adhesive-based techniques in Section 3.5.2.5. The key differences between them are the temperature ranges used (low for adhesive) and the ability to form hermetic bonds (which is not possible by adhesive bonding). Table 3.12 summarizes and compares the key features of common bonding methods.

3.5.2.1 Fusion Bonding

Silicon substrates are directly joined together without the use of intermediate layers by *fusion bonding*. Substrates must be very flat (<10 Å surface roughness and <5 µm bow over a 4" wafer [18]) and clean in order to form a robust bond. Bonding takes place at the interface between two joined silicon surfaces. First, the surfaces must be prepared by either a hydrophilic or hydration process. Clean silicon surfaces are rendered hydrophilic by boiling in nitric acid. Once surfaces are joined at room temperature, van der Waals forces bond the surfaces together. Stronger bonding is achieved by elevating the temperature (1100–1400°C). The bonded pieces exhibit no visible interface and form a hermetic seal.

The hydration process starts with oxidized wafers. The surfaces are modified through a soak in a mixture of hydrogen peroxide (H_2O_2) and sulfuric acid (H_2SO_4) to form OH groups. When substrates are brought into contact, bonding is achieved through chemical forces in which Si-O-Si bonds develop through heat dehydration. For both processes, high temperatures and flat substrates are required.

TABLE 3.12

Comparison of the Various Bonding Methods

Bonding Technique		Materials	Temperature	Comments
Direct	Anodic	Glass to Si	180–900°C	Hermetic, surface roughness <500 nm
	Fusion	Si to Si	>800°C	Hermetic, surface roughness <50 nm
Metal interlayer	Eutectic	Au to Si (or other metals to Si)	363°C	Hermetic, smooth surface not required
	Solder	AuSn, PbSn, InSn, or AlSi to Si and glass	>100°C	Hermetic, smooth surface not required
Insulating interlayer	Glass frit	Glass frit	<600°C	Hermetic, smooth surface not required
	Adhesive	SU-8, polyimide, benzocyclobutene, and so on	<300°C	Not hermetic, smooth surface not required

3.5.2.2 Anodic Bonding

The high temperatures of fusion bonding prohibit its use with lower-temperature materials. Instead, *anodic bonding*, also called *electrostatic bonding* or *field-assisted thermal bonding*, is used. This process was discovered in 1969 [19] and results in hermetic seals. In addition, minimal equipment is required making this a low-cost bonding method. Surfaces must be clean and smooth although greater roughness is tolerated compared to fusion bonding (<50 nm [18]).

Anodic bonding joins conductive silicon substrates to sodium-rich glass or quartz substrates. The bonding setup is shown in Figure 3.52. Silicon and glass substrates are pressed together, brought to elevated temperatures (450–900°C), and subjected to a DC bias (400–1000 V) with the glass side connected to the anode. The elevated temperature mobilizes the sodium ions present in the glass and the electric field attracts these ions to the negative cathode. The combination of these two effects depletes sodium ions within 1 μm of the interface and results in the formation of a large electrostatic pressure. The strong electric field allows bonding of surfaces that are not completely planar. This facilitates bond formation, although the exact bonding mechanism is not yet fully understood; it is hypothesized that covalent silicon oxygen bonds are formed at the interface.

Although bonding temperatures are lower than that of fusion bonding, the joining of two dissimilar materials necessitates matching of CTEs. Otherwise, significant thermal stress can develop at the interface that may result in warping or cracks. Pyrex 7740 is a glass formulated with a CTE closely matched to that of silicon. Thin layers of glass that are sputtered or evaporated onto silicon may also be joined by anodic bonding.

3.5.2.3 Low-Temperature Glass Bonding

The presence of high electric fields in anodic bonding may result in undesirable damage to the bonded materials. Instead, it is possible to join interfaces using thin LTO glass with pressure and heat but without high fields.

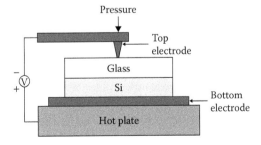

FIGURE 3.52
Anodic bonding setup.

Other types of glasses can also be used in similar processes, including PSG, borosilicate, spin-on glass (a methylsilsesquioxane polymer that flows at 150–210°C to fill grooves), and frits [9]. In general, the use of thin films is convenient; however, greater planarity is required compared to anodic bonding. Also, the bonds that are formed are also weaker. Frits can be used for nonplanar surfaces as the material is able to flow and fill in gaps created by roughness. However, this is achieved at the cost of elevated temperatures; the temperature must be raised above the glass melting temperature for bonding (<600°C).

3.5.2.4 Eutectic Bonding

Specific combinations of materials exhibit eutectic behavior such that when they are brought together at elevated temperatures, bonding occurs by diffusion between the two materials through the formation of an alloy. Stable hermetic bonds can be formed in this manner. To understand the mechanism of bonding, let us consider an example. If gold and silicon are brought together, the combination melts at 363°C, which is much lower than the melting point of pure gold (1064°C) and silicon (1410°C). This point is known as *eutectic temperature* and occurs for a specific mixture of gold and silicon (97.1% gold and 2.85% silicon [9]). No other combination of two materials or either of the pure forms possesses a lower fusion temperature. So, by heating gold and silicon to the eutectic temperature, diffusion of the elements occurs and causes a eutectic alloy to form. Gold and tin also exhibit this property (80% gold and 20% tin at 280°C with 1 MPa pressure [20]) and can form eutectic bonds. The combination of 60% tin and 40% lead has an even lower eutectic temperature of 183°C.

3.5.2.5 Adhesive Bonding

The simplest bonding method for joining parts is bonding by adhesives. Although adhesives are inexpensive, they cannot achieve hermetic seals due to their polymer nature. Bonds are typically not stable compared to other forms of bonding (e.g., with glass, metal, or silicon), and high yields may be difficult. Polymers may also age with time and exposure to environmental elements. Adhesion promoters are often required to chemically treat surfaces to prepare them for bonding.

Adhesive bonding is commonly used to attach individual dies to supports. Common adhesives are epoxy resins, silicone rubbers, and other organic materials. Epoxy resins exhibit good flexibility and form adequate seals following appropriate surface preparation. However, they are vulnerable to environment and temperature. Some formulations such as benzocyclobutene (BCB) and polyimides cure upon exposure to UV light. Both are photopatternable, allowing selective patterning of adhesive regions. Silicone rubbers have greater flexibility but also have low chemical resistance and strength. Adhesives are available in a number of formulations that exhibit different and sometimes specialized properties. For example, thermally conductive, electrically conductive, and optically transparent adhesives exist. For medical applications, adhesives suitable for skin contact and even chronic implantation are available.

3.5.3 Hermeticity and Sealing

The quality of the seal protecting a device determines its *hermeticity* or ability to protect the device from moisture and contamination. Moisture, in particular, must be minimized to reduce corrosion and unwanted electrolytic conduction. Regardless of the sealing method, all packages will leak to some degree. However, if the package provides a sufficiently small leak rate to maintain the desired conditions, then it is said to be hermetic.

Sealing by adhesives is the simplest method to protect devices; however, adhesives are vulnerable to moisture penetration, and thus the sealing is not hermetic. In fact, all polymer sealing methods (e.g., adhesives, o-rings, and gaskets) are permeable to moisture. Thus, these organic compounds are short-lived and inadequate for use under extreme conditions such as high temperatures and immersion in corrosive media. The gas permeability (P_{gas}) of a device package can be defined as follows (in $g/(cm \cdot s \cdot torr)$).

$$P_{gas} = \frac{Q_{gas} t_{barrier}}{A_{barrier} P} \tag{3.46}$$

where Q_{gas} is the gas flow rate, $t_{barrier}$ is the thickness of the barrier material, $A_{barrier}$ is the barrier area, and P is the pressure difference [21]. The moisture permeability of different sealants is shown in Figure 3.53. From the figure, it is clear that glasses and metals yield more hermetic packages than polymers when used alone.

Hermetic housings for conventional medical implants are made from titanium, tantalum, niobium, or stainless steel. However, it is not always possible to use metallic packages for MEMS devices. Alternatives that have been explored include micromachined ceramic packages [23], glass-sealed packages [24], and polymer encapsulation [25]. For glass-sealed packages formed by anodic bonding, a mean time to failure of 177 years at 37°C was obtained from accelerated lifetime saline soak tests [24]. In this same device, electronic feedthroughs for attaching to electrodes that are external to the package were achieved using glass-to-metal seals with sodium-free borosilicate glass.

Packages may experience high cavity pressures resulting from bonding processes. Metallic getters can be added into the package to lower the pressure. Common getter materials include pure metals and alloys of barium [Ba], aluminum, titanium, zirconium [Zr], vanadium [V], and iron [Fe] [21].

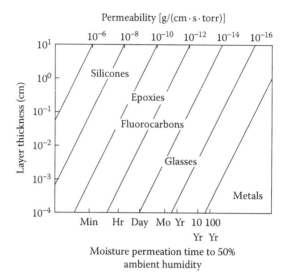

FIGURE 3.53
Calculated time for moisture permeation through various sealant materials to 50% of the exterior humidity. Calculations are performed for one particular geometry. (From Traeger, R. K. 1976. *IEEE City* 361. With permission.)

These reduce the moisture and pressure within packages by combining to form oxides and hydrides and through surface adsorption.

In addition to preventing moisture penetration, packages must also protect the mechanical features of microdevices. The moving structures are often fragile and may require sealing in a vacuum. This is done either by wafer bonding or by surface micromachining sealing techniques (Figure 3.44). For example, a cavity formed by a shell of material protecting a microstructure may be sealed by a chemical reaction occurring at the interfaces to be sealed. One form of this reactive sealing method is the growth of silicon dioxide in the shell-to-substrate gap. The shell is typically obtained by surface micromachining in which a sacrificial layer is first deposited on the die, followed by deposition of the shell material. The sacrificial layer is later removed and results in a shell-to-substrate gap. These processes occur at elevated temperatures, which must be factored into the overall process; the thermal properties of each material contribute to thermal stresses that must be managed.

3.5.4 Assembly

Bonding is only part of a number of processes involved in a device or system assembly. In research, assembly is performed manually on a few devices at a time. Consumer MEMS may be assembled using an automated approach with conventional robotics. For example, microassembly may include part feeding, part grasping, part mating, bonding, and fastening [17]. A common element in assembly is the alignment that occurs to ensure mating parts are properly registered. Novel self-assembly methods avoid manual or serial processes. For example, batch assembly may be performed by random movement of all parts in a bin that are selectively joined together through the presence of patterned bonding forces or geometrically matched receptacles.

3.5.5 Electrical Packaging

Devices that contain electrical components need to be connected to a chip package. In standard microelectronic packaging, wire bonding or flip-chip methods may be used. In wire bonding, bonding pads on the chip are connected to the internal lead of an electronic package with a thin wire. Popular wire materials include gold and aluminum. Gold is resistant to oxidation and corrosion; however, aluminum is less expensive and preferred in applications where its shortcomings can be tolerated. Other common wire materials are copper, silver, and palladium. To ensure high density connections, very thin wires ~20–80 µm in diameter are used.

To join the wires to bonding pad and lead connections, one of the following three methods may be used: (1) thermocompression, (2) ultrasonic, and (3) thermosonic. In thermocompression bonding, metal-to-metal bonds are formed by pressing at elevated temperatures (~400°C). A thin metal wire is fed through a capillary tool, and a ball is formed at the end by heating the wire with a spark or flame. This ball is bonded onto a pad by plastic deformation and atomic interdiffusion. Ultrasonic bonding may be used for delicate structures and applications requiring lower bonding temperatures. The presence of ultrasonic force provides enough energy for bonding without requiring higher temperatures. Usually, a wedge tool is used that vibrates (20–60 kHz) while compressing the wire onto the pad. Finally, thermosonic bonding combines features from thermocompression and ultrasonic bonding. Thermosonic bonding is performed at medium temperatures (100–150°C); it uses ultrasonic energy and forms ball joints with a capillary tool. Typical bonds formed by these processes are depicted in Figure 3.54.

Wire bonding forces the placement of bond pads at the periphery of a die. Also, long wires occupy space and add resistance. In flip-chip or bump bonding, balls or bumps are used to form connections and allow a more effective use of die real estate for the placement of electrical connections. Metallic balls are usually formed by electroplating and are arranged across the surface of a die. The assembly is flipped upside down and is brought into contact with mating pads in a package or a printed circuit board (Figure 3.55). The bump size is about 75 µm.

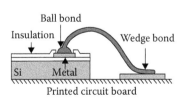

FIGURE 3.54
Ball and wedge wire bonds.

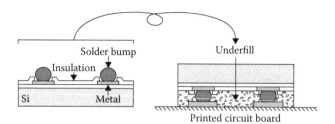

FIGURE 3.55
Flip-chip bonding.

3.5.6 Microfluidic Packaging

While electrical packaging and interconnects are standardized processes due to decades of development in the semiconductor industry, microfluidic packaging and interconnects are still a topic of much research. Reliable connects are needed to the channels within microfluidic devices to connect to macroscale inputs and outputs. The planar nature of microfabrication usually results in "out-of-plane" connections to microchannels. Access ports exist in either the back or top side of the die in which the microchannels exist. Fluid entering an interconnect must travel through a 90° bend to enter the microchannel. Also, a limited contact surface area exists in these connections forming a weak joint. This is not desirable, but few "in-plane" methods for connecting to microchannels exist [26,27].

Methods for interfacing to microfluidic devices are reviewed in references [27,28]. For compliant silicone rubber devices, the compression fitting of conventional microtubing into cored cylindrical access ports is commonly used [29, 30]. Devices constructed of other materials use a number of other methods for creating connections. The simplest method is adhesive bonding of conventional microtubing [31,32]. Tubing may also be attached by thermal bonding [31,33,34] or mechanical compression [32,35,36] methods. An alternative to direct bonding of tubes to access ports involves the use of microplumbing or connectors manufactured by microfabrication processes to match the geometry of fluidic access ports [35,37].

3.5.7 Medical Packaging Requirements

Microdevices intended for implantation in humans have additional packaging requirements. Primarily, the package must be biostable to survive the corrosive environment inside the human body and biocompatible in that it should not result in unwanted immunological or other adverse reactions. These objectives can be achieved by applying coatings that reside at the interface between the device and the body. These coatings include polymers, metals, and other materials. Silicone rubber [38,39], Parylene [40,41], and epoxy are commonly used polymers. Examples of metal packages are given in Section 3.5.3. Other interface materials include ultrananocrystalline diamond (UNCD) [42] and hydroxyapatite (HA)-titania (TiO_2). UNCD thin films, which are still being developed, may provide hermetic sealing properties. The HA-TiO_2 coating can be applied by micro-arc oxidation on titanium surfaces, which may improve cellular activity over titanium-only surfaces. These coatings are of interest in dental and orthopedic applications for the improvement of osseointegration of an implant with bones [43]. Other functional surface coatings are discussed in Section 3.6.

3.6 Surface Treatment

3.6.1 Stiction

When a sacrificial layer is removed by wet chemical etching, the structural layer may collapse and adhere to underlying layers (Figure 3.56). This phenomenon is called stiction (or static friction) and is an important cohesive force at the microscale. During sacrificial layer release, the adhesion of two surfaces is caused by capillary forces due to liquid surface tension. Stiction may also be caused by electrostatic forces, van der Waals forces, or hydrogen bonding. Regardless of the true nature of the cohesion, large forces are required to separate these layers.

Stiction is a significant failure mode in surface micromachining. Stiction can be avoided by avoiding the liquid phase altogether. Wet etching to release structural layers can be replaced with dry etching methods such as gas-phase or plasma etching. When wet etching cannot be avoided, the strategy to avoid stiction is to limit surface tension and capillary forces through freeze or supercritical drying. In freeze drying, liquids are frozen and then removed by sublimation in a vacuum chamber to avoid the liquid phase. Supercritical drying uses liquid carbon dioxide (CO_2) as the final rinsing solvent. To get to this point, water is replaced with methanol which is then replaced with liquid carbon dioxide. The liquid carbon dioxide is brought to critical-point pressure and temperature (7.39 MPa, 31°C) at which the surface tension is zero, and then into the supercritical region where the liquid is vaporized.

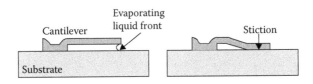

FIGURE 3.56
Stiction in a released cantilever beam.

Stiction also occurs during device operation. In-use stiction is avoided with appropriate low energy surface coatings to reduce electrostatic, capillary, and chemical bonding forces. Examples include fluorocarbon, hydrocarbon, or self-assembled monolayer (SAM) coatings that render a surface hydrophobic, which also prevents water condensation. The reduction of contact surface area by adding texture, such as small dimples, also prevents stiction (Figure 3.57).

3.6.2 Surface Modification

Fabricated surfaces typically lack the desired properties for the intended application, whether it be the prevention of nonspecific adsorption of proteins, prevention of immune response, or promotion of adhesion and integration with cells and tissue. Thus, surface modification is a necessary step prior to device use, and was alluded to in Section 3.5.7. This issue is revisited in specific examples throughout Chapters 5–9.

Bare silicon surfaces exhibit a native oxide layer when exposed to air or water. The resulting surface silanol groups acquire a negative charge when ionized in water and promote biofouling, an unwanted accumulation of biological matter on a wetted surface. A common example of biofouling is the attachment of barnacles and other marine organisms to a ship's hull. In microfluidic devices, biofouling influences device performance and lifetime. Surface modification through the application of thin film coatings can reduce biofouling.

Poly(ethylene glycol) (PEG) polymer coatings are effective in preventing protein and cellular adsorption [44]. Other polymers such as poly(hydroxyethyl methacrylate), poly(acrylamide), poly(N,N-dimethyl acrylamide), and dextran also exhibit this property, but to a lesser degree. Several PEG immobilization methods are available, including physical adsorption, covalent methods of attachment (e.g., grafting by photopolymerization or chemical surface attachment), vapor deposition, and gas-phase polymerization.

SAMs also prevent protein and cellular adsorption [45,46]. In general, surfaces to be treated with SAMs are immersed in a solution consisting of a surfactant and an organic solvent. Through a combination of adsorption and diffusion processes, a monolayer film spontaneously forms on the surfaces.

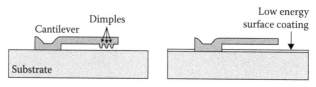

FIGURE 3.57
Stiction prevention by addition of dimples to a cantilever or a surface coating on the substrate.

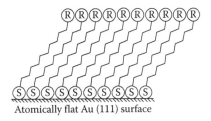

FIGURE 3.58
Alkanethiol self-assembled monolayer on a gold surface.

The films consist of a monolayer of well-ordered, regularly oriented, and closely packed molecules.

Many different types of SAMs exist, but perhaps the most studied are the alkanethiols ($R(CH_2)_nSH$, where R is a small functional group such as CH_3, CF_3, $CHCH_2$, CH_3OH, COOH), which are typically formed on gold surfaces. The density and regularity of the SAM highly depends on the crystallographic orientation of the surface to be modified. The best results are obtained using (111) gold deposited on atomically flat mica surfaces. SAMs consist of a head-and-tail group. For alkanethiols, the head group is the sulfur end and the tail group is the R part (Figure 3.58).

Many other SAMs have been explored, including thiols (RSH), disulfides (RSSR), sulfides (RSR), organosilanes ($RSiCl_3$, $RSi(OCH_3)_3$, $RSi(NH_2)_3$), alcohols (ROH), and amines (RNH_2). In addition to immersion processes, stamping and direct writing of SAMs have emerged as popular formation methods. Silicone rubber replicas made using casting techniques are patterned, immersed into SAM solutions, and held in contact with the surfaces to be modified. SAMs are also selectively assembled into the desired regions from inked AFM probes that are scanned across the surfaces. Common coatings used to entrap biomolecules and promote cell adhesion are discussed in Chapters 6–7.

In addition to the methods described, other common additive microfabrication processes, including spin and dip coating, evaporation, sputtering, and molecular beam epitaxy (with standard film thicknesses of 20 nm to 10 μm), may be used to selectively modify surfaces to achieve the desired properties. Other microfabrication techniques applied to achieve surface property control are described in reference [47].

3.7 Conversion Factors for Energy and Intensity Units

$1 \text{ mJ} = 1 \text{ mW} \cdot \text{s} = 10^4 \text{ ergs}$

$1 \text{ calorie} = 4.186 \text{ J}$

3.8 Problems

1. Design a simple experiment to determine the relationship between the spin speed and the thickness for a new positive photoresist (*i*-line). Assume that you have a profilometer available for you to measure the photoresist thickness. What other equipment would you need? How would you plan your experiment to obtain the desired data? Describe your experimental plan and methods.

2. You need to design a lithography process for a device that requires the use of a negative resist. You have an *i*-line UV source that provides 1000 W/m². From the data sheet for the photoresist, you find the graph in Figure 3.59. Based on this information, calculate the exposure time for (1) 25-, (2) 50-, and (3) 150-μm-thick photoresist films. Show your work.

3. The Rayleigh criterion describes the separation of two barely resolvable objects:

$$R = \frac{k_1 \lambda}{NA}$$

The Rayleigh DOF criterion tells us how much defocusing can be tolerated:

$$DOF = \pm \frac{k_2 \lambda}{(NA)^2}$$

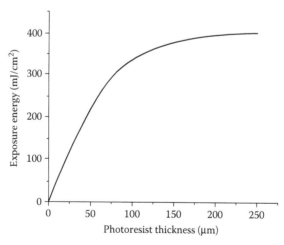

FIGURE 3.59
Graph from the data sheet for the photoresist.

Ideally, one would like the maximum DOF. Compare the above two equations and explain why high resolution and a large DOF are mutually exclusive.

4. Sketch a photomask design that will allow the letters "MICRO" to be etched into the surface of a silicon wafer using a negative resist etch mask. Sketch a photomask design that will result in a relief pattern of the letter "μ" to remain after etching with a positive-resist etch mask. For both photomask designs, clearly indicate which regions are transparent or opaque.

5. To align multiple photomasks used in the fabrication process to create a microdevice, registration, or alignment marks, are required. Create a simple set of alignment features for registration between two photomasks. Extend this strategy to a process that requires alignment of three photomasks. The marks you design should correct for x, y, and θ misalignment. Show the placement of the alignment marks on the photomask and briefly describe or indicate how the marks are intended to be used. Assume that the photomasks are intended for use in a contact mode lithography tool and a 100-mm (4 inch) silicon wafer.

6. Selectivity is an extremely important parameter that affects our ability to accurately transfer patterns in microfabrication processes. It is simply defined as a ratio of etch rates between (1) the material to be etched and (2) the masking material:

$$S = \frac{R_{substrate}}{R_{mask}}$$

Assume a selectivity of 0.8 and that the material you want to etch is 3000 Å thick. Using photoresist as a mask, what is the minimum photoresist thickness required to etch through this material?

7. Obtain a copy of the paper below. What materials are suitable as an etch mask for silicon during KOH etching based on etch selectivity?

Williams, K.R. and R.S. Muller, *Etch rates for micromachining processing.* Journal of Microelectromechanical Systems, 1996. 5(4): p. 256–269.

8. Suppose you want to build a hinge using an undoped polysilicon and surface micromachining techniques. Based on the data in table I from the study by Williams and Muller mentioned in Problem 7, which materials are suitable as sacrificial materials if you want to use 5:1 buffered hydrofluoric acid as the etchant? Choose at least three materials. Calculate the etch selectivity of the sacrificial materials versus the undoped polysilicon.

9. In designing both wet and dry etch processes, anisotropy needs to be considered. Let us say you want to etch a trench of width w into a film of thickness d. Another definition for the anisotropy for the film etch is

$$a = \frac{E_v}{E_h} = \text{anisotropy}$$

where v denotes vertical and h denotes horizontal.

(a) If the horizontal, or lateral, etch depth is δ, what is the expression for a? (b) Draw the cross section of the etched trench, showing both the masking material and the etched film. (c) If the original mask opening was w_m, what is the expression for w in terms of w_m and δ? This is the maximum width for the trench. (d) Solve this expression for δ and substitute it into your answer for part (a). This expression gives the anisotropy requirement.

10. Suppose you want to use KOH to etch a square through a hole in a (100) silicon wafer that measures no more than 200 μm on a side. Here, the 200-μm dimension refers to the smallest dimension measured in the x-y plane along the wafer surface. What mask dimensions should you use if the wafer measures (a) 300 μm or (b) 500 μm thick? What will the through hole dimensions be if you accidentally use the mask intended for the 300-μm thick wafer on the 500-μm thick wafer? Assume the mask is perfectly aligned and does not experience erosion or etching during the process. Show your work and draw images to accompany your numerical calculations.

11. You are preparing to etch (100) silicon wafers in KOH. To optimize the process, you consult a paper on anisotropic silicon etching. For safety reasons, you decide on a process temperature of 72°C. According to the information provided in Figure 3.60, what KOH concentration should be used to maximize the etch rate? You plan to use a silicon oxide etch mask. However, oxide has a nontrivial etch rate in KOH. What is the corresponding oxide etch rate at the selected KOH concentration based on the data given in the below figure? Show your work. These results will be used in the next problem.

Left: {100} silicon etch rate as a function of KOH concentration at a temperature of 72 °C. Right: Silicon dioxide etch rate for KOH and EDP type S as a function of temperature [48].

12. Suppose you want to use KOH to etch through holes in a 500-μm-thick (100) silicon wafer. What is the minimum thickness of the silicon dioxide etch mask that is needed to protect other regions of the wafer during KOH etching? If silicon nitride etches at 1.5 Å/min under the

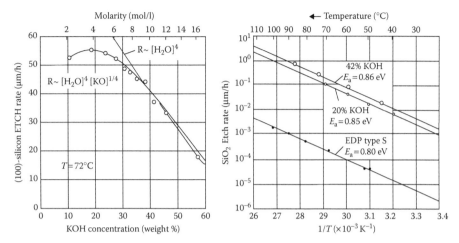

FIGURE 3.60
Graphs to determine KOH concentration.

same conditions, what is the minimum thickness the nitride layer would need to be in order to serve as a mask?

13. Draw the etch profile for the mask opening below for both isotropic (dry) and anisotropic (dry) silicon etches. Assume ideal conditions. Also, assume that etching is stopped as soon as the bottom of the silicon substrate is reached. Label any relevant features including planes, directions, and angles.

14. Of the following etch methods, indicate which one would be the most suitable for creating deep trenches with vertical sidewalls in a silicon (100) wafer and explain why. Etch methods: KOH wet etching, DRIE, XeF_2 gas-phase etching, sputtering.

15. What methods are available for the deposition of (1) metal and (2) silicon nitride films? List at least two for each material.

16. Suppose you need to deposit a uniform alloy thin film. Which additive process would you use and why?

17. Given the principles of evaporation and sputtering for deposition of thin films, which method would provide you with a more conformal coating? Which method is more suitable if the objective is to pattern the deposited film using a lift-off process? Justify your answers.

18. Devise a process flow with written steps and corresponding cross-section images for making a microchannel mixing device out of PDMS. Figure 3.61 shows the internal flow path through the microchannel; in other words, the device should contain a channel with the flow path drawn below. Also, draw each photomask necessary to produce the structure and indicate where each mask should be

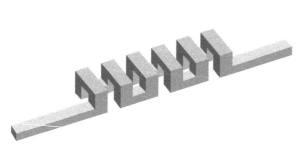

FIGURE 3.61
Internal flow path through the microchannel.

used in the process. Include any necessary alignment marks on the photomasks to achieve accurate registration between process layers. Briefly explain the function of each photomask.

19. Suppose you want to join two pieces of silicon together to form a microfluidic channel. One piece of silicon is etched and contains the channel structure. The second piece contains access holes and encloses the channel. What methods are available for you select from? What are the potential selection criteria for choosing one method over another?

20. Search the research literature and summarize at least two techniques used to avoid stiction. Provide the reference(s) using the IEEE format provided in Chapter 1.

3.9 Laboratory Exercises

Laboratory Exercise 1: Cleaving Silicon

Purpose: In this lab exercise, you will learn how to cleave silicon wafers into small dies by relying on the crystalline structure. After cleaving, a simple cleaning process will be used to remove debris and to prepare the dies for additional processing.

Materials and supplies
- Gloves
- Lint-free cleanroom cloths
- Single-side polished (100) silicon wafer
- Straight edge (glass slide, steel ruler, or similar rigid straight edge)
- Diamond-tipped scribe
- Tweezers
- Glass beakers

Chemicals
- Acetone
- Isopropyl alcohol
- Deionized water

Equipment
- Ultrasonic bath
- Chemical fume hood

Cleaving a wafer
1. Wear gloves and safety glasses. The wafer should be handled using gloves or tweezers.
2. Place the silicon wafer on a lint-free cleanroom cloth (shiny, polished side face down or up depending on your preferred cleaving method).
3. Align the straight edge, either perpendicular or parallel to the major flat, where you want to cleave the wafer.
4. Run the scribe tip along the straight edge while holding the scribe as vertically as possible to initiate a crack along the (100) crystal plane. Do not use excessive pressure.
5. Align a straight edge under the scribe mark. Align another straight edge above the scribe mark such that the silicon wafer is between the two aligned straight edges. Sandwich the wafer flat between the straight edges with gentle pressure. Then, press on the exposed wafer on the other side of the scribe mark to cleave the wafer into two pieces.
6. Alternatively, the wafer can be cleaved using the diamond scribe to press the scribe line near the edge of the wafer.
7. Yet another method is to hold the wafer at the edge using your hands with a grip on either side of the scribe line. A gentle pulling and snapping force will easily separate the two halves.
8. Repeat the scribe and cleaving process until the desired pieces are obtained.
9. Any sharp debris should be disposed of properly according to local safety guidelines.

Cleaning the cleaved dies
1. This process should be carried out in a chemical fume hood. To clean the cleaved pieces, a wet cleaning process with ultrasonic agitation will be used.
2. First, perform a simple rinse in filtered deionized water to remove large debris.
3. Place the dies to be cleaned in a beaker containing acetone.
4. Place the beaker in an ultrasonic bath for a few minutes.

5. Transfer the dies using chemical-resistant tweezers to a beaker containing isopropyl alcohol.

6. Place the beaker in an ultrasonic bath for a few minutes.

7. Finally, rinse the dies in a beaker of deionized water.

8. Gently blow each die dry with filtered compressed nitrogen.

9. When finished, dispose of the solutions appropriately.

Laboratory Exercise 2: Device and Process Flow Design

Purpose: In this lab exercise, you will learn how to create a process flow and mask set for a simple, single-level microfluidic device to be made using soft lithography techniques. The construction material will be PDMS.

Instructions: This exercise is to be completed in groups. Create a simple process flow for a simple, single-layer microfluidic device to be made out of PDMS by soft lithography. Show each step in writing. Illustrate the process flow using cross section images corresponding to key process steps. Draw the mask to scale using drawing software of your choice. Assume that your mold will be created on a 76.2-mm-diameter (3 inch) silicon wafer and your smallest feature size is 100 μm. Your design should occupy no more than one-fourth of the wafer and should be centered in one of the wafer quarters. Leave at least a 5 mm border around the periphery of the wafer. In other words, your design should not go all the way out to the edge. Make sure that you can output your design in postscript or the designated format.

The function and thus the design of the microfluidic device are up to you and your colleagues. For example, you may choose to design a mixer. To facilitate the design process, your group can search through the research literature for ideas. This device may have any function that you desire, and there is an adequate space to design more than one device. Leave a 1–2-mm gap between the devices.

Students will present their designs in class using electronic presentation software (~2–3 slides). The slides should include the mask design, process flow, cross-section drawings, and an explanation of what your device does.

Laboratory Exercise 3: Anodic Bonding

Purpose: In this lab exercise, you will learn how to join glass and silicon dies together using the anodic bonding process.

Background: A large voltage potential applied across the glass–silicon interface generates an electric field that drives Na^+ ions in the glass away from the interface region. The resulting Na^+ depletion zone leaves oxygen molecules at the interface, which diffuse into silicon to form silicon dioxide and a strong bond.

Materials and supplies

- Gloves
- Lint-free cleanroom cloths
- (100) silicon wafer pieces
- Pyrex 7740 glass pieces
- Tweezers
- Glass beakers
- Graduated glass cylinders
- Parafilm

Chemicals

- Ammonium hydroxide (NH_4OH)
- Hydrogen peroxide (H_2O_2)
- Deionized water

Equipment

- High voltage power supply
- Hot plate
- Stainless steel plate
- Metal probe
- Chemical fume hood

Cleaning the pieces using the Radio Corporation of America Standard Clean 1 (RCA SC1) process

1. This process should be carried out in a chemical fume hood. Based on the number of pieces you need to clean for anodic bonding and the glass beaker to be used for cleaning, determine the appropriate total volume of the cleaning solution that will cover the pieces. Then, calculate the volume of deionized water, NH_4OH, and H_2O_2 based on a 5:1:1 volumetric ratio.

2. First, place the deionized water into the beaker. Then, carefully add NH_4OH followed by H_2O_2.

3. Cover the solution with parafilm and heat to 50°C on a hot plate in a chemical fume hood.

4. Place the pieces into the beaker using chemical-resistant tweezers. Cover with parafilm and clean for at least 10 minutes. The tweezer tips can also be cleaned at the same time in the same beaker.

5. Remove items with RCA-cleaned chemical-resistant tweezers and rinse in a beaker of deionized water.

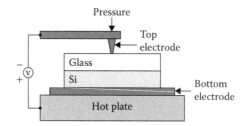

FIGURE 3.62
Anodic bonding setup.

6. Gently blow each die dry with filtered compressed nitrogen.

7. When finished, dispose of the solutions appropriately.

Anodic bonding

1. Follow the setup in Figure 3.62. Place a stainless steel plate on the hot plate. Then, place silicon and glass on top in the order indicated. Align the two pieces as desired.

2. Position the high-voltage probe above the stack and apply a slight pressure using the probe to the glass/silicon stack.

3. Plug the hot plate in, and turn it on to approximately 500°C.

4. Plug the power supply in and set the voltage to approximately 500 V. You will start to see a darker area near the probe tip. This is the bonded area.

5. When the bonded area covers the entire interface between the glass and silicon, the bonding process is complete. This may take several minutes.

6. Turn off the voltage and unplug the high voltage power supply. Turn off the hot plate and unplug the hot plate.

7. Carefully remove the probe. The hot plate will still be hot. Allow the bonded surfaces to cool slowly.

References

1. Helbert, J. N. 2000. *Handbook of VLSI Microlithography: Principles, Technology, and Applications.* 2nd ed. Norwich, NY: Noyes Publications/William Andrew Pub.
2. Moreau, W. M. 1988. *Semiconductor Lithography: Principles, Practices, and Materials.* New York: Plenum Press.

3. Wolf, S., and R. N. Tauber. 1986. *Silicon Processing for the VLSI Era*. Sunset Beach, CA: Lattice Press.

4. Runyan, W. R., and K. E. Bean. 1990. *Semiconductor Integrated Circuit Processing Technology*. Reading, MA: Addison-Wesley.

5. Brunner, T. 1997. Pushing the limits of lithography for IC production. *IEEE City* 9.

6. Manos, D. M., and D. L. Flamm. 1989. *Plasma Etching: An Introduction*. Boston: Academic Press.

7. Madou, M. J. 1997. *Fundamentals of Microfabrication*. Boca Raton, FL: CRC Press.

8. Deal, B. E., and A. S. Grove. 1965. General relationship for thermal oxidation of silicon. *J Appl Phys* 36:3770.

9. Sze, S. M. 1994. *Semiconductor Sensors*. New York: Wiley.

10. Robbins, H., and B. Schwartz. 1960. Chemical etching of silicon 2. The system HF, HNO_3, H_2O, and $HC_2H_3O_2$. *J Electrochem Soc* 107:108.

11. Williams, K. R., and R. S. Muller. 1996. Etch rates for micromachining processing. *J Microelectromech Syst* 5:256.

12. Petersen, K. E. 1982. Silicon as a mechanical material. *Proc IEEE* 70:420

13. Orthuber, R. K., J. E. Clemens, and B. B. Johnstone. 1956. *Method of Preparing Electrostatic Shutter Mosaics*. US Patent 2749598. June 12, 1956.

14. Nathanson, H. C., and R. A. Wickstrom. 1965. A resonant-gate silicon surface transistor with high-q bandpass properties. *Appl Phys Lett* 7:84.

15. Effenhauser, C. S., et al. 1997. Integrated capillary electrophoresis on flexible silicone microdevices: Analysis of DNA restriction fragments and detection of single DNA molecules on microchips. *Anal Chem* 69:3451.

16. Xia, Y. N., and G. M. Whitesides. 1998. Soft lithography. *Annu Rev Mater Sci* 28:153.

17. Hsu, T.-R. 2002. *MEMS and Microsystems: Design and Manufacture*. Boston: McGraw-Hill.

18. Bhushan, B. 2004. *Springer Handbook of Nanotechnology*. Berlin: Springer.

19. Wallis, G., and D. I. Pomerantz. 1969. Field assisted glass-metal sealing. *J Appl Phys* 40:3946.

20. Fung, C. D. 1985. *Micromachining and Micropackaging of Transducers*. New York: Elsevier.

21. Tummala, R. R., and M. Swaminathan. 2008. *Introduction to System-On-Package (SOP): Miniaturization of the Entire System*. New York: McGraw-Hill.

22. Traeger, R. K. 1976. Hermeticity of polymeric lid sealants. *IEEE City* 361.

23. Callewaert, L., et al. 1991. Programmable implantable device for investigating the adaptive response of skeletal-muscle to chronic electrical-stimulation. *Med Biol Eng Comput* 29:548.

24. Ziaie, B., et al. 1996. A hermetic glass-silicon micropackage with high-density on-chip feedthroughs for sensors and actuators. *J Microelectromech Syst* 5:166.

25. Claes, W., et al. 2002. A low power miniaturized autonomous data logger for dental implants. *Sens Actuators A Phys* 97–8:548.

26. Lo, R., and E. Meng. 2008. Integrated and reusable in-plane microfluidic interconnects. *Sens Actuators B Chem* 132:531.

27. Lo, R., and E. Meng. 2009. Macro-to-micro fluidic interfacing. In *Lab on a Chip Technology, Volume 1: Fabrication and Microfluidics*, Norfolk, UK: Caister Academic Press.

28. Fredrickson, C. K., and Z. H. Fan. 2004. Macro-to-micro interfaces for microfluidic devices. *Lab Chip* 4:526.

29. McDonald, J. C., et al. 2000. Fabrication of microfluidic systems in poly (dimethylsiloxane). *Electrophoresis* 21:27.
30. Liu, J., C. Hansen, and S. R. Quake. 2003. Solving the "World-to-chip" Interface problem with a microfluidic matrix. *Anal Chem* 75:4718.
31. Pattekar, A. V., and M. V. Kothare. 2003. Novel microfluidic interconnectors for high temperature and pressure applications. *J Micromech Microeng* 13:337.
32. Lee, E. S., et al. 2004. Removable tubing interconnects for glass-based microfluidic systems made using ECDM. *J Micromech Microeng* 14:535.
33. Puntambekar, A., and C. H. Ahn. 2002. Self-aligning microfluidic interconnects for glass- and plastic-based microfluidic systems. *J Micromech Microeng* 12:35.
34. Murphy, E. R., et al. 2007. Solder-based chip-to-tube and chip-to-chip packaging for microfluidic devices. *Lab Chip* 7:1309.
35. Meng, E., S. Wu, and Y.-C. Tai. 2001. Silicon couplers for microfluidic applications. *Fresenius J Anal Chem* 371:270.
36. Christensen, A. M., D. A. Chang-Yen, and B. K. Gale. 2005. Characterization of interconnects used in PDMS microfluidic systems. *J Micromech Microeng* 15:928.
37. González, C., S. D. Collins, and R. L. Smith. 1998. Fluidic interconnects for modular assembly of chemical microsystems. *Sens Actuators B* 49:40.
38. Hierold, C., et al. 1999. Low power integrated pressure sensor system for medical applications. *Sens Actuators A Phys* 73:58.
39. Flick, B. B., and R. Orglmeister. 2000. A portable microsystem-based telemetric pressure and temperature measurement unit. *IEEE Trans Biomed Eng* 47:12.
40. Takahata, K., Y. B. Gianchandani, and K. D. Wise. 2006. Micromachined antenna stents and cuffs for monitoring intraluminal pressure and flow. *J Microelectromech Syst* 15:1289.
41. Rodger, D. C., et al. 2008. Flexible parylene-based multielectrode array technology for high-density neural stimulation and recording. *Sens Actuators B Chem* 132:449.
42. Xiao, X., et al. 2004. Low temperature growth of ultrananocrystalline diamond. *J Appl Phys* 96:2232.
43. Lee, S. H., et al. 2006. Hydroxyapatite-tio2 hybrid coating on Ti implants. *J Biomater Appl* 20:195.
44. Sharma, S., R. W. Johnson, and T. A. Desai. 2004. Evaluation of the stability of nonfouling ultrathin poly (ethylene glycol) films for silicon-based microdevices. *Langmuir* 20:348.
45. Prime, K. L., and G. M. Whitesides. 1991. Self-assembled organic monolayers - model systems for studying adsorption of proteins at surfaces. *Science* 252:1164.
46. Ostuni, E., et al. 2001. A survey of structure-property relationships of surfaces that resist the adsorption of protein. *Langmuir* 17:5605.
47. Bohringer, K. F. 2003. Surface modification and modulation in microstructures: Controlling protein adsorption, monolayer desorption and micro-self-assembly. *J Micromech Microeng* 13:S1.
48. Seidel, H., L. Csepregi, A. Heuberger, et al. 1990. Anisotropic etching of crystalline silicon in alkaline-solutions: 1. Orientation dependence and behavior of passivation layers. *Journal of the Electrochemical Society* 137:3612-26.

4

Microfluidics

4.1 Introduction and Fluid Properties

4.1.1 Fluids

From the viewpoint of fluid mechanics, there are only two states of matter: solids and fluids. What distinguishes the two is their response to applied shear stress. Solids will not move in the presence of shear stress and instead undergo static deformation. Fluids, on the other hand, cannot resist shear stress and will move in response to it.

Both liquids and gases are considered fluids and are distinguished by the spacing of their constituent molecules. This intermolecular spacing has profound implications on the cohesive forces experienced by neighboring molecules, which in turn impacts the ability of the fluid to maintain a certain volume. Liquids are closely spaced and tend to retain their volume if left unconfined, whereas gases are composed of a collection of molecules separated by greater spacing and tend to expand to fill a volume.

4.1.2 Fluid as a Continuum

We normally do not consider the molecular spacing of molecules in a fluid because the length scales and volumes we deal with are large compared to the molecular spacing. At these large volumes, we consider that the number of molecules is constant and that the bulk fluid properties are therefore constant throughout the volume. For instance, we may need to know the density of the fluid. If the fluid can be considered a *continuum* and is well-defined everywhere in space, determining the density at a point is sufficient. This greatly simplifies the analysis of fluids. It is possible to consider fluids from a molecular point of view; however, this is not necessary unless we move to small scales. The continuum assumption holds until a length scale of ~1 μm for gases and ~10 nm for liquids. These are approximate values intended to provide a reference point and are calculated under the assumptions that at least 10^4 molecules (for <1% statistical variation in property measurements) are present and that the length scales are ten times the interaction-length scales of the constituent molecules [1]. For microfluidic systems, the

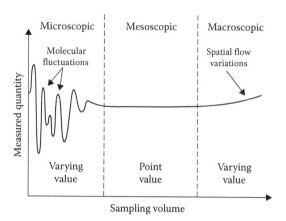

FIGURE 4.1
Continuum versus molecular fluid properties.

continuum assumption generally holds, but in nanofluidic systems, the assumption may need to be revisited. A visual representation of the validity of the continuum assumption over length scales is illustrated in Figure 4.1.

4.1.3 Properties of a Fluid

The primary thermodynamic properties of fluids include pressure, density, and temperature. The secondary properties relate to mechanical behavior and include viscosity, thermal conductivity, and surface tension.

4.1.3.1 Pressure

Pressure, P, is a dynamic variable in fluid mechanics and is defined as the normal force exerted per unit area over an area. In microfluidic systems, static pressure is often a driving force used to move fluids in microchannels. Pressure units are typically represented as pascals (Pa), 1 Pa being equivalent to 1 N/m^2 (N stands for newton).

4.1.3.2 Density

The *density* of a fluid is its mass per unit volume and is denoted as ρ (in kg/m^3). Density is a function of both temperature and pressure. Liquids are *incompressible* and possess constant densities even in the presence of pressure. The same is not true for most gases, which are said to be *compressible* because the density increases with pressure.

Specific weight, γ, is related to density and is defined as the weight per unit volume of the fluid. The relationship between specific weight and density is expressed as

$$\gamma = \rho g \tag{4.1}$$

where g is the gravitational constant ($g = 9.8$ m/s²). The unit for specific weight is N/m³.

Assuming that density is constant and that a fluid is stationary, we can define hydrostatic pressure as follows:

$$P = \rho g h \tag{4.2}$$

where h is the height measured in the direction of gravity (usually with reference to the surface of the earth).

4.1.3.3 Temperature

Temperature, T, quantifies a fluid's internal energy. The presence of a temperature gradient induces *heat transfer.*

4.1.3.4 Viscosity

The ability of a fluid to resist motion due to shear stress is expressed as the *absolute viscosity*, μ (in N • s/m² or Pa • s). Assuming that the shear stress, τ, is applied only in one dimension (y-direction), the shear stress and the velocity gradient, du/dy, are related by the absolute viscosity (Figure 4.2), as expressed in Equation 4.3

$$\tau = \mu \frac{du}{dy} \tag{4.3}$$

Kinematic viscosity, ν, is related to absolute viscosity through density as follows

$$\nu = \frac{\mu}{\rho} \tag{4.4}$$

The units of kinematic viscosity are in m²/s.

Fluids that behave in the manner described by Equation 4.3 are said to be Newtonian. Examples of Newtonian fluids, in which the viscosity is

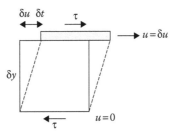

FIGURE 4.2
Shear stress acting on a fluid element through an adjacent plate moving at velocity δu causes continuous shear deformation (modified from [4]).

independent of the velocity gradient, are air, water, and alcohol. Viscosity does, however, vary with temperature and pressure.

An important implication of the relationship between fluid as a continuum and the effect of shear stress is that a fluid in contact with a moving surface must move with the surface. This is known as the *no-slip condition* and is so named because the fluid is considered to be stuck to the surface and does not slip across it.

When considering flows and the viscous behavior of Newtonian fluids, a dimensionless number known as the Reynolds number (Re) provides a convenient starting point to determine the relative importance of inertial effects compared to viscous effects and is represented as follows:

$$\text{Re} = \frac{\rho UL}{\mu} = \frac{UL}{\nu} \tag{4.5}$$

where U is the velocity and L is the characteristic length scale of the flow. A high Re suggests that flow is *turbulent* and that inertial effects on the flow are dominant. A low Re points to *laminar* flow, where viscous effects dominate. A clear boundary does not exist between these two types of flow; however, a general rule of thumb is that flows tend to be turbulent for Re > 2000. In microfluidics, the length scales are small and flows are nearly always laminar. Laminar and turbulent flows differ in their appearances. Laminar flow is smooth and steady; for example, a stream of viscous honey flowing through a small spout from a large bottle exhibits laminar flow. Turbulent flow, in contrast, is unsteady, irregular, and agitated in appearance, similar to the flow of water from a faucet.

Microfluidic channels assume a number of different geometries and their characteristic lengths are not always obvious. Instead, it is more convenient to substitute the *hydraulic diameter,* which is defined as follows:

$$D_{\text{hyd}} = \frac{4 \times \text{cross-sectional area}}{\text{wetted perimeter}} = \frac{4A}{P} \tag{4.6}$$

where A is the cross-sectional area of the channel and P is the wetted perimeter. The expression for the Re in this case becomes

$$\text{Re} = \frac{\rho U D_{\text{hyd}}}{\mu} \tag{4.7}$$

Hydraulic diameters for cross sections of common channels are listed in Table 4.1.

Not all fluids obey the linear relation. Blood is an example of a non-Newtonian fluid; it possesses a viscosity that is independent of the rate of shear. Instead, blood is a *pseudoplastic* non-Newtonian fluid that decreases in viscosity as the velocity gradient increases. There are also shear-thickening fluids, in which the viscosity increases with the velocity gradient (these are

TABLE 4.1

Hydraulic Diameters for Different Channel Geometries

Channel Cross Section	Hydraulic Diameter, D_{hyd}	Diagram
Circle	D	
Square	a	
Rectangle	$\dfrac{2ab}{a+b}$	
Equilateral triangle	$\dfrac{a}{\sqrt{3}}$	
Anistropic channel	$\dfrac{2h}{1+\dfrac{2h}{a+b}\sqrt{1+\left(\dfrac{a-b}{2h}\right)^2}}$	
Isotropic channel	$2h\,\dfrac{\dfrac{a}{h}+\dfrac{\pi}{2}}{\dfrac{a}{h}+\dfrac{\pi}{2}+1}$	

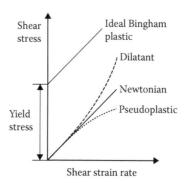

FIGURE 4.3
Stress-to-strain-rate relationship of different viscous fluids (modified from [4]).

called *dilatant* non-Newtonian fluids; e.g., corn starch in water). Other examples of non-Newtonian fluids include molasses, honey, ketchup, chocolate, and toothpaste. Ketchup, chocolate, and toothpaste fall into the category of *Bingham plastics* and possess the property whereby flow does not initiate until a finite amount of stress (yield stress) is applied. Ketchup does not come out of the bottle unless shaken. The stress-to-strain rate relationships of different viscous fluids, both Newtonian and non-Newtonian, are shown in Figure 4.3.

4.1.3.5 Thermal Conductivity

Heat transfer may also be present in microfluidic systems. The heat flow per unit area, q, is related to the temperature gradient, ∇T, through a proportionality constant similar to the case of stress and velocity gradient. This proportionality constant is the thermal conductivity, k, (in J/(s • m • K) or W/(m • K)). The relationship between these quantities is given by an empirical law known as Fourier's law of heat conduction, which, in one dimension, is expressed as

$$q_x = -k \frac{dT}{dx} \tag{4.8}$$

Heat flux is defined to be positive from higher to lower temperatures and is responsible for the minus sign in Equation 4.8. Similar to viscosity, thermal conductivity is also dependent on temperature and pressure.

4.1.3.6 Surface Tension

Liquid molecules experience cohesive forces that allow it to retain its volume. This also implies that liquids will form interfaces because they cannot expand as gases do. There are fewer molecules at the surface or interface of a liquid than in the interior. These molecules experience an inward pull arising from attraction forces associated with the interior molecules. Thus, it is possible to think of a liquid surface as an imaginary membrane that is in uniform tension. This phenomenon is called surface tension, Υ (expressed in N/m), which is an expression of the Gibbs free energy per unit area of the interface.

Surface tension exists at all interfaces due to the presence of discontinuities in the density pattern. For example, consider the boundary between water and oil. If a surface is curved, then a pressure difference exists and is the source of capillary effects. The Young-Laplace equation expresses this pressure difference between the interior and exterior of a general surface as

$$\Delta P = P_{interior} - P_{exterior} = \left(\frac{1}{R_1} + \frac{1}{R_2}\right)\Upsilon \tag{4.9}$$

where R_1 and R_2 are the two principal radii of curvature. Note that this quantity is positive and that higher pressure is associated with the side of the surface (liquid) with the smaller radius of curvature. Another way to think of this is that the pressure is highest in the convex medium in which the centers of the radii of curvature are placed. For a spherical droplet having radius R, the radii of curvature are equal and the pressure difference is given by

$$\Delta P = P_{interior} - P_{exterior} = \frac{2\Upsilon}{R} \tag{4.10}$$

4.1.3.7 Contact Angle

When a liquid droplet comes into contact with a solid, it forms an interface with the solid and surrounding medium. One can define an angle formed between these interfaces, as shown in Figure 4.4, which is called the *contact angle* or the *wetting angle,* θ. When the angle is <90°, the liquid is said to be *wetting* and the surface is *hydrophilic*. If the contact angle is >90°, then the liquid said to be *nonwetting* and the surface is *hydrophobic* (Figure 4.5). This angle arises due to the cohesive forces between the molecules of the three substances (gas, liquid, and solid) involved. Young's equation expresses the

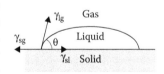

FIGURE 4.4
Contact angle definition between a liquid, solid, and gas.

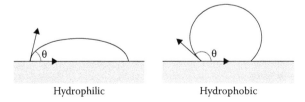

Hydrophilic   Hydrophobic

FIGURE 4.5
Hydrophilic ($\theta < 90°$) and hydrophobic ($\theta > 90°$) surfaces.

equilibrium contact angle as a function of the surface tensions of the three interfaces formed by the immiscible phases as follows

$$\cos\theta = \frac{\Upsilon_{sg} - \Upsilon_{sl}}{\Upsilon_{lg}} \qquad (4.11)$$

where Υ_{sg}, Υ_{sl}, and Υ_{lg} are the surface tensions of the solid/gas, the solid/liquid, and the liquid/gas interfaces, respectively.

We will revisit the concepts of surface tension and contact angle again in Section 4.2 in our discussion on capillarity.

4.2 Concepts in Microfluidics

4.2.1 Capillarity

The combined effects of surface tension, contact angle, and gravity can be seen when one end of a small diameter tube is dipped into a container of liquid. The liquid will rise within the tube; here, we assume the rise, H, is larger than the tube radius, a. The phenomenon is visually apparent if we choose a transparent tube. If we observe the interface between the liquid in the tube and the air above, a meniscus is observed (Figure 4.6). Assuming a spherical meniscus, the height of rise from the surface of the liquid in the container to the bottom of the meniscus (for $\theta < 90°$) is the capillary rise height, expressed as follows:

$$H = \frac{2\Upsilon}{\rho g a}\cos\theta \qquad (4.12)$$

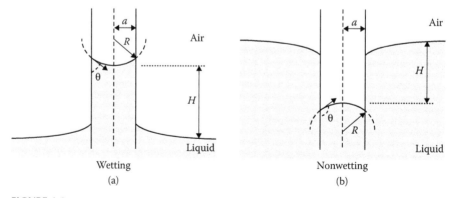

FIGURE 4.6
Capillary rise in a tube dipped in liquid for (a) wetting ($\theta < 90°$) and (b) nonwetting cases ($\theta > 90°$).

The radius of curvature, R, of the meniscus is given by

$$R = \frac{a}{\cos \theta} \qquad (4.13)$$

Based on the results of Equations 4.12 and 4.13, we can rearrange Equation 4.10 to derive the hydrostatic pressure present in this system (Figure 4.6a) due to the presence of a curved air/liquid interface.

$$\Delta p = p_{air} - p_{liquid} = \frac{2\Upsilon}{R} = \frac{2\Upsilon}{a} \cos \theta \qquad (4.14)$$

The pressure is greater above the interface on the air side (for $\theta < 90°$); the value of this pressure is simply the atmospheric pressure p_0. The capillary rise heights in microchannels can be quite large. The pressure developed due to capillary effects can also be used for pumping fluids (*capillary pumping*).

4.2.2 Fluid Flow

The motion of a viscous fluid flow is governed by the Navier-Stokes equation, which is derived from the fundamental equations for rate of change of flux densities of mass, momentum, and energy. Detailed derivations can be found in texts dealing with fluid mechanics and will not be repeated here [1,3–5]. Instead, we will examine a few simple cases of fluid flow that apply to microfluidics. While flows can be described using the velocity, U, it is often more convenient to use the volumetric flow rate, Q (expressed in m³/s) as shown in Equation 4.15:

$$Q = VA \qquad (4.15)$$

where A is the cross-sectional area of the flow. Flow rates are frequently given in µL/min in the literature on the subject.

4.2.3 Couette Flow

Couette flow is a classic problem in which the shear drives the fluid flow when applied to a liquid sandwiched between two infinite parallel plates (walls), one of which is moving (Figure 4.7). The plates are horizontally oriented such that gravity acts at an angle perpendicular to the plates. The plates are separated by a distance h and the top plate is moved at a constant speed v_0 in the x-direction. The flow at the stationary wall must be zero and, at the moving wall, it is equal to v_0 because the no-slip condition applies. Thus, assuming there is no pressure variation in the direction of the flow, the velocity profile is solely a function of y. From Equation 4.3, we can write

$$\frac{du_x}{dy} = \frac{\tau}{\mu} = \text{constant} \qquad (4.16)$$

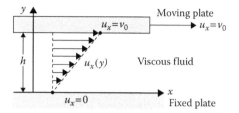

FIGURE 4.7
Couette flow between two infinite parallel plates: the bottom plate is fixed and the top plate is
moving at velocity v_0.

We could also start with the Navier-Stokes equation for this system and
write

$$\frac{d^2 u_x}{dy^2} = 0 \tag{4.17}$$

After integrating Equation 4.14, the velocity profile is given by

$$u_x(y) = a + by \tag{4.18}$$

If we apply the boundary conditions

$$\begin{aligned} u_x(0) &= 0 \\ u_x(h) &= v_0 \end{aligned} \tag{4.19}$$

we obtain the following solution in which the velocity follows a linear
profile

$$u_x(y) = v_0 \frac{y}{h} \tag{4.20}$$

4.2.4 Poiseuille Flow

Another classic problem is that of steady-state *Poiseuille flow* (or *Hagen-Poiseuille flow*; Poiseuille is pronounced "pwa-say") in which the driving force
is pressure. Most microfluidic systems rely on pressure-driven flow in which
a pressure difference exists between the ends of a microchannel. Although
this type of flow was originally studied in circular pipes, slight modifications can be made to account for the many microchannel geometries that
one might use.

Let us first consider a simple case in which Poiseuille flow is induced in
an infinite parallel-plate channel (Figure 4.8). A pressure difference ΔP is
applied across a channel of height h. To determine the velocity profile for

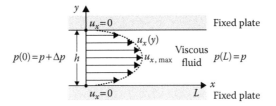

FIGURE 4.8
Poiseuille flow between two infinite parallel plates.

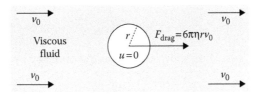

FIGURE 4.9
Stokes flow involving a rigid sphere at rest in an incompressible fluid moving with a constant velocity v_0 in the bulk.

this system, we assume that gravity can be neglected and thus obtain the following form of the Navier-Stokes equation:

$$\mu \frac{d^2 u_x}{dy^2} = -\frac{\Delta P}{L} \tag{4.21}$$

After integrating and applying the following boundary conditions (no-slip condition applies here)

$$\begin{aligned} u_x(0) &= 0 \\ u_x(h) &= 0 \end{aligned} \tag{4.22}$$

the solution is

$$u_x(y) = \frac{\Delta P}{2\mu L} h(h - y) \tag{4.23}$$

The maximum velocity occurs at the center ($y = h/2$):

$$u_{x,max} = \frac{\Delta P}{L} \frac{h^2}{8\mu} \tag{4.24}$$

We can also define a flow rate, Q, through a width w for this system as follows

$$Q = w \int_0^h u_x dy = w \int_0^h \frac{\Delta P}{2\mu L} h(h-y) dy = \frac{h^3 w}{12\mu L} \Delta P \qquad (4.25)$$

4.2.5 Hydraulic Resistance

In examining the expression in Equation 4.25 and drawing an analogy between fluid mechanics and electronics, one can define a parameter called the *hydraulic resistance*, R_{hyd} (expressed in Pa $\cdot$ s/m^3 or kg/(m^4 $\cdot$ s)) as follows

$$\Delta P = Q R_{hyd} \qquad (4.26)$$

If we compare this to Ohm's law, the following analogy becomes clear

$$V = IR \qquad (4.27)$$

The pressure difference is similar to a voltage drop and the flow rate is similar to the current. Therefore, the hydraulic resistance for a case of Poiseuille flow (where $h < w$) in a section of an infinite parallel plate system is

$$R_{hyd} = \frac{12\mu L}{h^3 w} \qquad (4.28)$$

This relationship allows one to quickly determine pressure drops or flow rates in microfluidic channels if the hydraulic resistance is known. Luckily, analytical solutions exist for many common channel cross-sectional geometries and are listed in Table 4.2.

Two additional solutions for the hydraulic resistance are those for a square cross-section channel (side of length h)

$$R_{hyd,square} = 28.4 \frac{\mu L}{h^4} \qquad (4.29)$$

and for a parabola cross-section channel (of width w and of height h in the thickest part; resembles an isotropically etched microchannel) [3]

$$R_{hyd,parabola} = \frac{105}{4} \frac{\mu L}{h^3 w} \qquad (4.30)$$

In analyzing microfluidic systems, the same basic circuit laws applying to series and parallel combinations of resistors also apply to series and parallel combinations of channels

$$R_{series} = R_1 + R_2 + \cdots + R_N \qquad (4.31)$$

TABLE 4.2

Flow Velocities and Hydraulic Resistance for Common Cross-Sectional Geometries of Channels

Channel Cross Section	Flow Velocity, Flow Rate, and Hydraulic Resistance	Diagram
Circular	$$u_x(y,z) = \frac{\Delta P}{4\mu L}\left(a^2 - y^2 - z^2\right)$$ $$Q = \frac{\pi a^4}{8\mu L}\Delta P$$ $$R_{hyd} = \frac{8\mu L}{\pi a^4}$$	
Elliptic	$$u_x(y,z) = \frac{\Delta P}{2\mu L}\frac{a^2 b^2}{a^2 + b^2}\left(1 - \frac{y^2}{a} - \frac{z^2}{b}\right)$$ $$Q = \frac{\pi}{4}\frac{1}{\mu L}\frac{a^3 b^3}{a^2 + b^2}\Delta P$$ $$R_{hyd} = \frac{4\mu L}{\pi}\frac{a^2 + b^2}{a^3 b^3}$$	
Equilateral triangle	$$u_x(y,z) = \frac{v_0}{a^3}\left(\frac{\sqrt{3}}{2}a - z\right)\left(z^2 - 3y^2\right)$$ $$Q = \frac{3}{160}v_0 a^2 = \frac{\sqrt{3}}{320}\frac{a^4}{\mu L}\Delta P$$ $$R_{hyd} = \frac{320}{\sqrt{3}}\frac{\mu L}{a^4}$$ $$v_0 = \frac{1}{2\sqrt{3}}\frac{\Delta P}{\mu L}a^2$$	
Rectangular	$$u_x(y,z) = \frac{4h^2 \Delta P}{\pi^3 \mu L}\sum_{n,\text{odd}}^{\infty}\frac{1}{n^3}\left[1 - \frac{\cos h\left(\frac{n\pi y}{h}\right)}{\cos h\left(\frac{n\pi w}{2h}\right)}\right]\sin h\left(\frac{n\pi z}{h}\right)$$ $$Q = \frac{h^3 w \Delta P}{12\mu L}\left[1 - \sum_{n,\text{odd}}^{\infty}\frac{1}{n^5}\frac{192}{\pi^5}\frac{h}{w}\tan h\left(\frac{n\pi w}{2h}\right)\right]$$ $$R_{hyd} = \frac{12\mu L}{h^3 w}\left[1 - \sum_{n,\text{odd}}^{\infty}\frac{1}{n^5}\frac{192}{\pi^5}\frac{h}{w}\tan h\left(\frac{n\pi w}{2h}\right)\right]^{-1}$$	

Source: Bruus, H. 2008. *Theoretical Microfluidics.* Oxford: Oxford University Press.

$$R_{parallel} = \left(\frac{1}{R_1} + \frac{1}{R_2} + \cdots + \frac{1}{R_N} \right)^{-1} \tag{4.32}$$

4.2.6 Hydrodynamic Capacitance

Continuing the analogy between electronics and fluid mechanics, it follows that a *hydraulic capacitance* (in m^5/N), or *compliance*, can also be defined as follows

$$C_{hyd} = \frac{dV}{d\Delta P} \tag{4.33}$$

where V is volume. In other words, fluidic capacitance is the pressure-dependent volume change. We can also write

$$Q = C_{hyd} \frac{d\Delta P}{dt} \tag{4.34}$$

4.2.7 Hydrodynamic Inductance

A *hydrodynamic inductance* (in kg/m^3) can also be defined as follows (but is rarely used)

$$L_{hyd} = \frac{\rho L}{A} \tag{4.35}$$

Fluidic inductance refers to the ability of a fluidic element to store kinetic energy as shown in the following equation:

$$\Delta P = L_{hyd} \frac{dQ}{dt} \tag{4.36}$$

4.2.8 Fluidic Circuit Theory

To complete the analogy between fluid flow and electrical circuits, fluidic circuit laws are necessary. From Kirchhoff's laws for electrical circuits, we can write a set of laws that applies to fluidic circuits as follows:

1. The sum of flow rates either entering or leaving a circuit node is zero.

2. The sum of pressure differences around a closed fluidic circuit loop is zero.

4.2.9 Stokes Drag

Many applications of microfluidics involve beads or cells in fluid flow. Consider the case of a rigid sphere in a moving incompressible fluid (Figure 4.9). A sphere at rest will experience a *drag force*, F_{drag}, which for low Re (<<1) is expressed as

$$F_{drag} = 6\pi\mu r v_0 \qquad (4.37)$$

where r is the radius of the sphere, and the fluid moves at a constant velocity v_0. This is known as *Stokes flow* or *creeping flow*. Drag is the force along the sphere in parallel to the direction of the flow, which has an associated *drag coefficient*, C_D, that is a function of the rigid body's shape and is expressed as follows:

$$C_D = \frac{F_{drag}}{\frac{1}{2}\rho U^2 A} \qquad (4.38)$$

For a sphere, the drag coefficient is

$$C_{D,sphere} = \frac{6\pi\mu r v_0}{\frac{1}{2}\rho v_0^2 \pi r^2} = \frac{24}{\rho v_0 d/\mu} = \frac{24}{Re} \qquad (4.39)$$

where d is the diameter of the sphere.

4.3 Fluid-Transport Phenomena and Pumping

4.3.1 Diffusion

Our focus thus far has primarily been on fluids in their bulk form. Fluid transport also takes place on a molecular scale through diffusion. Due to the low Re values associated with microfluidics, transport is largely laminar. Thus, mixing of fluids is largely achieved through diffusion. The same is true of micro- and nanoparticles in fluids at these scales. Fluid transport by diffusion is based on Brownian motion. Considering the case of point-source diffusion, a characteristic diffusion length, L, can be derived [3]. For the one-dimensional case,

$$L = \sqrt{2Dt} \qquad (4.40)$$

where D is the diffusion coefficient (in m²/s) and t is the time (in seconds). By rearranging, the characteristic diffusion time is derived as

$$t = \frac{L^2}{2D} \qquad (4.41)$$

The same can be done for the two-dimensional case

$$L = 2\sqrt{Dt} \qquad (4.42)$$

$$t = \frac{L^2}{4D} \qquad (4.43)$$

and the three-dimensional case

$$L = \sqrt{6Dt} \qquad (4.44)$$

$$t = \frac{L^2}{6D} \qquad (4.45)$$

The corresponding equation for diffusion from a point source in one dimension is

$$C(x,t) = \frac{C_0}{\sqrt{4\pi Dt}}\exp\left[-\frac{x^2}{4Dt}\right] \qquad (4.46)$$

where C_0 is the initial concentration of the point source. Previously, we have investigated diffusion from a planar source in Chapter 3 in the discussion on doping and have developed an equation (Equation 3.11) describing the concentration profiles over time.

The Stokes-Einstein equation allows the estimation of diffusion coefficients as follows:

$$D = \frac{kT}{6\pi r\mu} \qquad (4.47)$$

where k is the Boltzmann constant, T is the temperature, and r is the radius of the particle.

Let us consider a practical example of diffusion-based transport. Let us assume that we have spherical particles in water and that we want to determine their diffusion time to travel a certain distance. If $D = 3.33 \times 10^{-10}$ m²/s for these particles in water, then the diffusion times can be determined from Equation 4.45. The diffusion times for particles traveling through cubes of different lengths (and thus, volumes) are presented in Table 4.3.

TABLE 4.3

Calculated Diffusion Times for a Spherical Particle across Different Fluidic Lengths and Volumes

Side Length of Cube	Cube Volume	Diffusion Time over Cube Length
1 mm	1 μL	500 s
100 μm	1 nL	5 s
10 μm	1 pL	0.05 s
1 μm	1 fL	0.5 ms
100 nm	1 aL	0.05 ms

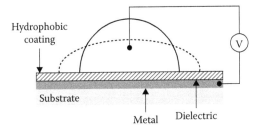

FIGURE 4.10

Electrowetting on dielectric setup in which the contact angle of a liquid droplet decreases on application of an electric field.

4.3.2 Electrowetting

Lippman discovered electrocapillarity in 1875 by observing that the capillary depression of mercury in contact with electrolyte solutions could be changed by applying a voltage. However, at sufficiently high voltages (a few hundred millivolts [mV]), electrolysis of water was induced. Much later, in the 1990s, Berge introduced a thin insulating film to separate the liquid from the electrode to eliminate electrolysis [6]. The ability to change the surface tension of a liquid by applying an electrical potential across the liquid/solid interface is called *electrowetting*. The systems that Berge introduced can be referred to as electrowetting on dielectric (EWOD) systems (Figure 4.10). Electrowetting combined with appropriately timed electric signals can be used to transport fluid droplets on surfaces. Other functions that have been demonstrated include droplet merging, splitting, and oscillation [6–11]. This principle can also be extended to the switching of fluid streams in channels [12].

Lippmann's law describes the change in surface tension of a liquid droplet on a solid surface, Υ_{sl}, in the presence of an electric field using the following equation:

$$\Upsilon_{sl} = \Upsilon_{sl,0} + \frac{1}{2}CV^2 \qquad (4.48)$$

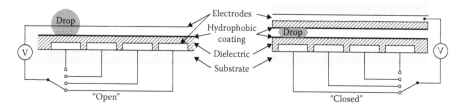

FIGURE 4.11
Electrowetting-on-dielectric droplet-manipulation system schemes.

where the 0 subscript refers to the state before application of the electric field, C is the capacitance, and V is the applied voltage. The Lippmann-Young law is obtained by combining Equation 4.48 with Young's law as follows:

$$\cos\theta = \cos\theta_0 + \frac{1}{2}\frac{C}{\Upsilon_{lg}}V^2 \qquad (4.49)$$

The implication is that the contact angle of the droplet can be changed, and by applying different electric potentials across the droplet, displacement is possible. This creates a wettability gradient and causes the droplet to move in the direction of the smaller contact angle. EWOD systems for droplet manipulation can be constructed such that the droplets are sandwiched between plates or in open systems without a cover (Figure 4.11).

4.3.3 Electrokinetics

The interaction of electric fields with ionic fluids or solid surfaces results in a number of phenomena classified as *electrokinetic* effects [13,14]. These phenomena are particularly useful at small scales for the pumping of fluids or manipulation of small objects without needing any moving parts [15,16]. There are four main electrokinetic phenomena, which are summarized in Table 4.4.

4.3.3.1 Electric Double Layers

The interaction of electrolytic solutions with solid surfaces results in the formation of an *electric double layer* (EDL), or *ionic double layer*. This phenomenon, which induces a surface charge on the solid, is central to electrokinetics. The surface charge develops because of a redistribution of charge, while the overall system maintains charge neutrality. The EDL is illustrated in Figure 4.12. Immediately adjacent to the charged surface, counter-ions are attracted and adsorb to the surface to form an immobile molecular film called the *Stern layer*. This in turn results in charge redistribution in the immediately adjacent liquid layer such that there is a net excess of counter-ions. In comparison to

TABLE 4.4

Electrokinetic Phenomena

Electrokinetic Phenomena	Movement Characteristics	Electrokinetic Coupling Mechanism
Electroosmosis	Moving phase: Liquid Stationary phase: Charged surface (usually a channel)	Movement induced by applied electric field
Electrophoresis	Moving phase: Charged surface (usually particles) Stationary phase: Liquid	
Streaming potential	Moving phase: Liquid Stationary phase: Charged surface (usually a channel)	Movement generates electric field
Sedimentation potential	Moving phase: Charged surface (usually particles) Stationary phase: Liquid	

the Stern layer, this *Gouy-Chapman layer*, or *diffuse layer*, is thicker and diffuse. The EDL consists of these two layers.

The net charge and mobility of the diffuse layer allow it to be manipulated by an electric field. Thus, the boundary between the two layers is referred to as a *shear surface* because of the mobility of the diffuse layer. The presence of excess charges produces potentials at the wall and shear surfaces, known as the *wall potential* (ϕ_w) and *zeta potential* (ζ), respectively. For example, polar liquids, such as water, placed into glass channels result in a negatively charged glass surface and a positively charged EDL. The negatively charged glass results primarily through the deprotonation of surface silanol groups (–Si–O–H) [13].

To determine both the profile of the electric

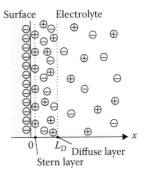

FIGURE 4.12
Electric double layer in relation to the interface between a charged solid surface and electrolyte solution.

potential developed and its impact on microfluidic systems, we start by examining the concentration of the ionic species present in the electrolyte (Figure 4.13). To simplify the analysis, we will assume a Boltzmann distribution of charge in the EDL on a flat plate (y-z plane):

$$C_i(x) = C_{i,0} \exp\left[-\frac{z_i q_e}{k_B T}\phi(x)\right] \tag{4.50}$$

where C_i is the concentration of the ionic species i, $C_{i,0}$ is the ionic concentration in the bulk solution ($C_i(\infty) = C_{i,0}$), z_i is the charge of the ionic species

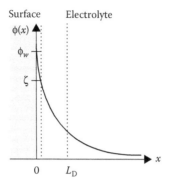

FIGURE 4.13
Profile of the electric potential from the charged solid surface to the bulk electrolyte solution in electrokinetic systems.

normalized to the electronic charge q_e, k_B is the Boltzmann constant, T is the temperature, and $\phi(x)$ is the local potential. The charge distribution coordinate x is obtained at a direction perpendicular to the plate, with the origin on the shear surface. Note that the potential in the bulk solution tends to zero $(\phi(\infty) = 0)$. The charge density of the EDL, ρ_e, for N species is given by

$$\rho_e(x) = \sum_{i=1}^{N} z_i q_e C_i(x) \tag{4.51}$$

To relate the charge density to the electrical potential near the plate, we use the Poisson equation as follows:

$$\nabla^2 \phi(x) = -\frac{\rho_e(x)}{\varepsilon} \tag{4.52}$$

where ε is permittivity of the liquid. Rearranging Equations 4.50 through 4.52, we obtain the Poisson–Boltzmann equation for the electrical potential as follows:

$$\nabla^2 \phi(x) = -\frac{1}{\varepsilon} \sum_{i=1}^{N} z_i q_e C_{i,0} \exp\left[-\frac{z_i q_e}{k_B T} \phi(x) \right] \tag{4.53}$$

This expression simplifies to the following for the case of an electrolyte with two monovalent ions:

$$\nabla^2 \phi(x) = 2 \frac{q_e z C_0}{\varepsilon} \sin h\left(\frac{z q_e}{k_B T} \phi(x) \right) \tag{4.54}$$

A useful approximation is the Debye-Hückel limit, which applies when the electrical energy is small compared to the thermal energy of the ions in the EDL, or when

$$zq_e\zeta \ll k_BT \qquad (4.55)$$

This condition holds true when ζ is less than 26 mV at room temperature. By applying a Taylor expansion, the following solution is obtained:

$$\nabla^2\phi(x) = \frac{\phi(x)}{L_D^2} = 2\frac{z^2q_e^2C_0}{\varepsilon k_B T}\phi(x) \qquad (4.56)$$

From this expression, we can extract the Debye length, L_D, as

$$L_D = \sqrt{\frac{\varepsilon k_B T}{2z^2q_e^2C_0}} \qquad (4.57)$$

which is the characteristic length of the EDL and corresponds to the distance at which ζ decays by $1/e$. The Debye lengths for pure water and 1M KCl are 1 μm and 0.3 nm, respectively. Microfluidic channels are typically greater than 10 μm in width; hence, the Debye lengths are small in comparison.

Finally, the potential distribution for the flat plate away from the wall ($x > 0$) is expressed as

$$\phi(x) = \zeta\exp\left[-\frac{x}{L_D}\right] \qquad (4.58)$$

4.3.3.2 Electroosmosis

Let us consider the case of a solution in the presence of a charged surface, such as a microchannel. The channel's cross section is constant over the length, and an electric field is applied across the ends of the channel (Figure 4.14). The ions in the fluid beyond the shear surface respond to the field through an induced electrostatic body force and drag the fluid as they move. For glass channels, the surface charge is negative and the diffuse layer is positively charged; it follows that the net flow is toward the cathode in the direction of the electric field. This is called the *electroosmotic flow* (EOF).

To understand the impact of the EDL potential on flow, we combine it with the Navier-Stokes equation to obtain the following expression

$$\rho\frac{DU_x}{dt} = -\frac{dP}{dx} + \mu\nabla^2 U_x + \rho_e E_x \qquad (4.59)$$

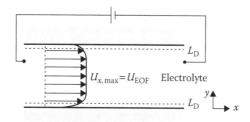

FIGURE 4.14
Electroosmotic flow induced in a glass channel.

If we assume (1) a fully developed flow, (2) no pressure gradient across the channel, and (3) a uniform channel cross section, the equation simplifies (after combining with the Poisson equation (Equation 4.52)) to

$$\mu \nabla^2 U_x = \varepsilon E_x \nabla^2 \phi(x) \tag{4.60}$$

Flow is solely due to electroosmosis and the only gradients are along the x-direction, represented as follows:

$$\mu \frac{d^2 U_{eof}}{dx^2} = \varepsilon E \frac{d^2 \phi}{dx^2} \tag{4.61}$$

The final solution to this equation is the Helmholtz-Smoluchowski equation, which is written as follows:

$$U_{eof} = \frac{\varepsilon \zeta}{\mu} E \tag{4.62}$$

in which the wall potential is equal to the zeta potential. Thus, the net result of EOF is a constant flow velocity across the bulk of the microchannel and a corresponding planar plug flow profile. We can also rewrite Equation 4.62 in the following manner:

$$\mu_{eof} \equiv \frac{U_{eof}}{E} = \frac{\varepsilon \zeta}{\mu} \tag{4.63}$$

where μ_{eof} is the *electroosmotic mobility* (not to be confused with viscosity, which is represented by μ).

In contrast, Poiseuille pressure-driven flow has a parabolic flow profile. The two flow profiles are compared in Figure 4.15.

4.3.3.3 Electrophoresis

Electrokinetics also applies to charged surfaces in the form of particles or molecules. The application of an electric field induces forces on these

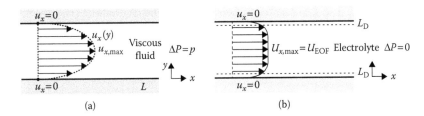

FIGURE 4.15
(a) Comparison of pressure-driven and (b) electroosmotic flow profiles in a microchannel.

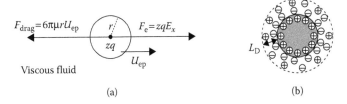

FIGURE 4.16
(a) Electrophoretic forces acting on a spherical particle. (b) Charge distribution on the particle.

surfaces and allows them to be manipulated with respect to the solution containing them. This phenomenon is called *electrophoresis*. Consider the case of a spherical particle in a stationary polar solution (Figure 4.16). The sphere has a radius r and a charge zq (where z is the valence number and q is the electron charge). Under an applied electric field E_x, the particle experiences a force expressed as

$$F_e = zqE_x \qquad (4.64)$$

We also know from our discussion of Stokes drag that the particle will experience a drag force acting in the opposite direction. The force balance for this system is given by

$$F_e = F_{drag} = zqE_x = 6\pi\mu r U_{ep} \qquad (4.65)$$

Solving, we obtain the *electrophoretic drift velocity*, U_{ep}, on the particle and a corresponding *electrophoretic mobility*, μ_{ep}, as shown in Equations 4.66 and 4.67:

$$U_{ep} = \frac{zq}{6\pi\mu r} E_x \equiv \mu_{ep} E_x \qquad (4.66)$$

$$\mu_{ep} = \frac{zq}{6\pi\mu r} \qquad (4.67)$$

Equations 4.66 and 4.67 apply to cases in which the diameter of the particle is much smaller than the Debye length ($d \ll L_D$ or ~100 nm). For much larger particles such as cells and beads ($d \gg L_D$ or ~100–10,000 nm), the electrophoretic velocity reduces to an expression resembling the Helmholtz-Smoluchowski equation for a flat plate:

$$U_{ep} = \frac{\varepsilon \zeta}{\mu} E_x \qquad (4.68)$$

$$\mu_{ep} = \frac{\varepsilon \zeta}{\mu} \qquad (4.69)$$

The electrophoretic mobility of different species is a convenient parameter by which to achieve their *separation*, a Nobel Prize–winning concept (1948) that was introduced by Tiselius in 1937. *Capillary electrophoresis* (CE) systems are examples of lab-on-a-chip microfluidic systems that can separate samples such as amino acids, proteins, DNA, and RNA. These systems will be discussed in more detail in Chapters 5 and 7.

4.3.3.4 Dielectrophoresis

When particles in an electrolyte are subject to a nonuniform electric field, they experience a force associated with *dielectrophoresis* (DEP). The field may be either alternating current (AC) or direct current (DC). Here, the force is useful as a tool to manipulate and position the particles as opposed to transporting them.

Consider the case illustrated in Figure 4.17, where a neutral particle of radius r is subjected to an inhomogeneous field produced by a pair of electrodes, one of which is flat and the other is spherical. The particle and the solution possess different dielectric constants (ε_p and ε_l, respectively) and

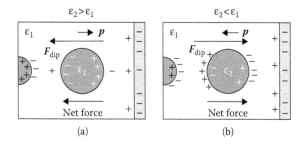

(a)   (b)

FIGURE 4.17
Dielectric sphere subjected to an inhomogeneous electric field: (a) a case where the particle is more polarizable than the fluid and (b) a case where the particle is less polarizable than the fluid In each figure, the particle is centered between a pair of electrodes—a spherical electrode on the left and a flat one on the right.

hence different polarizabilities. In the case of a DC field E, the polarization P of a dielectric fluid is given by

$$P = \varepsilon_0 \chi E \qquad (4.70)$$

where ε_0 is the permittivity of vacuum and χ is the susceptibility. A dipole moment p is induced on the dielectric particle, represented as follows:

$$p = \alpha E \qquad (4.71)$$

where α is the polarizability. A dielectric force F_{dip} acts on the dipole moment, as shown in Equation 4.72:

$$F_{dip} = (p \cdot \nabla) E \qquad (4.72)$$

If the particle has a larger dielectric constant than the fluid ($\varepsilon_p > \varepsilon_l$), then the particle acquires more charges at its surface. The net dielectric force then acts toward the spherical electrode. The reverse is true if the magnitude of the relative dielectric constants flips (i.e., $\varepsilon_p < \varepsilon_l$); the fluid acquires more charges at its surface and the net force is toward the flat electrode.

The dielectrophoretic force can be expressed as follows (time-averaged in the AC case):

$$DC: \quad F_{dep} = 2\pi r^3 \varepsilon_1 K \nabla E^2 \qquad (4.73)$$

$$AC: \quad \langle F_{dep} \rangle = 2\pi r^3 \varepsilon_1 \, Re(K(\omega)) \nabla E_{rms}^2 \qquad (4.74)$$

These expressions can be rewritten in terms of the Clausius-Mossotti factor as

$$DC: \quad K = \frac{\varepsilon_p - \varepsilon_1}{\varepsilon_p + 2\varepsilon_1} \qquad (4.75)$$

$$AC: \quad K(\omega) = \frac{\varepsilon_p^* - \varepsilon_1^*}{\varepsilon_p^* + 2\varepsilon_1^*} \qquad (4.76)$$

where $E_{rms} = E/\sqrt{2}$ and ε_p^* and ε_1^* are the complex permittivities of the particle and fluid, respectively. The complex permittivity for an isotropic homogeneous dielectric is given by

$$\varepsilon^* = \varepsilon - j\frac{\sigma}{\omega} = \varepsilon_0 \varepsilon_r - j\frac{\sigma}{\omega} \qquad (4.77)$$

where σ is the dielectric conductivity and ω is the frequency of the field.

The DEP force sign is given by the Clausius-Mossotti factor. For a positive DEP force, particles collect at the electrode edges, where the field is the highest (positive DEP); and for a negative DEP force, the particles are repelled from the electrode edges (negative DEP). This assumes, of course, that in both cases the DEP force is greater than the viscous drag force.

4.3.4 Magnetophoresis

DEP is nonspecific in that forces are induced on all of the dielectric particles. Fortunately, where specificity is desired, biological samples can be labeled with magnetic beads on which magnetic fields preferentially act. The vast majority of biological materials are not affected by weak magnetic fields. Therefore, although one might argue that the magnetic forces used in *magnetophoresis* are weaker than those used in DEP, the ability to isolate a specific species through bead tagging is a definite advantage.

Magnetic beads are used as vehicles through which specific bound biological species can be manipulated. Following this manipulation, the linkage between the beads and the biological species is typically broken to allow further processing of the biological species. Two types of magnetic particles can be isolated in this technique: paramagnetic particles experience an induced magnetization in the presence of a magnetic field, whereas ferromagnetic particles possess a permanent magnetization. Paramagnetic particles (or even superparamagnetic) are preferred as they disperse in the carrier fluid by Brownian motion following the removal of the field, whereas ferromagnetic particles form aggregates, which is undesirable in many applications.

Magnetic beads (~50 nm–2 μm) are constructed by embedding magnetic nanoparticles in a polymer bead. Other less common bead types are also available, and the exact type to be used is the user's decision. Beads having pre functionalized surfaces ready to bind specific species are available, in addition to fluorescently tagged beads that facilitate optical viewing.

Beads acquire a magnetization, M, in the presence of a magnetic field, H, as follows:

$$M = \chi H \qquad (4.78)$$

where χ is the magnetic susceptibility. The magnetic induction, B, obtained through Maxwell's equations is

$$B = \mu_0 (H + M) = \mu_0 (1 + \chi) H \equiv \mu_0 \mu_r H = \mu H \qquad (4.79)$$

where μ_0 is the magnetic permeability of vacuum, μ_r is the relative permeability, and μ is just the permeability ($\mu = \mu_0 \mu_r$). A nonconducting spherical magnetic particle of radius a and permeability μ exposed to a homogeneous external magnetic field experiences a magnetic force *F*. The force is expressed as

$$F = 2\pi\mu_0 \frac{\mu - \mu_0}{\mu + 2\mu_0} a^3 \nabla H^2 \qquad (4.80)$$

Note that Equation 4.80 is very similar to Equation 4.73 for a DEP force.

4.4 Flow Control

The combination of fluid phenomena at small scales and microfabrication enables microfluidic flow control. Precise manipulation of flow is the basis of many lab-on-a-chip devices, which are introduced in Chapter 5.

4.4.1 Microchannels

Although it is possible to transport fluids in droplet form through mechanisms such as electrowetting, it is far more common to confine laminar streams of flow to microfabricated channels on flat substrates. A number of additive and subtractive, bulk and surface micromachining techniques exist to fabricate microchannels. One can either carve part of a channel into a substrate or build a channel from the surface using multiple structural layers. Several common strategies are illustrated in Figure 4.18.

Due to the planar nature of microfabrication processes, most channels bear rectangular cross sections (round channels are possible but require precise

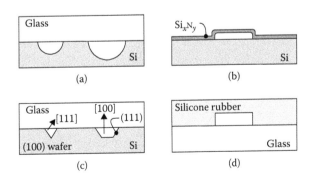

FIGURE 4.18
Cross-sectional images of channels fabricated using common strategies: (a) bulk micromachined (isotropic wet etch), (b) surface micromachined, (c) bulk micromachined (anisotropic wet etch), and (d) soft lithography. The channel flow path is drawn such that it is perpendicular to the page.

process control to achieve the desired cross-sectional profile). In subtractive or bulk micromachining processes, the bottom and sidewalls of a channel are formed by etching. Then, a separate top plate is bonded to the etched base to seal the channel. Access to the channel is commonly obtained by drilling inlet/outlet ports into either the substrate or the top plate. The bottoms of these channels may have differing profiles depending on the technique used to remove the material (e.g., isotropic or anisotropic wet etching or deep reactive ion etching; Figure 4.19). Surface micromachined channels use a patterned sacrificial layer to define all four sides. For example, photoresist is used as a sacrificial layer for constructing Parylene surface micromachined channels on a substrate. To remove the sacrificial layer, the photoresist is dissolved through the channel's inlet and outlet ports etched into the substrate or through the structural material. However, for long channels, this process is extremely slow. To speed up dissolution of the sacrificial material, side access ports are etched and later sealed in a subsequent processing step. Complete channels without the need for additional substrates are produced by surface micromachining.

Soft lithography or casting methods may also be used to produce microchannels in a rapid manner from a master mold. The master contains information on the channel sidewalls and either the top or the bottom surface. Polydimethylsiloxane (PDMS) prepolymer is poured onto the master and allowed to cure. As in bulk micromachined channels, a separate substrate is required to seal and complete the channel structure. In PDMS devices, this substrate may be one of a number of materials including PDMS, glass, or silicon. Inlet/outlet connections in PDMS are made simply by coring the polymer with a needle and inserting a tubing of slightly larger diameter into the hole, where it is held by compressive forces.

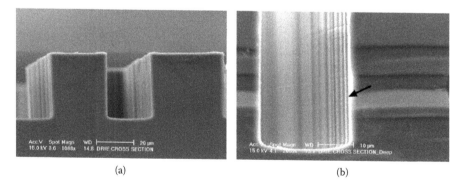

(a)	(b)

FIGURE 4.19

(a) Scanning electron microscopy images of microchannels formed using deep reactive ion etching. (b) Closeup showing scallops (indicated by arrow) formed on the sidewalls arising from the cyclical nature of the process. (Images: Courtesy of Po-Ying Li and Christian Gutierrez).

4.4.2 Laminar Flow in Microchannels

A major consequence of laminar flow in microchannels is that mixing between fluid streams in long, smooth channels occurs only by diffusion. Streams of flow that empty into a single channel are observed to maintain distinct borders with neighboring streams (Figure 4.20). Only after the streams remain in contact over large distances does significant diffusion occur.

This behavior is particularly useful in focusing streams of flow aerodynamically or hydrodynamically. In the examples shown in Figure 4.21, parallel streams of gas or liquid driven at higher pressures focus the central stream. The parallel focusing streams are known as *sheaths* (this type of flow

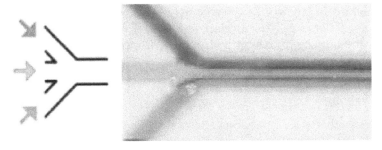

FIGURE 4.20
Three laminar fluid streams (colored with different dyes) are pumped and converge into a single channel without turbulent mixing (flow direction is from left to right). (From Gu, W., et al. 2004. *Proceedings of the National Academy of Sciences of the United States of America* 101:15861-6. With permission.)

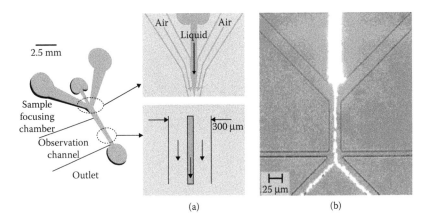

FIGURE 4.21
Flow focusing using (a) aerodynamics or (b) hydrodynamic methods. In the hydrodynamically focused stream, fluorescent beads are present, which enter both sides of the split outlet channel with equal probability. Beads enter one channel preferentially when a perpendicular electroosomtic flow is induced in the intersecting channel above the split outlet. (From Huh, D., et al. 2002. *Biomed Microdevices* 4:141; and Dittrich, P. S., and P. Schwille. 2003. *Anal Chem* 75:5767. With permission.)

is also called *sheath flow*). The central stream may contain particles or cells. For example, in *flow cytometry*, it is desirable to count small objects as they pass an optical detector in single file. Flow focusing can be used to force cells and particles to travel in the flow in single file. Flow confinement of a single stream can also be achieved using only one additional parallel sheath stream.

This concept can be extended to multiple-channel systems and used to switch flows into the desired outlet. An example of flow switching is shown in Figure 4.22, in which the relative sheath and sample flow rates are adjusted precisely to guide the focused flows into the desired outlet port. Electrowetting methods may also be applied to streams of flow. Aerodynamically focused flows of water are switched by adjusting the surface energy of the bottom of a channel. In Figure 4.23, the channel floor is rendered hydrophobic by a coating of Teflon. Applying an electric potential to the underlying electrodes causes the surface to exhibit hydrophilic properties and the fluid stream switches toward this region in response.

4.4.3 Valving

Gating of otherwise continuous streams of flow is achieved with microvalves. In general, valves and pumps may be categorized by their electrical power requirements (passive versus active). Passive valves operate without electrical power and are used extensively as parts of certain pumps. Active valves rely on actuation to control flow that requires power. Many actuation methods have been explored for microelectromechanical systems devices and will be briefly introduced in the context of their use in flow control components.

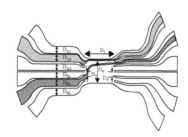

FIGURE 4.22
Flow switching of laminar streams using hydrodynamic methods. (From Lee, G. B., B. H. Hwei, and G. R. Huang. 2001. *J Micromech Microeng* 11:654. With permission.)

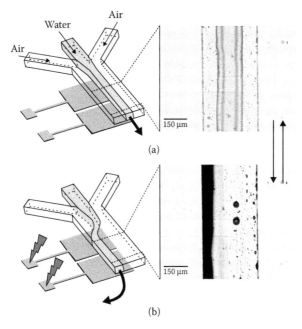

FIGURE 4.23
Flow switching of laminar streams using electrowetting principles: (a) an illustration and photograph of focused flow in a 100-μm-high channel and (b) an illustration and photograph of flow switching by electrowetting. (From Huh, D., et al. 2003. *J Am Chem Soc* 125:14678. With permission.).

Valves may further be classified on the basis of their initial state. For example, valves may initially be in a *normally open* or a *normally closed* mode. Normally closed valves may switch to being open and remain open or may close again upon application of appropriate conditions. Valves may also exhibit bistable properties in which the valve switches back and forth between open and closed but does not have a well defined "off" state. Key valve characteristics include the flow capacity, resistance against leakage, power consumption, pressure range, and temperature range. In-depth reviews of valves and their characteristics are found in references [1,21,22].

4.4.3.1 Passive Valves

Passive valves are pressure sensitive and only require a pneumatic source to operate. Simple normally open or closed valves can be made using multilayer soft lithography (Figure 4.24). They consist of a flow channel and an under-lying pneumatic control channel separated by a thin membrane. In the normally open valve, a positive pressure applied to the control channel inflates the thin membrane to shut off flow in the channel above it [23]. In contrast, a vacuum applied to the control channel in the normally closed valve deflects the membrane to allow a flow path to open across the valve [24].

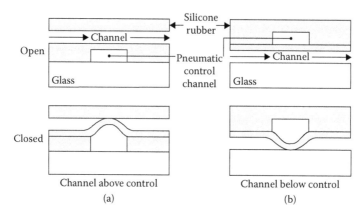

FIGURE 4.24

Two types of normally open valves fabricated using multilayer soft lithography in which the flow channel is either (a) above or (b) below the pneumatic control channel.

Check valves are a common type of pressure-sensitive passive valve. These valves allow flow in only one direction and most are normally closed. Positive pressure deflects the valving element away from the valve seat; once the pressure is removed, a mechanical restoring force returns the element to the valve seat. Reverse pressure applied across the valve serves to further tighten the seal between the valving element and the valve seat; there is no flow in this direction. Common types of check valves are shown in Figure 4.25.

Finally, in-channel valves may also be created without the need for any moving parts. Hydrophobic channels with sudden changes in their cross-sectional areas exhibit a pressure drop across this region [26]. The flow past this junction is regulated by the inlet pressure. The same concept can be implemented in a rotational format using centrifugal pressure to initiate flow in capillary channels [27,28]. Selectively patterned hydrophobic patches in hydrophilic channels also serve as pressure-controlled valves [29,30].

4.4.3.2 Actuation Methods and Active Valves

Actuators convert electrical energy into mechanical energy. This generates the desired movement to displace valve elements such that flow is either stopped or allowed to pass. In general, actuators are selected for their ability to generate appropriate forces and displacements, in addition to their power consumption, size, response time, and ease of integration.

Coulombic attraction between oppositely charged metal plates forms the basis of *electrostatic* actuation. The force generated between parallel plates is captured by Coulomb's Law and is represented as

$$F = \frac{1}{2} \frac{\varepsilon_0 \varepsilon_r A V^2}{d^2} \tag{4.81}$$

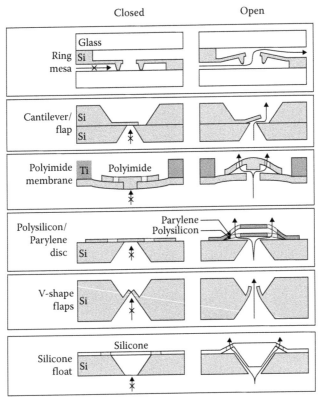

FIGURE 4.25
Common normally closed check-valve configurations. The forward-flow path is indicated using arrows.

where A is the electrode area, V is the applied voltage, and d is the electrode separation. While their construction is simple, electrostatic actuators are nonlinear and require large voltages to produce small forces and strokes (Figure 4.26).

Electromagnetic forces useful for actuation are present when a current-carrying wire is subjected to a magnetic force. The Lorenz force that the wire experiences is given by

$$F = (I \times B)L \qquad (4.82)$$

where I is the current passing through the wire, B is the magnetic field, and L is the wire length. Large forces are possible in this system, but an external magnetic field is necessary (e.g., a permanent magnet), which adds significant bulk to a microsystem (Figure 4.27). Another useful form of electromagnetic actuation may be used in a magnetic circuit with an air gap.

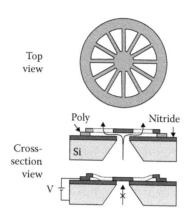

Top
view

Poly   Nitride

Cross-
section
view

Si

V

FIGURE 4.26
Example of a normally open electrostatic valve. Application of voltage attracts the top polysilicon plate toward the substrate and shuts off the bottom-to-top flow path.

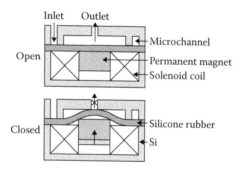

Inlet   Outlet

Microchannel

Open

Permanent magnet
Solenoid coil

Closed

Silicone rubber

Si

FIGURE 4.27
Example of a normally closed electromagnetic valve. A permanent magnet is deflected by electromagnetic forces generated by energizing the solenoid coil; this in turn deflects a silicone rubber membrane to block the outflow path. The flow path is complex and not completely revealed in the cross-sectional drawings.

The magnetic circuit consists of a magnetic core with an air gap wrapped with a coil. The force developed across the air gap is given by

$$F = \frac{\phi_g^2}{2\mu_0 A_g} = \frac{\left. NI \middle/ \left(\mathfrak{R}_c + \mathfrak{R}_g\right)\right.}{2\mu_0 A_g} \tag{4.83}$$

where ϕ_g is the magnetic flux in the circuit, A_g is the cross-sectional area in the air gap, N is the number of coil turns, I is the coil current, $\mathfrak{R}_c$ is the magnetic reluctance of the magnetic core, and $\mathfrak{R}_g$ is the magnetic reluctance of the air gap.

Special piezoelectric materials undergo strain in the presence of an applied electric field. As a result, large forces can be produced but with small displacements (Figure 4.28). This phenomena and related equations are given in Section 2.1.5.

Electrolysis is the process by which application of electricity splits water into hydrogen and oxygen gas bubbles. Very large volume expansion, and thus large forces and deflections, are possible (Figure 4.29). The pressure generated is a function of the volume of gas produced compared to the volume of water consumed by electrolysis, and the maximum pressure possible (~200 MPa [34]) is calculated using the following expression:

$$P = \frac{3}{2} \frac{RT}{M_{H_2O}/\rho_{H_2O}} \tag{4.84}$$

where R is the gas constant, M_{H_2O} is the molecular weight of water, and ρ_{H_2O} is the density of water.

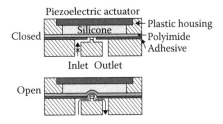

FIGURE 4.28
Example of a normally closed piezoelectric valve. The valve is sealed shut at the interface between the thin film silicone rubber–covered valve seat and the polyimide valve membrane. When the piezoelectric actuator is energized, the silicone and polyimide composite membrane deflects upward to open the flow path indicated by the arrow.

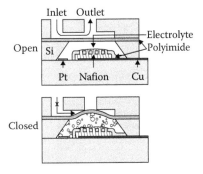

FIGURE 4.29
Example of a normally open electrolysis valve. The thin film polyimide membrane deflects when current is passed between the electrolysis actuator electrodes to induce gas formation. The pressure increase in the electrolyte chamber pushes the polyimide membrane upward and eventually seals the flow path when sufficient actuation pressure is reached.

Thermal energy can be harnessed to produce a volume and/or phase change that results in thermopneumatic actuation. This actuation method has been demonstrated with waxes and specially formulated fluids (such as Fluorinert™ in Figure 4.30). For liquids stored in a closed system (one compliant wall is necessary for actuation), force can be extracted from the pressure increase as follows:

$$P = E\left(\alpha_T \Delta T - \frac{\Delta V}{V}\right) \tag{4.85}$$

where E is Young's modulus, α_T is the thermal expansion coefficient, ΔT is the temperature increase, and $\Delta V/V$ is the volume change. Large pressures can be generated for large deflections and forces but at the cost of significant power consumption. In addition, the thermal processes are slow during both pressure generation and cooling to restore the initial state.

Materials having dissimilar thermal expansion coefficients exhibit thermally induced stresses that result in small deflections. These bimetallic materials require the presence of a temperature gradient for actuation, which must be large because thermal expansion coefficients are fixed and small (Figure 4.31). The force produced is proportional to the difference in thermal expansion coefficients and the temperature difference as follows:

$$F \sim \left(\alpha_{T,1} - \alpha_{T,2}\right)\Delta T \tag{4.86}$$

Another special class of materials that provide actuation in the presence of a temperature gradient are shape memory alloys (e.g., gold/copper, indium/titanium, and nickel/titanium). These alloys undergo reversible temperature-dependent phase transitions, but as with most thermal actuation

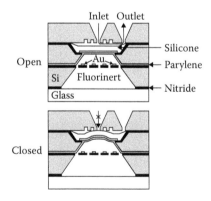

FIGURE 4.30
Example of a normally open thermopneumatic valve. Au-resistive heaters are suspended in the cavity, which heat the surrounding Fluorinert to initiate a phase change. This leads to a pressure increase that deflects the silicone rubber-Parylene composite valve and shuts off the flow path.

methods, the response time is slow and the power consumption is high (Figure 4.32).

4.4.4 Pumping

Pumps can be active or passive; it is sometimes more useful to categorize pumps based on the source of kinetic energy for driving fluid flow. In other words, pumps may derive their driving force from mechanical or nonmechanical means. Pump performance is measured in terms of flow rate range, sustainable back pressure, and pump efficiency. Back pressure opposes pumping and takes the form of a positive pressure gradient from the output to the input. In-depth reviews of micropumps and their characteristics are found in references [1,22,39,40].

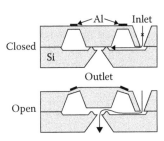

FIGURE 4.31
Example of a normally closed bimetallic valve. The force to open the flow path is generated by the presence of an electrically induced temperature gradient and dissimilar thermal expansion coefficients between silicon and aluminum.

4.4.4.1 Passive Pumps

A simple passive pump can be assembled from three passive normally open valve elements (Figure 4.33). Each valve element is pneumatically actuated in a sequential manner to displace small fluid packets. The progression of valve activation dictates the net fluid motion; the pump is symmetrical and may be operated in forward or reverse. A typical valve activation sequence is 1, 1–2, and 2–3, where the numbers refer to the valve or valves that are turned on to trap or displace a fluid packet. The frequency of valve switching determines the flow rate. Due to the packet-by-packet nature of fluid displacement, flow may appear to be pulsatile (as opposed to continuous).

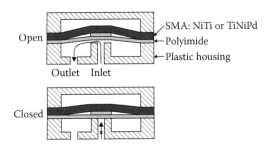

FIGURE 4.32
Example of a normally open shape memory alloy valve. The spacer between the polyimide membrane and the shape memory alloy device can be made thicker or thinner to adjust the maximum back pressure. Heating the shape memory alloy device above the phase transition temperature regains the undeflected shape and causes the valve to close.

4.4.4.2 Active Pumps: Mechanical

The most common mechanical pump is the diaphragm type, which consists of a pumping chamber bounded by a compliant diaphragm and two check valves (Figure 4.34). The check valves are oriented in opposite directions such that when an actuator deflects the compliant diaphragm, fluid from within the chamber is forced out through one valve due to an increase in the chamber pressure (pump or discharge mode). When the diaphragm returns to the resting position, the other valve pulls in fluid to replace what was pumped out (supply mode). To operate the check valves, the pressure produced must exceed their cracking pressures.

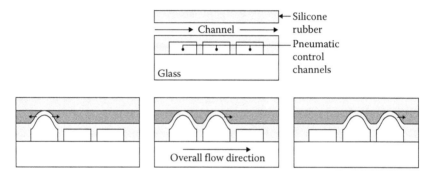

FIGURE 4.33
Top: Layout of a passive peristaltic pump fabricated using multilayer soft lithography. The pump is simply a combination of three pneumatic valves, similar to the ones in Figure 4.24. Bottom: Valves are pneumatically actuated in a sequential manner to selectively displace a packet of fluid. An example of a pneumatic pumping scheme is also shown. Pumping may occur in either direction due to the symmetry of the device.

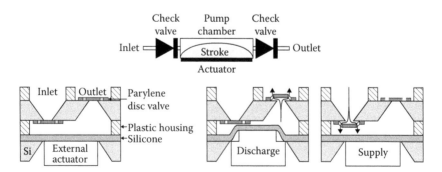

FIGURE 4.34
Top: Schematic representation of a typical diaphragm pump structure. Bottom: Example of a diaphragm pump. The two operational modes (discharge and supply) are shown. This pump is sometimes referred to as a reciprocating pump due to the cycling between modes to achieve pumping.

The check valves in diaphragm pumps are necessary to rectify flow. However, in valveless pumps, inlet and outlet channels in the shape of nozzles (converging duct) or diffusers (expanding duct), respectively, are used instead (Figure 4.35). In all other respects, the layout is quite similar to that of a diaphragm pump. Flow rectification occurs due to the pressure drop developed across the rectification structures, with lower pressure loss in the diffuser direction. This pressure drop is sensitive to the flow direction such that during the pumping mode, fluid is pushed out through both the inlet and outlet but with a greater magnitude at the outlet. In other words, the outlet behaves as a diffuser with lower flow resistance compared to the inlet, which behaves as a nozzle. The opposite is true in the supply mode. Other common active mechanical pumps include the peristaltic, rotary, ultrasonic, and centrifugal types.

4.4.4.3 Active Pumps: Nonmechanical

Electrokinetic principles can be used to create pumps that drive fluids using electrical fields but without any moving parts; the only limitation is that the fluids must be conductive. Both electroosmosis and electrophoresis are used in pumping applications. A simple electroosmotic pump was previously introduced in Figure 4.14. Electrokinetic pumps function best when glass is used to construct the microchannel due to the high voltage and immobilized surface charge requirements.

Surface tension is another nonmechanical means to drive fluid flow. For example, we have already noted that fluid is displaced by capillarity (Figure 4.6). Surface tension is also a function of temperature, and surface-tension gradients give rise to fluid flow (*Marangoni effect*). The temperature gradient and the resulting surface tension gradient across a vapor bubble in a microchannel may be used to drive flow in one of two ways (Figure 4.36) [43]. First, the pressure difference across a single bubble heated asymmetrically

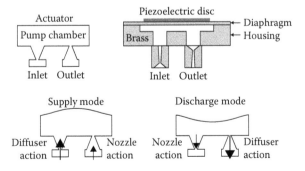

FIGURE 4.35
Top left: A schematic representation of a valveless pump. Top right: Example of a valveless pump. Bottom: The two operational modes of the pump.

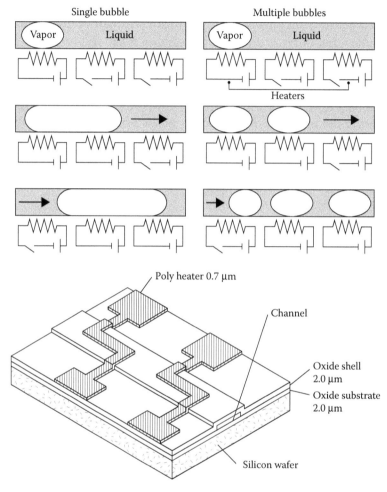

FIGURE 4.36
Top: Two methods of thermocapillary pumping. Bottom: Schematic representation of a thermocapillary pump. (From Jun, T. J., and C.-J. Kim. 1998. *J Appl Phys* 83:5658. With permission.)

at either end drives the flow in a microchannel toward the hotter side. Alternatively, multiple bubbles generated sequentially may be used, in which a previously generated vapor bubble acts as a valve.

Osmosis is a chemical process by which a pressure gradient can be achieved for pumping. A solvent moves across a semipermeable membrane from a region having low solute concentration to one with high concentration. The osmotic pressure that develops is expressed using the van't Hoff equation as follows:

$$P_{\text{osmotic}} = cRT \qquad (4.87)$$

where c is the molar concentration of the solute, R is the gas constant, and T is the absolute temperature. This pressure acts on a flexible membrane to displace fluid. A pump that uses an osmotic actuator to drive fluid is shown in Figure 4.37. In this example, the semipermeable membrane is formed from cellulose acetate and the actuator membrane is formed from vinylidene chloride and acrylonitrile copolymer [44]. Sodium chloride is used as the osmotic driving agent [45]. Other common active nonmechanical pumps include magnetohydrodynamic, ferrofluidic, evaporative, and electrolytic types.

4.4.5 Mixing

A low Re results in laminar flow for the vast majority of microfluidic devices. Thus, mixing cannot be achieved by turbulent processes as in macroscale flows. Instead, diffusion is the predominant method and efficient mixing by diffusion is determined by the process time. From Equation 4.41, diffusion time (for the one-dimensional case) is proportional to the square of the diffusion length, or mixing path. This suggests that diffusion time is decreased in microscale systems by virtue of both shorter mixing paths and large surface area-to-volume ratios. In practice, efficient mixing by diffusion at microscale levels occurs only in limited cases. In addition, diffusion in a single channel is ultimately limited by length; long channels consuming significant chip real estate are often needed to achieve adequate mixing. For example, a Y-mixer consists

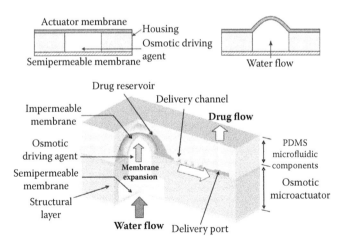

FIGURE 4.37
Top: Osmotic actuator. Bottom: Diagram of an osmotic pump intended for drug-delivery applications. (From Su, Y. C., and L. W. Lin. 2004. *J Microelectromech Syst* 13:75. With permission.)

of three channels arranged in a Y-shape. Two streams join into a single stream and eventually mix. To increase the length of the mixing channel while minimizing the real estate occupied, the straight channel can be replaced with a winding serpentine channel. Many alternative solutions for efficient mixing have been explored.

Splitting and recombining fluid streams is a mixing strategy called lamination. A simple example involves taking two fluid streams and subdividing each stream into two laminae (Figure 4.38) [46]. By doing so, the mixing speed is increased by a factor of n^2, in which n is the number of laminae. The diffusion time is now written as

$$t = \frac{L^2}{2n^2D} \qquad (4.88)$$

Many forms of lamination in two and three dimensions have been explored. A similar concept involves focusing a single stream composed of one or more laminar streams into a smaller flow channel. This action, termed hydrodynamic focusing, creates thinner flow streams for faster mixing.

A serpentine mixing channel not only conserves chip real estate but also takes advantage of another microscale mixing principle called *chaotic advection*. The corners present in the twisting channel induce turbulence on the flow through folding and stretching or local chaotic regimes. Texturing straight channels can also result in mixing by chaotic advection. Adding staggered herringbone-shaped ridges along the length of a microchannel also results in forced fluid circulation around the axis of fluid flow. The folding and stretching of fluid is visualized with a dye, as in Figure 4.39. The methods described thus far are all passive mixing schemes. Mixing can also be achieved through active means, as in valving and pumping. Ultrasonic energy or other means of energetic agitation are possible.

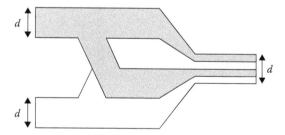

FIGURE 4.38
A simple mixer that functions on the principle of lamination of successively thinner streams. In this example, two fluid streams are each split into two laminae that are then rejoined downstream.

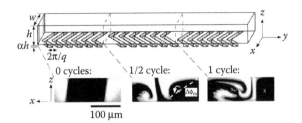

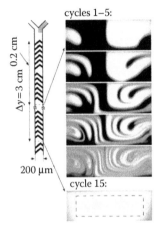

FIGURE 4.39

Chaotic mixing in a straight polydimethylsiloxane channel using staggered herringbone structures. Top: A full mixing cycle is achieved when fluid passes through two sequences of herringbone-shaped ridges. Images taken at a perpendicular angle to the direction of flow show the progress of mixing between a fluorescent and a nonfluorescent solution. Bottom: Mixing results after 15 cycles. (From Stroock, A. D., et al. 2002. *Science* 295:647. With permission.)

4.5 Problems

1. Consider a capillary tube with one end dipped in a liquid. What happens to the capillary rise height if $\theta > 90°$? Explain your answer using Equation 4.12. Similarly, what happens to the pressure difference across the air/liquid interface?

2. Calculate the capillary rise height for a microchannel of radius 50 µm for water in a glass channel ($\rho_{water} = 1 \times 10^3 \, kg/m^3$, $\Upsilon_{water} = .0729 \, N/m$, $\theta_{water/glass} = 25°$). What is the rise height for mercury in the same system ($\rho_{mercury} = 1.36 \times 10^4 \, kg/m^3$, $\Upsilon_{mercury} = .4865 \, N/m$, $\theta_{mercury/glass} = 140°$)?

3. What is the Reynolds number for an isotropically etched channel with a hemicylindrical cross section? The channel diameter is 50 µm, the fluid velocity is 1 mm/s, and the liquid is water at room temperature ($\rho_{water} = 1 \times 10^3 \, kg/m^3$ and $\mu_{water} = 1.002 \times 10^{-3} \, Pa \cdot s$).

4. You are asked to design a microfluidic channel of adequate length for a particular application but need to balance length with flow resistance and, thus, the pressure drop. The channel will be used for liquids having a viscosity equal to that of water (@20°C, μ_{water} = 1.002×10^{-3} Pa • s). The channel width and height are predetermined and cannot be changed (100 μm and 10 μm, respectively). In addition, you will use a flow rate of 0.1 μL/min. Using Equations 4.26 and 4.28, determine the pressure drops for channel lengths of 10, 5, and 1 mm.

5. What is the Debye length for the following system: a microchannel filled with a monovalent salt dissolved in water (ε_{water} = 78.5 and C_0 = .1 mol/L)? Assume that the system is at room temperature (25°C). The electron charge q_e is 1.6×10^{-19} C, the permittivity of vacuum ε_0 is 8.85×10^{-12} C/V • m, and the Boltzmann constant k_B is 1.38×10^{-23} J/K. Avogadro's number is 6.022×10^{23}/mol.

6. In a case where the electrolyte solution is asymmetrical (the anion valency is different from the cation valency), then the term (z^2C_0) in Equation 4.57 is replaced with the following expression for the ionic strength (I) of the solution (a function of all ions present in the solution):

$$I = \frac{1}{2}\sum_i m_i z_i^2$$

where for the ionic species i, m_i is the molar concentration, and z_i is the valency. The sum is taken over all ions in the solution. What is the Debye length for the system in Problem 5 if the monovalent salt solution is replaced by a 1:3 asymmetric electrolyte (0.1 mol/L)? How does the Debye length change as a result of the increased ionic strength (even though the concentrations are the same)?

7. For a zeta potential of 100 mV and an applied potential of 100 V, what is the electroosmotic flow velocity for water (μ_{water} = 1.002×10^{-3} Pa • s, ε_{water} = 78.5) in a channel measuring 1 mm long? If the channel measures 10 μm high and 100 μm wide, what is the volumetric flow rate (in μL/min)?

8. What are the advantages of driving dielectrophoresis using an AC field compared to using a DC electric field?

9. There are many ways to make microfluidic channel structures apart from the few examples provided in the text. Suggest another method to make an enclosed microchannel (one having walls throughout the cross section). Briefly describe the fabrication process and draw the cross-sectional images that correspond to the major steps in your process. The fabrication process should include all the major steps required to form the microchannel.

10. How would you make a normally closed valve using a multilayer soft lithography approach? Briefly describe the materials needed and draw the master molds you will need. Be sure to include a description of how the layers will align on top of one another. Draw the top view and cross-sectional images to show the valve structure. Be sure to indicate where the cross-sectional images are taken in the top view. You may use multiple cross-sectional images if necessary. Explain in your own words how your valve design operates. Be sure to include descriptions of how the valve changes from the normally closed state to the open state.

11. Select and read a journal paper published either this year or last regarding a microvalve device. Provide the reference in the Institute of Electrical and Electronics Engineers (IEEE) format given previously in Chapter 1. Answer the following questions in your own words. Is the valve a normally closed or normally open valve? Is the valve active or passive? How does the valve work?

12. Select and read a journal paper published either this year or last regarding a micropump device. Provide the reference in the IEEE format given previously in Chapter 1. Answer the following questions in your own words. Is the pump active or passive? How does the pump work?

13. Let us say you want to mix water with nitric acid at room temperature. You have chosen to design a simple micro Y-mixer. If you set the flow rates of the two solutions to be 5 μL/min and use a channel cross section ($w \times h$) of 50 μm × 150 μm, what should the length of your channel be to provide adequate mixing? The diffusion coefficient of nitric acid in water at room temperature is 2.60×10^{-5} cm^2/s. For simplicity, assume that mixing occurs only in one dimension across the width of the channel.

4.6 Laboratory Exercise

Laboratory Exercise 4: Soft Lithography and Microchannels

Purpose: In this laboratory exercise, you will learn how to create a simple silicone rubber (also known as PDMS) microfluidic device using soft lithography methods. This exercise assumes that a master mold containing microfluidic features is already prepared and available.

Background: A two-part PDMS resin system (Sylgard® 184, Dow Corning, Midland, MI) will be used to create microfluidic devices. Part A (elastomer base) contains vinyl groups, and Part B (curing agent) contains hydrosiloxane groups. When mixed together, a prepolymer consisting of a cross-linked network of dimethylsiloxanes is formed. The prepolymer mixture is poured over a master mold and allowed to cure. The replica can be removed to form part of a microfluidic channel system. Here, a single-layer process involving only one replica layer will be used.

Materials and supplies
- Gloves
- Lint-free cleanroom cloths
- Master mold (i.e., photoresist patterned on a silicon or glass substrate)
- Glass microscopic slides
- Tweezers
- Plastic beakers or containers (for mixing prepolymer)
- Glass beakers
- Glass graduated cylinders
- Glass mixing rod
- Stainless steel spatula
- Aluminum foil
- Blunt-tip syringe needle (~20 G)
- Syringe

Chemicals
- Sylgard® 184 two-part silicone rubber kit (or another comparable type of silicone rubber)
- Ethanol
- Isopropyl alcohol
- Deionized water

Equipment
- Digital balance
- Hot plate
- Vacuum chamber and pump (optional, but recommended)
- Oven (optional)

Calculating the volume of PDMS prepolymer to mix
1. Based on the geometry of your master mold and the desired thickness of your replica, estimate the total volume of PDMS prepolymer required. Note that the final replica will shrink slightly during the curing process.

2. The prepolymer mixture will contain 10 parts elastomer (Part A) to 1 part curing agent (Part B) by weight. Approximate the desired volume of prepolymer and calculate the weights of Parts A and B. Make sure the balance you are using possesses adequate accuracy to obtain the desired amounts of Parts A and B.

Preparing PDMS prepolymer

1. Obtain a plastic beaker or container for mixing the prepolymer.
2. Place the vessel on a balance and tare the balance so that the reading is zeroed.
3. Using the reading on the balance, pour the desired amount of Part A into the vessel. Record the weight and tare the balance again.
4. Then add the desired amount of Part B into the vessel with Part A.
5. Thoroughly mix Parts A and B using a glass mixing rod or stainless steel spatula. An electronic mixer can also be used if available. Parts covered in prepolymer can be cleaned using a cloth and isopropyl alcohol.

Forming the replica

1. Place the master mold into a suitable plastic container (e.g., a polystyrene Petri dish) or a container fashioned from a piece of aluminum foil (you can fold excess aluminum foil up around the edges of the master mold).
2. Slowly pour the PDMS onto the master mold in the container. Try not to trap air bubbles in the polymer during the pouring process.
3. Degas the PDMS (remove trapped air bubbles) in a vacuum chamber. Ensure that the container holding the PDMS has high walls because the PDMS can expand to occupy a space many times its original volume as the trapped gases are removed.
4. Cure the PDMS at room temperature (for approximately 24 hours) or at 65°C using an oven/hot plate (for about 2 hours).

Bonding the PDMS replica to a glass substrate

1. First, the surface of the PDMS replica will be treated to allow bonding to a glass microscope slide. When you are ready to perform the surface treatment process, remove the replica from the mold in a clean, dust-free environment using tweezers.
2. In a glass beaker, make a 1000:1 (by volume) solution of deionized water and hydrochloric acid. Calculate the volume necessary to completely submerge the replica. Then add water to the beaker first, followed by the acid. Stir the solution with a glass stirring rod to mix it.
3. Place the beaker on a hot plate and set the temperature to 60–80°C.
4. Soak the replica in the heated solution for 30 minutes.

5. Remove the replica and rinse in deionized water.

6. Rinse the replica in ethanol and blow-dry the chip using compressed nitrogen gas.

7. Place the replica with the molded side face up on a clean glass microscope slide. Use caution when handling the treated replica. Only hold the replica on the edges and do not handle or touch the side to be bonded. The glass microscope slide can be cleaned in ethanol and dried using compressed nitrogen gas. Ensure that the surface is free of dust and other contamination.

8. Pick up the replica and invert it such that the molded side is facing the glass slide. After aligning the replica with the slide, place the two in contact starting on one edge and gently allow the rest of the replica to come into contact with the slide. Press gently to squeeze out trapped air bubbles but not hard enough to collapse any channels.

9. Place the stack in an oven or on a hot plate set at 80°C for at least 3 hours.

Preparing microfluidic connections

1. Examine the bonded microfluidic chip to identify the location of the inlet and outlet ports.

2. Using a blunt-tip syringe needle (~20 G), core a hole through the PDMS at the site of the desired inlet/outlet port.

3. Remove the needle along with the cored PDMS.

4. Insert a new syringe needle or rigid tubing having the same outer diameter as the coring needle into the cored hole. Do not insert the needle/tubing up to the glass substrate—leave a small gap to allow fluid to pass through and enter the microfluidic system.

5. Repeat this process for all inlet and outlet ports.

Check connections

1. Attach a syringe to one of the inlet needle/tube ports created and inject deionized water (could be dyed to facilitate viewing) to ensure that a fluidic connection was successfully established.

2. Repeat this process as necessary for all the inlet ports.

3. The microfluidic chip is now ready to perform its intended function.

References

1. Nguyen, N.-T., and S. T. Wereley. 2002. *Fundamentals and Applications of Microfluidics*. Boston: Artech House.

2. Batchelor, G. K. 1967. *An Introduction to Fluid Dynamics*. Cambridge: Cambridge University Press.
3. Bruus, H. 2008. *Theoretical Microfluidics*. Oxford: Oxford University Press.
4. White, F. M. 2008. *Fluid Mechanics*. 6th ed. New York: McGraw-Hill.
5. Sabersky, R. H., A. J. Acosta, and E. G. Hauptmann. 1989. *Fluid Flow: A First Course in Fluid Mechanics*. 3rd ed. New York: Macmillan.
6. Mugele, F., and J. C. Baret. 2005. Electrowetting: From basics to applications. *J Phys Condens Matter* 17:R705.
7. Pollack, M. G., R. B. Fair, and A. D. Shenderov. 2000. Electrowetting-based actuation of liquid droplets for microfluidic applications. *Appl Phys Lett* 77:1725.
8. Moon, H., et al. 2002. Low voltage electrowetting-on-dielectric. *J Appl Phys* 92:4080.
9. Pollack, M. G., A. D. Shenderov, and R. B. Fair. 2002. Electrowetting-based actuation of droplets for integrated microfluidics. *Lab Chip* 2:96.
10. Cho, S. K., H. J. Moon, and C. J. Kim. 2003. Creating, transporting, cutting, and merging liquid droplets by electrowetting-based actuation for digital microfluidic circuits. *J Microelectromech Syst* 12:70.
11. Paik, P., et al. 2003. Electrowetting-based droplet mixers for microfluidic systems. *Lab Chip* 3:28.
12. Huh, D., et al. 2003. Reversible switching of high-speed air-liquid two-phase flows using electrowetting-assisted flow-pattern change. *J Am Chem Soc* 125:14678.
13. Hunter, R. J. 1981. *Zeta Potential in Colloid Science: Principles and Applications*. London: Academic Press.
14. Probstein, R. F. 1994. *Physicochemical Hydrodynamics: An Introduction*. 2nd ed. New York: Wiley.
15. Sharp, K. V., et al. 2002. Liquid flows in microchannels. In *The MEMS Handbook*. Boca Raton, FL: CRC Press.
16. Devasenathipathy, S., and J. G. Santiago. 2005. Electrokinetic flow diagnostics. In *Microscale Diagnostic Techniques*. Berlin: Springer.
17. Gu, W., X. Y. Zhu, N. Futai, et al. 2004. Computerized microfluidic cell culture using elastomeric channels and braille displays. *Proceedings of the National Academy of Sciences of the United States of America* 101:15861-6.
18. Huh, D., et al. 2002. Use of air-liquid two-phase flow in hydrophobic microfluidic channels for disposable flow cytometers. *Biomed Microdevices* 4:141.
19. Dittrich, P. S., and P. Schwille. 2003. An integrated microfluidic system for reaction, high-sensitivity detection, and sorting of fluorescent cells and particles. *Anal Chem* 75:5767.
20. Lee, G. B., B. H. Hwei, and G. R. Huang. 2001. Micromachined pre-focused m x n flow switches for continuous multisample injection. *J Micromech Microeng* 11:654.
21. Oh, K. W., and C. H. Ahn. 2006. A review of microvalves. *J Micromech Microeng* 16:R13.
22. Zhang, C. S., D. Xing, and Y. Y. Li. 2007. Micropumps, microvalves, and micromixers within PCR microfluidic chips: Advances and trends. *Biotechnol Adv* 25:483.
23. Unger, M. A., et al. 2000. Monolithic microfabricated valves and pumps by multilayer soft lithography. *Science* 288:113.
24. Grover, W. H., et al. 2006. Development and multiplexed control of latching pneumatic valves using microfluidic logical structures. *Lab Chip* 6:623.

25. Shoji, S., and M. Esashi. 1994. Microflow devices and systems. *J Micromech Microeng* 4:157.
26. Ahn, C. H., et al. 2004. Disposable smart lab on a chip for point-of-care clinical diagnostics. *Proc IEEE* 92:154.
27. Duffy, D. C., et al. 1999. Microfabricated centrifugal microfluidic systems: Characterization and multiple enzymatic assays. *Anal Chem* 71:4669.
28. Johnson, R. D., et al. 2001. Development of a fully integrated analysis system for ions based on ion-selective optodes and centrifugal microfluidics. *Anal Chem* 73:3940.
29. Andersson, H., et al. 2001. Hydrophobic valves of plasma deposited octafluoro-cyclobutane in DRIE channels. *Sens Actuators B Chem* 75:136.
30. Andersson, H., W. van der Wijngaart, and G. Stemme. 2001. Micromachined filter-chamber array with passive valves for biochemical assays on beads. *Electrophoresis* 22:249.
31. Yobas, L., et al. 2003. A novel integrable microvalve for refreshable Braille display system. *J Microelectromech Syst* 12:252.
32. Bae, B., et al. 2002. Feasibility test of an electromagnetically driven valve actuator for glaucoma treatment. *J Microelectromech Syst* 11:344.
33. Shao, P. G., Z. Rummler, and W. K. Schomburg. 2004. Polymer micro piezo valve with a small dead volume. *J Micromech Microeng* 14:305.
34. Cameron, C. G., and M. S. Freund. 2002. Electrolytic actuators: Alternative, high-performance, material-based devices. *Proc Natl Acad Sci U S A* 99:7827.
35. Neagu, C. R., et al. 1997. An electrochemical active valve. *Electrochim Acta* 42:3367.
36. Yang, X., C. Grosjean, and Y. C. Tai. 1999. Design, fabrication, and testing of micromachined silicone rubber membrane valves. *J Microelectromech Syst* 8:393.
37. Jerman, H. 1994. Electrically activated normally closed diaphragm valves. *J Micromech Microeng* 4:210.
38. Kohl, M., et al. 2000. Thin film shape memory microvalves with adjustable operation temperature. *Sens Actuators A Phys* 83:214.
39. Nguyen, N. T., X. Y. Huang, and T. K. Chuan. 2002. MEMS-Micropumps: A review. *J Fluids Eng Trans ASME* 124:384.
40. Laser, D. J., and J. G. Santiago. 2004. A review of micropumps. *J Micromech Microeng* 14:R35.
41. Meng, E., et al. 2000. A check-valved silicone diaphragm pump. In *MEMS 2000*. Japan: Miyazaki.
42. Stemme, E., and G. Stemme. 1993. A valveless diffuser/nozzle-based fluid pump. *Sens Actuators A Phys* 39:159.
43. Jun, T. J., and C.-J. Kim. 1998. Valveless pumping using traversing vapor bubbles in microchannels. *J Appl Phys* 83:5658.
44. Su, Y. C., and L. W. Lin. 2004. A water-powered micro drug delivery system. *J Microelectromech Syst* 13:75.
45. Su, Y. C., L. W. Lin, and A. P. Pisano. 2002. A water-powered osmotic microactuator. *J Microelectromech Syst* 11:736.
46. Erbacher, C., et al. 1999. Towards integrated continuous-flow chemical reactors. *Mikrochim Acta* 131:19.
47. Stroock, A. D., et al. 2002. Chaotic mixer for microchannels. *Science* 295:647.

5

Lab-on-a-Chip and Micro Total Analysis Systems

5.1 Microanalytical Systems in Chemistry and Biology

The term "lab-on-a-chip" (LOC) and micro total analysis systems (µTAS) are a subset of microelectromechanical systems (MEMS) dedicated to chemical and biological analyses and discoveries. The first example of a miniaturized LOC was created in the late 1970s by S. Terry at Stanford University [1]. The device was a gas chromatographic analyzer that combined an injection valve, a separation column, and a thermal-conductivity detector on the surface of a single silicon wafer (5 cm in diameter). It could separate gaseous mixtures of hydrocarbons in less than 10 seconds (Figure 5.1).

This amazing achievement was unappreciated until the 1990s when researchers worldwide recognized the potential and benefits of miniaturization in chemical and biological analyses. It was at this time that the term "micro total analysis systems" was first introduced [3] with the vision that these systems could perform sample handling, analysis, and detection in compact systems that were enabled by microfabrication. The basic elements of such a system are shown in Figure 5.2. Early work in the field included the development of microfluidic components (e.g., pumps and valves), followed by demonstration of planar sample-handling processes (e.g., injection and separation) on tiny chips. Many new microfabrication technologies that explored inexpensive techniques to incorporate polymers into the materials toolbox developed simultaneously. Following decades of development, many practical LOC systems that integrate a multitude of functions now exist [4–8]. Today, there is a scientific conference that shares the µTAS name and a journal named *Lab on a Chip* [9]. Industry (e.g., Agilent, Aclara, Caliper, and Fluidigm) has also taken notice of the many exciting developments and has successfully converted research laboratory curiosities into commercial products.

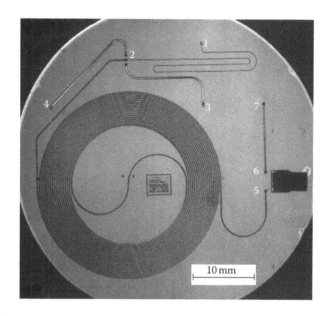

FIGURE 5.1
Gas chromatograph on a silicon wafer. The numbers correspond to (1) carrier gas input, (2) through holes to a backside valve, (3 and 4) sample gas input and output, respectively, (5 and 6) through holes to a thermal conductivity detector, and (7) a carrier gas vent. (From Terry, S. C., J. H. Jerman, and J. B. Angell. 1979. *IEEE Trans Electron Devices* 26:1880; and modified from de Mello, A. 2002. *Lab Chip* 2:48N. With permission.)

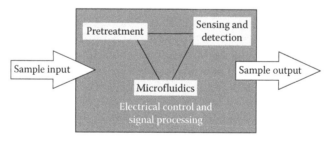

FIGURE 5.2
The basic elements of an LOC system.

This chapter focuses on the sample handling and treatment aspects of LOC. In fact, sample handling and processing are often the most difficult in LOC and in macroscale analytical methods. Microfluidic components were introduced in the Chapter 4, and sensing and detection are covered in Chapter 6. Examples of applications of LOC and other biomedical systems are covered in Chapters 7–9.

5.2 Sample Pretreatment

Biological and chemical samples are typically mixtures that contain diluted amounts of the species of interest. If the amount of the target species is below the detectable limit, the sample must be preconcentrated, or enriched, prior to any additional processing or detection. For example, some proteins exist only in low concentrations. Circulating tumor cells (CTCs) play a role in cancer metastasis but evade detailed investigation because they are difficult to isolate and identify due to their low concentrations in blood. Airborne toxins may be detectable in trace amounts prior to the onset of a major catastrophe. All these applications require significant sample pretreatment.

Sample pretreatment is usually performed in preparation for subsequent sample separation or detection. Not all pretreatment processes need to be miniaturized and, in some cases, it is not possible to outperform macroscale processes by scaling. For example, processing of crude samples by grinding, dissolving, or removing large particles is better performed off-chip because these processes require significant power input or may otherwise be impractical in a small form factor device or process [10]. The processes and methods involved in sample pretreatment are outlined in Figure 5.3. Additional reviews of pretreatment methods are found in references [10–12].

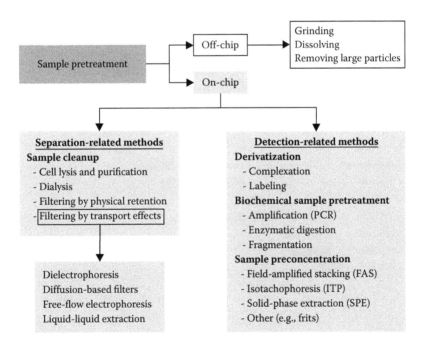

FIGURE 5.3
Categorization of sample pretreatment methods into on- and off-chip processes.

5.2.1 Filtration

Microchannels are particularly susceptible to clogging by particles in sample solutions due to their small sizes. Samples may include interfering species that require removal prior to further processing. Filtration is performed to clean up samples by removing unwanted particles or other species present in fluids (Figure 5.4). Filters may also be used to collect materials for further analysis. Some filtration steps are easily performed off-chip, whereas others require the use of discrete microfabricated filters or filters that are integrated into a microfluidic system. Filtration devices can also be used for retention. For example, it may be necessary to retain particles within a microchannel that are intended for interaction with the sample flow (as in chromatography). The focus of this section is predominantly on filtration; however, some examples of such devices are given here and later in Chapter 7 in the context of retaining cells and particles.

Microfabricated filtration devices have predefined pores that are dimensionally controlled either by lithographic methods or by the thickness of the sacrificial layer. Lithographic methods result in relatively large pores, whereas sacrificial layer processes create tortuous filtration paths. Several researchers have devised clever modifications to these standard processes, which allow even finer dimension control. Regardless of the fabrication method, a key design parameter for filters is the opening factor (β), which

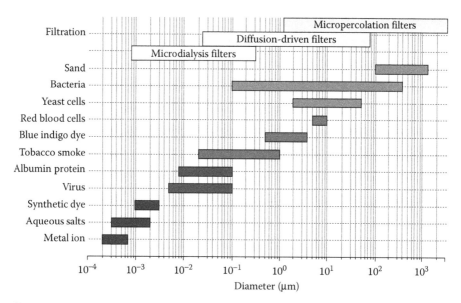

FIGURE 5.4
Dimensions of particles and analytes as related to microfabricated filtration devices. (From Lichtenberg, J., N. F. de Rooij, and E. Verpoorte. 2002. *Talanta* 56:233. With permission.)

compares the area of the pores in the filter ($A_{opening}$) to the overall filter area (A_{total}) as follows:

$$\beta = \frac{A_{opening}}{A_{total}} \qquad (5.1)$$

Naturally, the opening factor impacts the pressure drop across the filter, and a high opening factor is desired. However, a high opening factor and an adequate mechanical robustness of the filter are opposing requirements, and a trade-off must therefore be made. Filters fabricated using sacrificial layer technology are generally more robust due to their horizontal gaps.

5.2.1.1 Filtration Structures

Lithography, in conjunction with etching, allows direct control of the geometry of a filtration structure. Let us consider filtration and retention in the axial direction in relation to fluid flow. Pillars, posts, dams, and other features are etched into a silicon substrate by deep reactive ion etching (DRIE) [13] or using potassium hydroxide (KOH) [14]. Other researchers have used thick photoresist (SU-8) [15] or polymer replication [16] to produce these structures directly in polymer films or substrates.

In one study, silicon pillars with high aspect ratio were etched using DRIE to form filters that would retain beads within a reaction chamber while still allowing fluid to flow through (Figure 5.5). The bead surfaces were used for both chemical reactions and detection; thus, the beads needed to be retained in the vicinity of the defined reaction chamber. Linear arrangements of rectangular silicon pillars (3–5 µm long, 5–20 µm wide, and 50 µm high) were etched into the silicon substrate in groups containing between 20 and 792 pillars, and situated in the flow path. The channel was sealed with an anodically bonded Pyrex plate and was used to retain both magnetic and nonmagnetic beads (2.8-mm magnetic beads and 5.5-mm polystyrene beads) [13].

An alternative approach to axial filtration is the use of weir-type filters in which a small gap is defined in a portion of a microchannel. These filters were used in a study to separate white blood cells (WBCs) from red blood cells to isolate DNA from the WBCs [14]. The dimensions of the gap (3.5 µm wide) were based on the diameters of the different types of WBCs (6–20 µm); cells larger than the gap were either lodged in the gap or collected next to it as the blood sample flowed through the microsystem. An isolation yield of 6.8–15% of the entire population of WBCs was obtained. The gap was formed using a silicon substrate containing a microchannel with arrays of etched dam structures (7.2 mm long 560 µm wide by 23 µm long) bonded to a glass cover. Although not all the WBCs present were collected (some cells squeezed through the gap and escaped), the yield was adequate for the intended application (Figure 5.6).

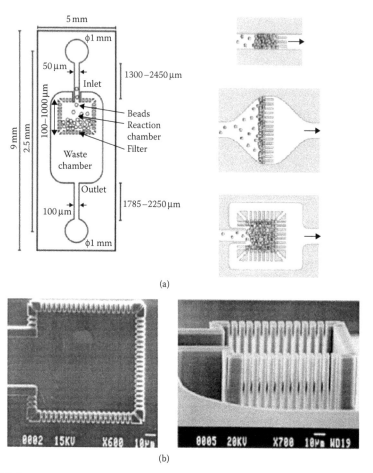

(a)

FIGURE 5.5
(a) Flow-through reactor with integrated filter fabricated using DRIE of silicon with different filter designs. (b) Top and side views of a fabricated silicon filter structure. (From Andersson, H., et al. 2000. *Sens Actuators B Chem* 67:203. With permission.)

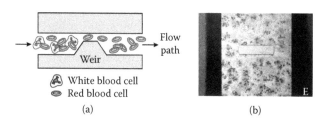

FIGURE 5.6
(a) Weir-type filter for separating white from red blood cells and (b) stained white blood cells trapped in the 3.5-mm gap above the dam. A few red blood cells (darker spots) remain among the white blood cells, but none are present in the gap. (From Wilding, P., et al. 1998. *Anal Biochem* 257:95. With permission.)

5.2.1.2 Filtration Membranes

Discrete filters may also be fabricated by etching lithographically defined perforations (typically a few microns in dimension) into thin membranes suspended over a cavity etched into the substrate. Classic examples of such membrane filters were constructed from anisotropically etched (using KOH) silicon substrates and thin nitride membranes [17,18]. Many perforation geometries and layouts are possible; however, the particle cutoff size is typically limited by the lithography resolution and the fidelity of the etching process. In one study, perforation dimensions were reduced by coating nitride membranes with a conformal Parylene coating. While this reduced the opening factor, the Parylene coating improved the mechanical strength and reduced the cutoff size (Figure 5.7) [18].

It is also possible to escape the limits of photolithography by using sacrificial layers to define a particle cutoff size in the range of 10s of nanometers. For example, by carefully controlling the dry thermal oxidation process, silicon dioxide layers as thin as 5–100 nm are possible. By depositing these layers over patterned step structures made of doped silicon and/or polysilicon, both straight and tortuous filtration paths are created (tortuous filtration paths involve a combination of 90° turns to pass from one side to the other). These devices require a combination of surface and bulk micromachining to pattern the filtration path

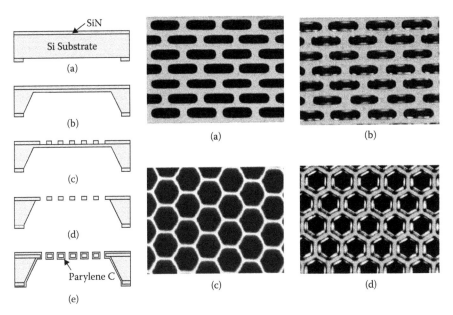

FIGURE 5.7
Left: Fabrication process for Parylene-coated silicon nitride membrane filters: (a) SiN deposition, (b) KOH etching, (c) SiN patterning, (d) Si etching, and (e) Parylene C deposition. Right: Images showing the two types of hole geometries—rectangular and hexagonal—before and after Parylene coating: (a) a SiN membrane filter with rectangular holes, (b) the filter after Parylene coating, (c) a SiN membrane filter with hexagonal holes, and (d) that filter after Parylene coating. (From Yang, X., et al. 1999. *Sens Actuators A Phys* 73:184. With permission.)

and membrane, respectively [19–24]. Examples of fabrication processes that produce both straight and tortuous filtration paths are provided in Figures 5.8 and 5.9. These exemplify the creativity needed to overcome fabrication limitations through a combination of otherwise unlikely partnered processes.

5.2.1.3 Dialysis

Dialysis is a process that permits the removal of small molecules from a solution by exploiting the differences in diffusion rates in the presence of a concentration gradient through a semipermeable membrane. A continuous flow of sample containing the target (called the *perfusate* or *perfusion fluid*) is perfused across the membrane. The target having a molecular weight below the *molecular weight cutoff* (MWCO) diffuses across the membrane and is collected on the opposite side of the membrane in the *dialysate*. Common membrane materials include cellophane, polysulfone, cellulose acetate, polycarbonate, polyether sulfone, polyamide, and polyacetal.

The performance metric for dialysis systems is expressed as the *relative recovery* (RR), which is a measure of the extraction fraction, as follows:

$$RR = \frac{C_d - C_0}{C_s - C_0} \tag{5.2}$$

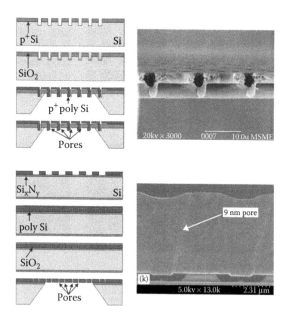

FIGURE 5.8

Two examples of straight, vertical pore filtration membranes. The fabrication process is shown on the left side (from Chu, W. H., et al. 1999. *J Microelectromech Syst* 8:34. With permission.) and scanning electron microscope images of the fabricated devices are shown on the right (from Lopez, C. A., et al. 2006. *Biomaterials* 27:3075. With permission.)

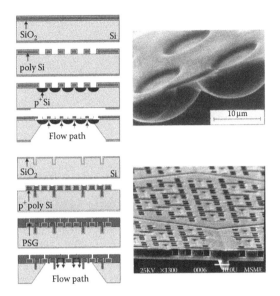

FIGURE 5.9
Two examples of tortuous path filtration membranes. The fabrication process is shown on the left side (from Kittilsland, G., G. Stemme, and B. Norden. 1990. *Sens Actuators A Phys* 23:904. With permission.) and scanning electron microscope images of the fabricated devices are shown on the right (from Chu, W. H., et al. 1999. *J Microelectromech Syst* 8:34. With permission.)

where C_d is the concentration of the analyte in the dialysate, C_s is the concentration of the analyte in the sampling region, and C_0 is the concentration of the analyte in the perfusate. A high extraction fraction is desired. Dialysis may be used to transfer a sample of interest or to removeunwanted species in either direction across the membrane; regardless of the operation mode, the appropriate concentration difference must exist. For example, C_s must be higher than C_0 for diffusion into the perfusate to occur. Because RR is highly and inversely dependent on flow rates, high recovery necessitates low flow rates that increase the contact time of the target with the membrane. Typical flow rates are of the order of ~µL/min. The trade-off, however, is increased time and a departure from real-time monitoring if coupled with a detection system.

Conventional cylindrical microdialysis probes possess large bores (~1 mm), consume large volumes of perfusate, and require long sampling times (minutes to hours). MEMS-based devices benefit from a reduction in size, thus affording smaller sample volumes, faster equilibrium times, and low dead volumes. An example of a microdialysis chip is given in Figure 5.10. Here, semipermeable membranes (7–50 µm thick) possessing different MWCOs were formed in a fused silica microchannel by *in situ* polymerization of a monomer solution—2-(N-3-sulfopropyl-N,N-dimethylammonium)ethyl methacrylate and deionized water with N,N′-methylenebisacrylamide—with an ultraviolet (UV) laser [25]. The membrane is supported by periodically spaced posts along the center of the microchannel. In a counterflow system with

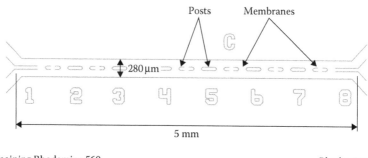

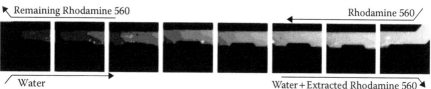

FIGURE 5.10

Top: Fused silica microchannel (20 μm deep) with an in-channel dialysis membrane. Bottom: extraction of Rhodamine 560 with a counterflow dialysis over 4.8 mm of the channel. (From Song, S., et al. 2004. *Anal Chem* 76:2367. With permission.)

the perfusate and dialysate flowing in opposite directions, the lower-MWCO membrane extracted Rhodamine 560 but not other proteins such as insulin, lactalbumin, bovine serum albumin, and antibiotin (all of which have molecular weights greater than 5700 Da). By varying the solvent mixture, a higher-MWCO membrane was formed that allowed the dialysis of lactalbumin (14,000 Da).

5.2.2 Extraction

5.2.2.1 Liquid-Phase Extraction

Transferring a sample across a liquid–liquid interface is used for concentrating, isolating, or separating a species of interest. The liquid–liquid interface usually involves an aqueous phase and an organic solvent phase, in which the aqueous phase is extracted with the solvent. Liquid-phase extraction (LPE) occurs due to the difference in solubility of the target in the two immiscible liquids. The solute exchanges across the liquid–liquid interface until the equilibrium concentration is reached, which indicates the end of the extraction process. This is expressed as follows:

$$K_p = \frac{[Z]_{\text{org}}}{[Z]_{\text{aq}}} \tag{5.3}$$

where K_p is the partition constant or distribution constant, $[Z]_{\text{org}}$ is the analyte concentration in the organic phase, and $[Z]_{\text{aq}}$ is the analyte concentration in the aqueous phase. The extraction process can be alternatively quantified using the following distribution ratio or distribution coefficient, D

$$D = \frac{[\text{total concentration of all forms of Z}]_{\text{org}}}{[\text{total concentration of all forms of Z}]_{\text{aq}}} \qquad (5.4)$$

The major difference between the two expressions is that the expression for D also accounts for the presence of the analyte in multiple forms (i.e., neutral or ionic). Many factors, including the nature and concentration of the analyte, the type of organic solvent, the pH of the aqueous phase, the effect of masking agents in the sample, and the temperature, play a role in determining the efficiency and selectivity of the extraction process [12].

A simple method to establish a liquid–liquid interface for extraction is to use a Y-microchannel laminar flow system. In an earlier work, an aqueous phase that contained the target (iron-bathophenanthrolinedisulfonic acid complex) and an organic phase (tri-*n*-octylmethylammonium chloride) was introduced at the system inlets, as shown in Figure 5.11 [26]. The iron complex was extracted into chloroform 45 seconds after the flow was stopped using a glass microchannel chip with a 10-mm-long extraction region (100 µm deep and 250 µm wide). However, the complex adsorbed to the channel walls, resulting in a reduced extraction efficiency compared to conventional systems.

More recently, a modified form of LPE was demonstrated in which aqueous droplets were used [27]. A mass exchange between water (or water/glycerol) droplets and octanol (octan-1-ol) allowed either the extraction of fluorescein from the solvent or the rejection of rhodamine into the solvent. Extractions were performed in four different configurations of polydimethylsiloxane channels (30–95 µm high, 205 µm wide, and 4–12 cm long; Figure 5.12). The droplets formed in the channels ranged in size between 165 and 195 µm.

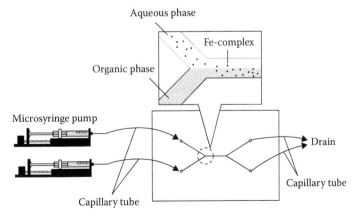

FIGURE 5.11
Microfabricated liquid–liquid extraction system. (From Tokeshi, M., T. Minagawa, and T. Kitamori. 2000. *Anal Chem* 72:1711. With permission.)

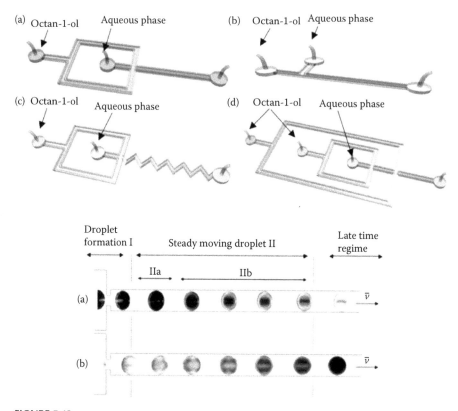

FIGURE 5.12
Top: Schematic representation of a microfluidic chip for liquid-phase extraction in droplets:
(a) hydrofocusing, (b) T-geometry, (c) winding channel flow-focusing, and (d) double flow-
focusing devices. Bottom: Fluorescent microscopy images showing (a) extraction of fluorescein
into a water/glycerol (60% wt) droplet and (b) purification of rhodamine from a water droplet. Both operations were performed using hydrofocusing device geometry. (From Mary, P., V.
Studer, and P. Tabeling. 2008. *Anal Chem* 80:2680. With permission.)

5.2.2.2 Solid-Phase Extraction

In solid-phase extraction (SPE), the sample of interest is retained by a stationary phase to preconcentrate or separate it from a matrix. The extraction
material is selected such that the analyte is preferentially adsorbed. Later, the
extracted analyte is eluted from the system (typically in a more concentrated
state) by washing the extraction material with an appropriate solvent. In some
cases, it is also possible to separate the analyte from the extraction material
by increasing the temperature. A standard SPE system consists of a compart-
ment with the extraction material through which a sample-carrying fluid is
either introduced for a predetermined duration or is flowed continuously.

Several adsorption mechanisms may be employed, including dipole–
dipole interactions and hydrogen bonding (normal-phase SPE), van der

Waals or nonpolar–hydrophobic interactions (reversed-phase SPE), ionic interactions (ion-exchange SPE), and heterogeneous immunoaffinity interactions. The immunoaffinity interactions possess greater specificity than the other methods. Factors that play a role in the retention of analytes in SPE include flow rate, particle size and porosity, column length, surface functional groups, temperature, and extraction material [12]. Several examples of chip-based SPE are presented next.

A simple soda lime glass channel (Figure 5.13) was coated with octadecyltrimethoxysilane (ODS; the solid phase) for enriching neutral Coumarin dye (C460) diluted in a 15% (v/v) acetonitrile solution (8.7 nM) [28]. The solution was electrokinetically introduced into the coated enrichment channel, and after 160 seconds, an 80-fold enrichment of the dye was accomplished. This is an example of reversed-phase SPE.

Alternatively, the stationary extraction phase may be in the form of trapped beads [29]. Glass channels were etched to form a flow through system containing a packed bead bed formed by weir structures (Figure 5.14). Then, the ODS-coated silica beads (1.5–4 um) were electrokinetically pumped through a bead introduction channel into an etched cavity (10 µm high) bounded by two 9-µm-high weir structures (a form of reversed-phase SPE) to trap the beads in the chamber. A 1-nM solution of boron dipyrromethane (BODIPY) 493/503, a nonpolar analyte, was introduced through the bead bed by electroosmotic flow and was retained on the beads with a 500-fold concentration enhancement.

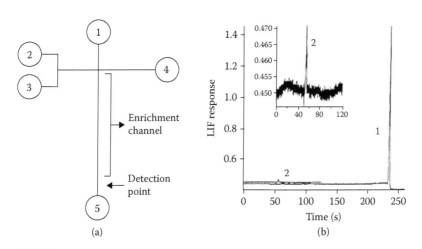

FIGURE 5.13
(a) Lateral solid-phase extraction microchip layout (channel depth/width = 5.1/53 µm and enrichment channel length = 33 mm), with labeled reservoirs corresponding to (1) sample, (2) buffer with 60% acetonitrile, (3) buffer with 15% acetonitrile, (4) sample waste, and (5) waste. (b) Detection of the eluted C460 dye by laser-induced fluorescence following (1) 160 seconds of enrichment and (2) 1 second of gated electrokinetic injection into the enrichment channel. (From Kutter, J. P., S. C. Jacobson, and J. M. Ramsey. 2000. *J Microcolumn Sep* 12:93. With permission.)

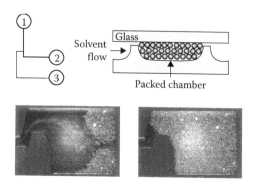

FIGURE 5.14

Top left: Layout of a bead-based solid-phase extraction device (1: outlet buffer channel, 2: inlet sample/buffer channel, 3: bead introduction channel); the channels are 580 μm wide and 10 μm deep, whereas the bead introduction channel is 30 μm wide with 9-μm-high weirs (After [29]). Top right: Cross section of the device through the packed detection chamber for solid-phase extraction of boron dipyrromethane. Bottom: Images showing the microfluidic bead-packing process. (From Oleschuk, R. D., et al. 2000. *Anal Chem* 72:585. With permission.)

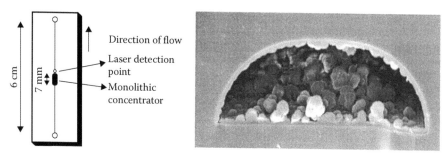

FIGURE 5.15

Left: Layout of a solid-phase extraction microchannel device (100 μm wide, 40 μm deep, and 6 cm long). Right: Scanning electron micrograph of a porous polymer monolithic concentrator. (From Yu, C., et al. 2001. *Anal Chem* 73:5088. With permission.)

Micro ion-exchange and hydrophobic SPE was demonstrated using a porous polymer monolith formed by *in situ* photoinitiated polymerization in an etched glass microchannel (Figure 5.15) [30]. The polymer monoliths were formed by UV-initiated polymerization of the following precursor systems: (1) butyl methacrylate (BMA) and ethylene dimethacrylate (EDMA) or (2) 2-hydroxyethyl methacrylate (HEMA), [2-(methacryloyloxy)ethyl] trimethylammonium chloride (META), and EDMA. The first and second systems formed a hydrophobic concentrator and an ion-exchange concentrator, respectively. Pore size (average pore sizes of 13.2 and 19.5 μm) was controlled in a separate processing step by introducing porogenic mixtures of hexane and methanol. (A large pore size reduces the flow resistance and allows high flow rates of sample solution.) Dilute solutions of Coumarin dye (C519; 10 or 100 nmol/L), a tetrapeptide (Phe-Gly-Phe-Gly; 10 nmol/L), and green fluorescent protein (18.5 nmol/L) were introduced under pressure-driven flow into the extraction region, and a concentration increase of greater than 10^3 times was achieved. Compared to the other extraction materials, the porous monoliths were reported to have much greater concentration capacities.

5.3 Sample Introduction

5.3.1 Electrokinetic Injection

In electrokinetic flow, an analogy between electrical circuits and microfluidic systems is convenient. Fluid filled channels may be considered resistors, and Kirchoff's voltage and current laws apply for the analysis of these electrofluidic circuits. If voltages are applied at points a and b of a channel over a length L, the resulting electric field strength in the channel segment is given by

$$E = \frac{|V_a - V_b|}{L} \qquad (5.5)$$

For our discussion on sample introduction using electrokinetic injection, we will use a simple network of channels that intersect in a cross pattern, as shown in Figure 5.16. The inlets and outlets for each channel are connected to reservoirs; the channels and reservoirs are named according to their purpose: buffer (B), sample (S), waste (W), and analysis (A). This simple microfluidic device is used for injection of sample mixtures and separation, typically by electrophoresis.

Application of the appropriate combination of (1) voltages at each reservoir with respect to a reference (typically established at the cross intersection) or (2) electric fields along and between the channel end and the intersection reference point results in the desired flow. For example, E_B refers to a field established between the buffer reservoir and the intersection.

Here, we are interested in methods that achieve sample introduction. For separations, introduction of a small sample volume (called a plug) into the longer analysis channel initiates the process. In earlier work, methods to dispense both sample stacks and plugs were developed (Figure 5.17) [32]. Higher resolution is possible with plugs, which have smaller volumes but also reduced intensities, thus requiring detection techniques that are more sensitive. Minimizing the injection plug width improves the performance

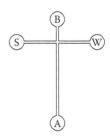

FIGURE 5.16
Diagram of a simple cross-intersection microchannel system with sections corresponding to the buffer (B), sample (S), waste (W), and analysis (A) reservoirs.

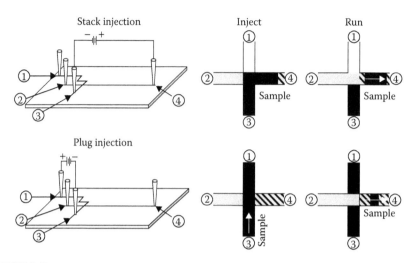

FIGURE 5.17
A comparison between stack (top) versus plug (bottom) injection. The numbers correspond to fluid reservoirs and points at which to apply driving voltages for injection. (From Woolley, A. T., and R. A. Mathies. 1994. *Proc Natl Acad Sci USA* 91:11348. With permission.)

of separations (i.e., increases the number of theoretical plates, a parameter defined in Section 5.4). Therefore, we will focus on plug injection methods.

Plugs are shaped by combining the appropriate biasing schemes and channel geometries. The most commonly used plug dispensing methods include *pinched*, *double tee*, and *gated*. The first two methods yield the shortest plugs (measured in the axial direction) but need more time due to the sample plug reloading process. Gated dispensing is faster because reloading is obviated, but requires larger electrical biases.

Pinched injection is so named because it uses electrokinetic flows to avoid leakage of the sample plug into the analysis channel prior to injection (Figure 5.18). First, the sample is loaded into the intersection under the following conditions:

$$E_B + E_S + E_A = E_W \tag{5.6}$$

where the subscripts correspond to the electric field strength in the buffer (B), sample (S), analysis (A), and waste (W) channels. The intersection in part defines the plug volume. The buffer and analysis fields confine the sample to the intersection, but also further narrow the sample stream and, thus, the resulting plug. Without these pinching voltages, *floating injection* results. Then, the biasing conditions are changed as follows to initiate plug injection:

$$E_B = E_S + E_A + E_W \tag{5.7}$$

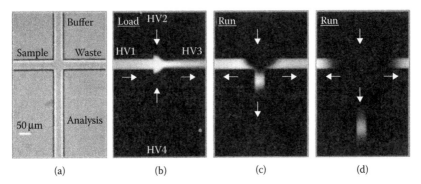

(a)   (b)   (c)   (d)

FIGURE 5.18
Visualization of pinched injection where the leftmost image (a) is a white light image and the remaining images are a sequence of fluorescent images corresponding to (b) sample loading and (c and d) the run mode. The arrows show the flow direction. (From Jacobson, S. C., and C. T. Culbertson. 2006. Microfluidics: Some basics. In *Separation Methods in Microanalytical Systems*, 575. Boca Raton, FL: Taylor & Francis. With permission.)

These conditions cause the movement of flow from the buffer region to the analysis region. A secondary flow also exists from the buffer region to the sample and waste regions to prevent any additional sample from entering the analysis channel. To achieve these conditions, the following field relations must be applied:

$$E_B > E_S, E_A, E_W \tag{5.8}$$

Overall, a constant volume plug is injected. Immediately after injection, the system returns to the sample loading condition.

A double tee injection system rearranges and offsets the intersections of the sample and waste channels. The benefit is that no pinching voltage is necessary for sample confinement. Thus, floating injection is performed without any sample leakage. However, a larger plug volume results in reduced resolution. A triple tee structure has also been investigated, which allows adjustment of the sample volume but requires an additional waste channel and intersection (Figure 5.19) [33]. The sample flows from the sample reservoir to the two waste reservoirs; the distance between the two waste channel intersections defines the sample plug volume.

If one considers pinched and double tee to be discretized methods of injection, gated injection can be thought of as a continuous method. Instead of switching between loading and injection, the throughput is increased by allowing the sample to flow continuously to the waste. Periodic injection events interrupt this flow (Figure 5.20). The obvious drawback is that more sample is needed in this method. The plug length is adjusted using injection time, and the plug length is generally increased by the faster migrating species. The latter factor results in a sample bias that is much greater than that found in the discretized methods.

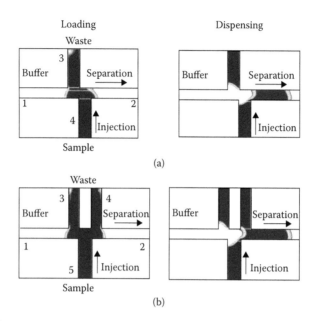

FIGURE 5.19
Sample injection by (a) double tee and (b) triple tee systems. (From Fu, L. M., et al. 2002. *Anal Chem* 74:5084. With permission.)

Under standard operating conditions, the sample flows continuously between the sample and waste reservoirs as follows:

$$E_B + E_S = E_A + E_W \tag{5.9}$$

To prevent the sample from entering the analysis channel, the following conditions are maintained:

$$E_S \geq E_W \quad \text{or} \quad E_B \leq E_A \tag{5.10}$$

Continuous flow is interrupted only for a brief period to dispense a small plug into the analysis channel. The following field strength relationships are established for this to occur:

$$E_S \leq E_W \quad \text{or} \quad E_B \geq E_A \tag{5.11}$$

The result is a plug that may vary in volume as a function of the injection time. The plug is cut off by raising the buffer potential or reapplying the initial loading conditions. Using these methods and microfabricated channels, injection of picoliter volume plugs is possible.

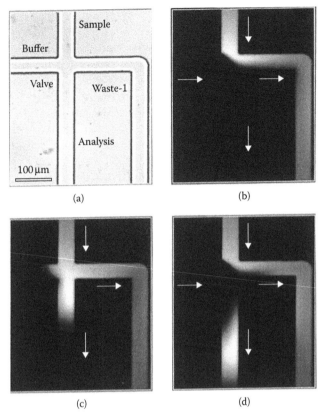

(a)   (b)

(c)   (d)

FIGURE 5.20
Visualization of gated injection where the leftmost image (a) is a white light image and the remaining are a sequence of fluorescent images corresponding to (b) and (d) the run and (c) the dispensing modes. The arrows show the flow direction. (From Jacobson, S. C., S. V. Ermakov, and J. M. Ramsey. 1999. *Anal Chem* 71:3273. With permission.)

5.3.2 Hydrodynamic Injection

Sample bias is always present when electrokinetic injection methods are used. Hydrodynamic injections using pressure-driven flows avoid this problem. Although these concepts are introduced separately here, in practice, both methods may be employed in a device.

Currently, fewer systems employ hydrodynamic injections; nevertheless, it has several advantages. First, the flow is independent of fluid composition, whereas in electrokinetic flow, fluid properties such as pH and electrolyte concentration are important. In addition, hydrodynamic systems may be constructed from a wide variety of materials including silicon. Silicon cannot be used for electrokinetic systems which require channel wall surfaces made of materials with low electrical conductivity. The wide choice of materials also enables the use of more types of solvents. The lack of an electric field in the flow prevents interference with electrical detectors. The primary disadvantages are the system pressure requirements and the parabolic flow profile. The available inlet pressure to drive flow is dependent upon the available pumping technologies and their interface with the chip. Whether an on- or off-chip pump is used, the pressure head provided is limited, which can be problematic in separation applications. A vacuum can also be used to drive the hydrodynamic flow.

Hydrodynamic injection is also used in separation methods; the parabolic flow profile is an asset in both field flow fractionation (FFF) and hydrodynamic chromatography. These two methods do not require an exchange with a second phase during the separation process and are introduced in the following section on separation techniques.

5.4 Separations

Separations are central to many chemical and biological analyses because they enable the isolation, recovery, and detection of individual analytes in a mixture. The physical separation of mixture components is achieved by differential mass transport mechanisms, as opposed to dispersive modes. The primary separation methods are chromatography and electrophoresis, which are distinguished by the physical parameters used for the separation of analytes. In chromatography, the solute is distributed differentially over two phases of matter, whereas in electrophoresis, ions possessing varying mobilities are separated in the presence of an electric field.

Scaling of a separation system not only minimizes sample usage, but also results in major performance improvements. The separation efficiency tends to be comparable to the counterpart macroscale systems, and smaller sample size allows for fast separations to be performed in shorter periods. Although many applications for separations exist, medical diagnostics perhaps best motivate the need for a high-throughput screening of multiple samples to produce rapid results. However, a trade-off exists: smaller sample volumes contain fewer molecules and require better detection methods. The principles that govern common separation techniques are discussed next, along with examples of their implementation in LOC devices.

5.4.1 Separation Performance

Let us assume that we start with a sample plug and that the separated product appears as a sequence of bands. To quantify the performance of our separation process, we start by calculating the efficiency. Naturally, narrow bands are preferred; however, a separated band is not distinct due to dispersive mass transport. Instead, bands possess a Gaussian profile, with the variance expressed as follows:

$$\sigma^2 = 2Dt \tag{5.12}$$

where D is the diffusion coefficient and t is the time. Because $t = L/u$ where L is the separation length and u is the velocity, the equation can be rearranged as follows to obtain the plate height:

$$H = \frac{\sigma^2}{L} = \frac{2D}{u} \tag{5.13}$$

To account for additional sources of band broadening, the van Deemter equation is used:

$$H = A + \frac{B}{u} + C \tag{5.14}$$

where the terms A, B, and C correspond to contributions from constant sources, diffusion, and mass transfer, respectively [35]. Another quantification of efficiency is the plate number, the number of plate heights per separation length, derived as follows:

$$N = \frac{L}{H} = \frac{L^2}{\sigma^2} \tag{5.15}$$

Plate number and height, however, do not characterize the ability of a separation to distinguish between two components. For example, the bands may overlap or separation may be incomplete. The resolution between the peaks of two bands having a Gaussian distribution is given by

$$R = \frac{\sqrt{N}}{4} \frac{\Delta u}{u} \tag{5.16}$$

Here, the base width of each band is assumed to be 4σ. In chromatography, the resolution of one species over a second is expressed in terms of the selectivity and capacity factor of the second component, as follows:

$$R_{chrom} = \frac{\sqrt{N}}{4} \frac{\alpha - 1}{\alpha} \frac{k_2}{k_2 + 1} \tag{5.17}$$

In electrophoresis, the resolution between two components is expressed as

$$R_{ep} = \frac{\sqrt{N}}{4} \frac{\Delta\mu_{1,2}}{\mu_{1,2} + \mu_{eo}} \tag{5.18}$$

where $\Delta\mu_{1,2}$ is the difference in the mobilities of the components, $\mu_{1,2}$ is their average mobility, and μ_{eo} is the electroosmotic mobility.

Ultimately, a separation system is limited by the achievable peak capacity or number of peaks that can fit within a separation length, as expressed in the following equation:

$$n = \frac{L}{w} = \frac{L}{4\sigma R} \tag{5.19}$$

where w is the average peak width and is assumed to be 4σ.

5.4.2 Chromatography

Chromatography is a general term used to describe the separation of a mixture to isolate the solutes dissolved in the mobile phase (i.e., a solvent) through the stationary phase (i.e., a packed particle bed, silica gel, or cellulose). Isolation may be performed simply (1) for isolation of the solute for subsequent analysis or (2) in preparation for additional processing. Separation is visualized with a chromatogram and the peaks correspond to the separated analytes that are present in the mixture. A typical chromatogram is shown in Figure 5.21.

The solute is distributed over the two phases, which is quantified by the following distribution coefficient:

$$K = \frac{c_s}{c_m} \tag{5.20}$$

where c is the concentration and the subscripts s and m correspond to the stationary and mobile phases, respectively. This type of separation relies on the ability of one phase to retain the solute. The capacity factor for the solute of interest compares the amount of solute present in each phase as follows

$$k = \frac{c_s V_s}{c_m V_m} = K \frac{V_s}{V_m} \tag{5.21}$$

where V_s and V_m are the volumes of the stationary and mobile phases, respectively. The retention ratio is related to the capacity factor and is the fraction of the solute retained by the mobile phase compared to the overall solute amount, as follows

$$R = \frac{c_m V_m}{c_m V_m + c_s V_s} = \frac{1}{1+k} \tag{5.22}$$

For complex mixtures of more than one solute, it is useful to define the selectivity of the separation, as shown in the following equation

$$\alpha = \frac{k_2}{k_1} \tag{5.23}$$

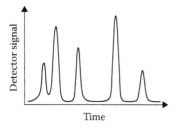

FIGURE 5.21
A chromatogram with well-isolated peaks.

where the numerical subscripts correspond to the different species present in solution and the selectivity is defined such that it is greater than 1.

Many forms of chromatography have been developed for separating liquids and gases (hence the classifications of liquid chromatography and gas chromatography) [36]. A few chromatographic methods, which possess enhanced performance afforded by miniaturization, are introduced in this section, with an emphasis on liquid separations; a few examples of microfabricated devices are also provided.

5.4.2.1 High-Performance Liquid Chromatography

High-performance liquid chromatography (HPLC) is a modern liquid chromatographic method and is distinguished by the use of small packing particles in a separation column coupled with high driving pressures. The smaller packing particles improve resolution by increasing the velocity and thus minimizing the diffusion times of the analytes. The first demonstration of an on-chip HPLC integrated (1) an injector, (2) a column with retaining frit, and (3) an optical detector in an anisotropically etched silicon substrate (overall size of 4.5- × 25- × .75 mm) [37]. Pumping was performed off-chip. Separation of two dyes (fluorescein and acridine orange) was performed with minimal *dead volume* (less than 2.5 nL; Figure 5.22).

Dead volume is an important concept in both microfluidics and LOC. The general definition is "the inaccessible fluid volume within a device." For example, a reciprocating diaphragm pump chamber may have dead volume in its corners and crevices, where fluid can enter but is not displaced by the actuated diaphragm. In microfluidic systems, fluid packaging, including interconnects and tubing, is often the source of additional dead volume. In chromatography, the definition is more subtle: dead volume refers to the volume of a completely unretained solvent (mobile phase) in the separation column or channel (also known as the thermodynamic dead volume). Regardless of the specific definition or the source, dead volume is an undesired parameter and is minimized greatly in microfluidic systems when compared to their macroscopic counterparts.

Electrokinetic methods have dominated microscale separations thus far. This is in part due to the technological challenges faced by HPLC [38]. One example involves generating and supporting the high pressures required for separation. A clever method was devised in references [39,40] to strengthen surface micromachined polymer channels. These devices were demonstrated using an HPLC method that uses temperature gradients to improve the separation of derivatized amino acids (temperature-gradient interaction chromatography, or TGIC). The system featured a separation channel (8 mm long, 100 μm wide, and 25 μm high) with integrated electrochemical detection in the form of interdigitated microelectrodes. The channels were strengthened by anchoring the Parylene into the depth of the silicon substrate by deposition into narrow etched trenches. This allowed the channels to sustain pressures

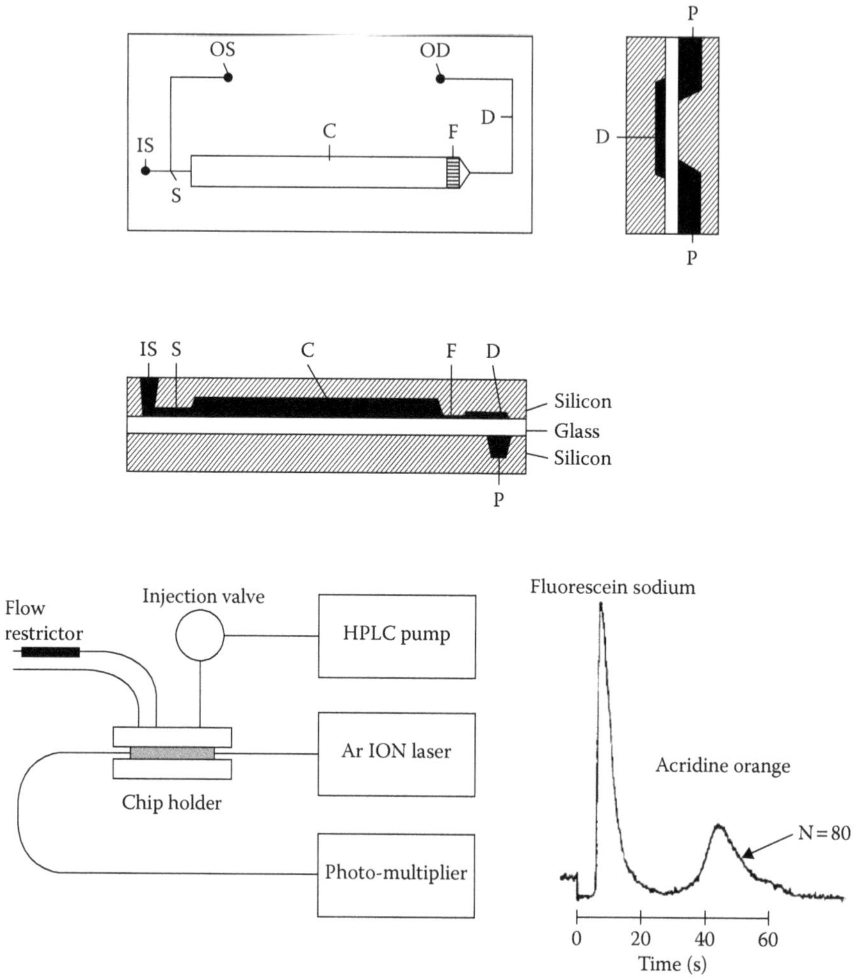

FIGURE 5.22
Top: Layout of a high-performance liquid chromatography chip (IS: inlet for the sample and the mobile phases; S: split injector; OS: outlet for the rejected portions of the sample and mobile phases; C: separation channel; F: frit; D: optical detector cell; OD: waste outlet; P: grooves for positioning the optical fiber). Bottom left: A schematic representation of a high-performance liquid chromatography system. Bottom right: Separation of fluorescein and acridine orange. (From Ocvirk, G., et al. 1995. *Anal Methods Instrum* 2:74. With permission.)

up to 600 psi without delaminating. To minimize power dissipation to the substrate during thermal cycling, an air gap was etched around the periphery of the separation channel. The channel and surrounding substrate were bridged across the air gap with a thin film of Parylene that was also anchored into the substrate with cross-linking stitches. Following fabrication, the channels were packed using a conventional slurry method with 5-μm C18-coated silica microbeads (C18 is a type of hydrophobic alkyl chain; Figure 5.23).

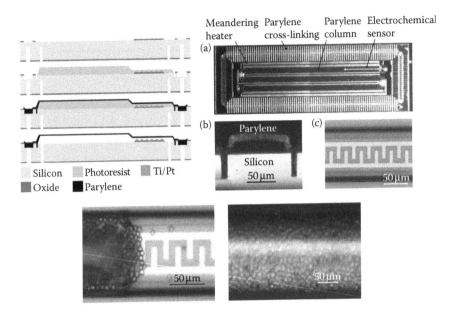

FIGURE 5.23
Top left: Fabrication process. Top right: (a) Top view of the chip showing the layout, (b) a cross section of the chip showing the anchoring of thin film Parylene channels to the substrate to sustain the high driving pressures, and (c) an interdigitated electrochemical detector. Bottom: Two views of the 5-μm C18 silica beads packed into the Parylene column. (From Shih, C. Y., et al. 2006. *Sens Actuators A Phys* 127:207; and Shih, C. Y., et al. 2006. *J Chromatogr A* 1111:272. With permission.)

5.4.2.2 Hydrodynamic Chromatography

Hydrodynamic chromatography (HDC) is a type of HPLC that is sometimes implemented with a characteristic packed bed of particles in a separation channel. The purpose of the packed bed is to assist in the size separation process; however, no exchange occurs between the particles and the analyte flow. Thus, this method is faster than other HPLC methods, such as size-exclusion chromatography, which requires an exchange with the matrix and operates based on steric exclusion effects.

HDC uses the velocity gradient of the parabolic flow profile to separate analytes based on size; analytes that are large compared to the width of the flow channel have a center of mass away from the edge of the channel where flow velocity is low. Thus, these larger analytes must necessarily move faster (Figure 5.24). In fact, the main application of HDC is to separate out larger species, such as larger molecules (polymers and macromolecules), organelles, and cells. The flat flow profile for electrokinetic flow cannot reproduce this velocity-gradient separation effect and thus there exists no electrokinetic analog to this hydrodynamic flow separation method.

In references [41,42], silicon and glass microfluidic devices were used for separating fluorescent nanospheres and macromolecules by HDC (Figure 5.25). An 8-cm-long shallow separation channel with a large aspect ratio (1 μm deep and

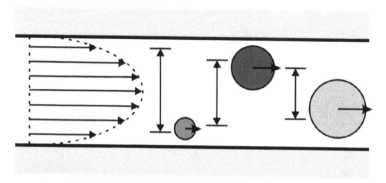

FIGURE 5.24

Principle of separation using hydrodynamic chromatography, in which only large analytes experience the velocities in the central zone of the channel, as a direct result of their size.

1000 μm wide) was created and fed by a 300-pL injector. The separation channel was created by growing a layer of oxide (1 μm thick), patterning the channel by photolithography, and etching the oxide with buffered hydrofluoric acid. Next, injection slits were defined by photolithography and etched by DRIE. The channels were enclosed by bonding to a Pyrex wafer. The sample-injection paradigm is shown in Figure 5.25. The separation channel length between the second and third injection slits—the sample-in and the sample-out slits—defines the injected volume. The first slit supplied a buffer to push the sample plug into the separation channel. Valving and pumping were performed with off-chip components. The characteristic parabolic flow profile is evident in the separated bands and became more pronounced as the flow profile developed over time.

5.4.2.3 Affinity Chromatography

In affinity chromatography, an immobilized affinity moiety is used to separate the analytes. The immobilized species may be deposited on the channel walls, bound to beads packed in the column, or bound to a porous slab in the column. A novel immobilized affinity moiety in the form of nanoparticles modified with a stimulus-responsive polymer was adapted for an affinity chromatography chip [43]. Latex particles coated with poly(N-isopropylacrylamide), or PNIPAAm, could be reversibly adhered to polyethylene terephthalate (known as PET) channel walls by controlling the temperature. PNIPAAm undergoes a reversible phase transition from a hydrophilic to hydrophobic form as the temperature is increased above 26°C (the lower critical solution temperature of the polymer). This allows selective attachment, removal, and replacement of the immobilized affinity matrix (Figure 5.26).

5.4.2.4 Field Flow Fractionation

FFF uses both axial and transverse transport to achieve separation. Axial transport is pressure-driven and takes advantage of the parabolic flow profile

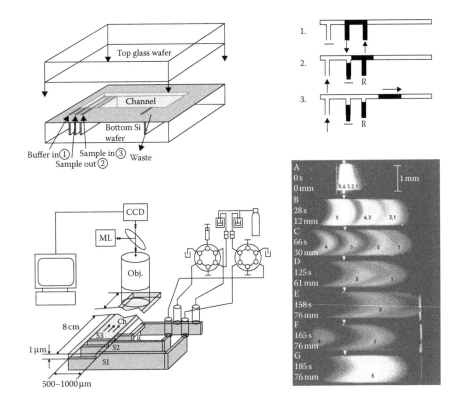

FIGURE 5.25
Top left: Layout of a hydrodynamic chromatography chip. Top right: Injection scheme in which (1) the sample is filled between the second and third slits, (2) the carrier flow sweeps the sample plug into the separation channel, and (3) sample tailing is prevented by valving and appropriate fluidic resistance. Bottom left: Experimental setup for hydrodynamic chromatography, including an external valve manifold for distributing the pressurized fluids and an imaging microscope (Ch: separation channel; R: resistance channel; S1: carrier liquid; S2: sample inlet; and S3: sample outlet). Bottom right: Sequence of fluorescent images showing the separation achieved by the chip over time under a pressure of 4 bars (the numbers 1–3 indicate 110-, 44-, and 26-nm diameter fluorescent polystyrene particles, respectively; number 4: fluorescently labeled anionic dextran [10 kDa]; and number 5: fluorescein [0.05 mg/mL]). (From Blom, M. T., et al. 2002. *Sens Actuators B Chem* 82:111; and Chmela, E., et al. 2002. *Anal Chem* 74:3470. With permission.)

to provide different elution times of analytes. A field established transversely across the flow channel redistributes the analytes based on differences in some physical property. This spatial elution variation is what separates the component mixtures. A key feature of FFF systems is that the ratio of the channel width to the height is large (>10; Figure 5.27). FFF is used for larger species such as macromolecules, particles, vesicles, and cells. The transverse, spatially varying field may be a function of temperature, electric potential, gravity, or another controllable parameter. For example, a temperature gradient separates the analytes based on their thermal diffusivities, and an electrical gradient separates them based on differences in their electrophoretic

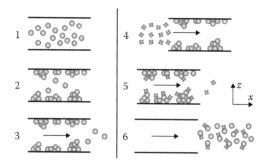

FIGURE 5.26
A schematic representation of the steps in affinity chromatography: (1) the channel fills with biotinylated, poly(N-isopropylacrylamide)-coated beads; (2) the beads aggregate following a temperature increase to 37°C; (3) the buffer washes away the unbound beads; (4) fluorescently tagged streptavidin is introduced into the channel; (5) streptavidin binding occurs, followed by a wash to remove the unbound molecules; and (6) bead aggregates detach and elute from the channel following the cooling of the system to room temperature. (From Malmstadt, N., et al. 2003. *Anal Chem* 75:2943. With permission.)

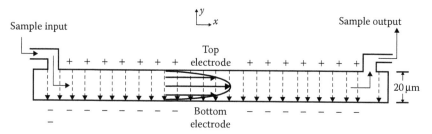

FIGURE 5.27
A schematic representation of the electrical field flow fractionation system. (From Gale, B. K., K. D. Caldwell, and A. B. Frazier. 2001. *Anal Chem* 73:2345. With permission.)

mobilities. However, not all forms of FFF benefit from a reduction in scale. Clearly, gravity-based methods (sedimentation FFF) do not consider our discussion in Chapter 1 regarding the scaling of forces on the microscale level. However, electrical, thermal, and magnetic FFF have been shown to achieve performance improvements with scale [9,44].

The advantages of miniaturizing electrical FFF (eFFF) were quantified and compared to a standard macroscale device in one study [9]. The microfabricated separation channel (5.4 cm long, 6 mm wide, and 28 μm high) included on-chip gold electrodes. The system was capable of separating (1) polystyrene particles based on size and (2) carboxylated polystyrene particles of identical size by varying the surface charge. The injected sample volumes were 0.1–0.3 μL. The performance was compared to a conventional eFFF separation channel (64 cm long, 2 cm wide, and 176 μm high) with external graphite electrodes. The latter system not only required larger injection volumes (5–10 µL) but also more carrier fluid (approximately 1000 times more). The size-based separation was performed in only 2 minutes in the micro-eFFF system, compared to 2 hours for the macroscale system. In addition to improvements in sample/carrier volumes and analysis time, the micro-eFFF system demonstrated increased retention and resolution while decreasing peak broadening and power consumption. Although the separation column was miniaturized, the supporting equipment (i.e., power supplies, data-acquisition hardware, and external pumps) are still quite bulky and occupy a nontrivial footprint.

5.4.3 Electrophoresis

Electrophoretic phenomena were covered in Chapter 4. In the presence of an electric field, a particle experiences both an electrical force and a drag resistance that balance one another to yield an electrophoretic force. An electrophoretic mobility, μ_{ep}, is associated with the resulting velocity of the particle in the electric field (Equation 4.66). Sometimes, a system may include ions that can be transported by electrophoretic and electroosmotic forces. Thus, both mobility terms are present, and it is possible to define an overall electrokinetic mobility as follows:

$$\mu_{ek} = (\mu_{ep} + \mu_{eof})$$
(5.24)

In this case, separation only results due to selective transport processes; electroosmostic flow is nonselective and therefore does not play a role in separation.

Electrophoretic separation is generally used when the analytes in a mixture possess a difference in their charge-to-mass ratios resulting in a difference in their mobilities. For example, although proteins, peptides, and amino acids are charged, their ratios are similar and, therefore, they are only amenable to separation by chromatographic techniques. Electrophoretic separations are visualized using an electropherogram or by an image of the separation channel. An electropherogram is very similar in appearance to a chromatogram in that the detected peaks are plotted against time (Figure 5.21).

Electrophoretic separation efficiency may be measured by the number of theoretical plates, derived as follows:

$$N_{ep} = \frac{\mu_{ek}V}{2D}$$
(5.25)

where V is the applied voltage and D is the diffusion coefficient. To improve efficiency and N_{ep}, the separation voltage may be increased. However, this approach leads to increased heat evolution, which may not be tolerated by the species present, leading to thermal band broadening mechanisms. Instead, the number of theoretical plates is improved by scaling. Additional benefits include reduction in sample size and rapid throughput.

A few chip-based electrophoresis methods are introduced in this section; for reviews of the many devices that have been developed, the reader is referred to [35,45,46].

5.4.3.1 Capillary Electrophoresis and Capillary Gel Electrophoresis

Capillary electrophoresis (CE), or capillary zone electrophoresis (CZE), can be performed with simple microfluidic devices having channels with a cross intersection (Figure 5.16). This structure also integrates sample injection. Conventional CE systems use fused-silica capillaries; thus, the first microfabricated CE devices were made in glass or quartz because of their similar electroosmotic flow (EOF) properties and known surface chemistries. In addition, glass and quartz are conveniently transparent, chemically inert, and electrically insulating. More recently, polymer-based systems have been recognized to offer a reduction in both cost and fabrication time. This is advantageous for disposable, single-use applications and in the rapid prototyping of preliminary designs. Naturally, the introduction of new materials necessitates the study of new surface chemistries and characterizations of EOFs.

Standard CE is useful for the separation of analytes having varying charge-to-mass ratios (e.g., proteins). However, analytes having similar charge-to-mass ratios cannot be separated by this technique. A variant of capillary electrophoresis, called capillary gel electrophoresis (CGE), is used for the separation of nucleic acids and DNA by size. In CGE, the separation column is filled with a polymer gel sieving matrix, such as polyacrylamide or agarose, which hinders the passage of molecules in relation to their size.

A large number of simultaneous separations can be performed with capillary array electrophoresis (CAE) devices, in which multiple serpentine parallel CGE lanes exist on the same substrate. In references [47,48], 96 lanes were isotropically etched into a glass substrate and covered with a second thermally bonded glass substrate (Figure 5.28). A sample was loaded into individual sample reservoirs for each lane. In this system, adjacent lanes share common cathode and waste reservoirs. At the center of the round substrate, the 96 lanes meet and share a central anode reservoir.

First, immobilization and preconcentration of DNA were performed on-chip. The desalted sample was then thermally released and injected for CGE separation. Finally, the device was loaded on a rotary confocal scanner for rapid detection, intended for DNA sequencing, a concept that is introduced in Chapter 7. Devices having up to 384 lanes have been demonstrated [49]. These systems feature serpentine channels that have been optimized to minimize band broadening that usually results due to turns in the separation channels [50].

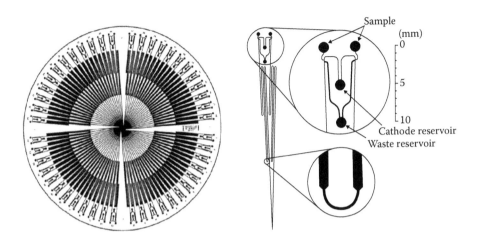

FIGURE 5.28
Left: Layout of a 96-lane capillary array electrophoresis device intended for DNA sequencing applications. Right: View of an injector pair with common cathode and waste reservoirs but individual sample reservoirs. The bottom inset shows a close-up of a turn. (From Paegel, B. M., et al. 2002. *Proc Natl Acad Sci USA* 99:574. With permission.)

5.4.3.2 Isoelectric Focusing

Isoelectric focusing (IEF) utilizes the isoelectric point (the pH value where the molecule does not possess a surface charge) of ampholytes for electrophoretic separation in a pH gradient. Ampholytes are charged molecules with both acidic and basic groups. A channel filled with ampholytes separates into distinct pH zones when a voltage is applied across it, in which the solution near the anode is acidic and basic near the cathode. The separation is based on differences in the isoelectric points of the ampholytes, and the analytes are focused into zones where they possess no net charge. Once the analytes are collected at their corresponding isoelectric points, a steady-state condition is reached. IEF is often used for high-resolution separation of peptides and proteins.

An interesting application of IEF was devised in a study in which resolution was improved by coupling it to a second separation process using CE [51]. This method is referred to as a two-dimensional separation because the sample is initially separated by IEF and then later processed by CE one band at a time (Figure 5.29). This two-step separation concentrates the initially dilute sample and eliminates dispersion of the sample during injection. A simple cross-intersection channel network is sufficient to perform both operations in a single chip. Channels (200 μm wide and 20 μm high) were imprinted in acrylic from electroplated metal masters. The IEF section measured 2.54 cm, and the CE section was formed from intersecting segments measuring 2.5 and 2.8 cm. With this system, peaks that could not be detected in one-dimensional separations under similar conditions were identified due to the increased resolution of this IEF–CE approach.

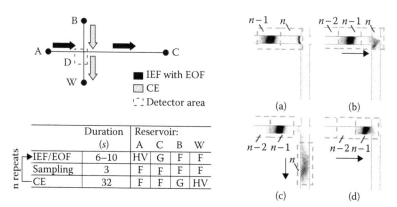

	Duration (s)	Reservoir:			
		A	C	B	W
IEF/EOF	6–10	HV	G	F	F
Sampling	3	F	F	F	F
CE	32	F	F	G	HV

(n repeats)

FIGURE 5.29
Left: Layout of a chip showing the sample movement path and the algorithm for the two-dimensional isoelectric focusing–capillary electrophoresis separation (where HV = high voltage, G = ground, and F = float). Right: Images showing the sampling process. The horizontal channel is for isoelectric focusing and the vertical channel is for capillary electrophoresis. (a) Samples are focused and induced to move toward reservoir C under the influence of an electric field. (b) The sample of interest reaches the intersection and the electrodes are set as floating. (c) A high voltage is applied between B and W for capillary electrophoresis separation. (d) The sample is refocused by isoelectric focusing after completion of capillary electrophoresis. The process is repeated until all the individual samples are obtained. (From Herr, A. E., et al. 2003. *Anal Chem* 75:1180. With permission.)

5.5 Problems

1. What other fabrication methods or physical principles have been employed to perform filtration? Find two examples in peer-reviewed journal articles. For each article, describe the filtration method and discuss the merits and specific applications of the selected approach. Cite your sources.

2. Figure 5.30 shows a simple cross channel for the electrokinetic injection of a small sample plug. Using the notation provided in the figure, describe the complete voltage-biasing conditions for each of the following cases (include relations for V_1, V_2, V_3, V_4, and V_J). For example, one possible condition is $V_1 > V_2$.

   a. Initiation of sample flow from the sample well (1) to the sample waste (2).

   b. Pinching of the sample flow from the sample well (1) to the sample waste (2).

   c. Rapid reversal of the flow from the sample waste (2) to the sample well (1) to align a thin sample plug from the pinched stream with the injection/separation channel.

   d. Sample injection into the analysis/buffer waste well (4). The sample (1) and the sample waste (2) are pulled back to minimize the injected plug volume.

3. What is the injected sample-plug volume needed for the two electrokinetic sample-injection systems shown in Figure 5.31? What phenomena might increase the actual injected sample-plug volume? Note that the drawings are not to scale.

4. For the cross-intersection electrokinetic sample-injection system shown in Problem 3, what is the voltage at the junction? Use a simple resistor model to find the voltage. What is the electric field in the separation channel?

   a. Buffer reservoir at 5 kV
   b. Analysis reservoir at 0 kV
   c. Sample reservoir at 1 kV
   d. Sample waste reservoir at 1 kV

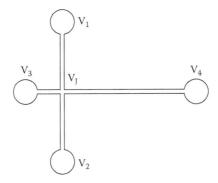

FIGURE 5.30
Simple cross channel for the electrokinetic injection of a small sample plug.

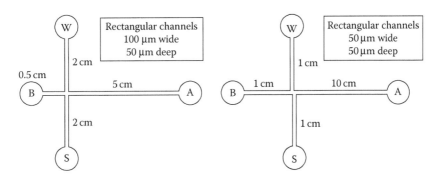

FIGURE 5.31
Two electrokinetic sample-injection systems.

5. Find an example of a microfabricated electrokinetic separation system in a peer-reviewed journal article. Briefly describe the device and include a picture. What measures were taken to improve the separation efficiency? Provide a citation for the article.

6. Find an example of a microfabricated hydrodynamic separation system in a peer-reviewed journal article. Briefly describe the device and include a picture. Why were hydrodynamic separation principles employed in this application? Provide a citation for the article.

7. C18 beads were used in the HPLC device shown in Figure 5.23. What is the composition of these beads and why are they commonly used for retention in HPLC systems? In addition, explain reversed-phase liquid chromatography. Cite your source.

8. What is sample bias and why should it be avoided? Cite any references used.

9. Band broadening may also be caused by turns in a separation channel. These turns are often necessary to allow the separation channel to fit within the small footprint of a chip. To minimize band broadening, is it better to use a channel with a single turn or two turns? Assume the channel width through each turn does not change and that the turns are oriented at 180°. Also assume that diffusion is insignificant within the turns.

References

1. Terry, S. C., J. H. Jerman, and J. B. Angell. 1979. Gas-chromatographic air analyzer fabricated on a silicon wafer. *IEEE Trans Electron Devices* 26:1880.
2. de Mello, A. 2002. On-chip chromatography: The last twenty years. *Lab Chip* 2:48N.
3. Manz, A., N. Graber, and H. M. Widmer. 1990. Miniaturized total chemical analysis systems: A novel concept for chemical sensing. *Sens Actuators B Chem* B1:244.
4. van den Berg, A., and T. S. J. Lammerink. 1998. Micro total analysis systems: Microfluidic aspects, integration concept and applications. *Topics in Current Chemistry* 194:21–49.
5. Auroux, P. A., et al. 2002. Micro total analysis systems. 2. Analytical standard operations and applications. *Anal Chem* 74:2637.
6. Reyes, D. R., et al. 2002. Micro total analysis systems. 1. Introduction, theory, and technology. *Anal Chem* 74:2623.
7. Vilkner, T., D. Janasek, and A. Manz. 2004. Micro total analysis systems. Recent developments. *Anal Chem* 76:3373.
8. Haeberle, S., and R. Zengerle. 2007. Microfluidic platforms for lab-on-a-chip applications. *Lab Chip* 7:1094.
9. Gale, B. K., K. D. Caldwell, and A. B. Frazier. 2002. Geometric scaling effects in electrical field flow fractionation. 2. Experimental results. *Anal Chem* 74:1024.

10. Lichtenberg, J., et al. 2006. Sample preparation on microchips. In *Separation Methods in Microanalytical Systems*, 575. Boca Raton, FL: Taylor & Francis.
11. Huang, Y., et al. 2002. MEMS-based sample preparation for molecular diagnostics. *Anal Bioanal Chem* 372:49.
12. Lichtenberg, J., N. F. de Rooij, and E. Verpoorte. 2002. Sample pretreatment on microfabricated devices. *Talanta* 56:233.
13. Andersson, H., et al. 2000. Micromachined flow-through filter-chamber for chemical reactions on beads. *Sens Actuators B Chem* 67:203.
14. Wilding, P., et al. 1998. Integrated cell isolation and polymerase chain reaction analysis using silicon microfilter chambers. *Anal Biochem* 257:95.
15. L'Hostis, E., et al. 2000. Microreactor and electrochemical detectors fabricated using Si and EPON SU-8. *Sens Actuators B Chem* 64:156.
16. Russo, A. P., et al. 2002. Microfabricated plastic devices from silicon using soft intermediates. *Biomed Microdevices* 4:277.
17. van Rijn, C., et al. 1997. Deflection and maximum load of microfiltration membrane sieves made with silicon micromachining. *J Microelectromech Syst* 6:48.
18. Yang, X., et al. 1999. Micromachined membrane particle filters. *Sens Actuators A Phys* 73:184.
19. Kittilsland, G., G. Stemme, and B. Norden. 1990. A submicron particle filter in silicon. *Sens Actuators A Phys* 23:904.
20. Chu, W. H., et al. 1999. Silicon membrane nanofilters from sacrificial oxide removal. *J Microelectromech Syst* 8:34.
21. Desai, T. A., D. Hansford, and M. Ferrari. 1999. Characterization of micromachined silicon membranes for immunoisolation and bioseparation applications. *J Memb Sci* 159:221.
22. Desai, T. A., D. J. Hansford, and M. Ferrari. 2000. Micromachined interfaces: New approaches in cell immunoisolation and biomolecular separation. *Biomol Eng* 17:23.
23. Lopez, C. A., et al. 2006. Evaluation of silicon nanoporous membranes and ECM-based microenvironments on neurosecretory cells. *Biomaterials* 27:3075.
24. Fissell, W. H., et al. 2009. High-performance silicon nanopore hemofiltration membranes. *J Memb Sci* 326:58.
25. Song, S., et al. 2004. Microchip dialysis of proteins using in situ photopatterned nanoporous polymer membranes. *Anal Chem* 76:2367.
26. Tokeshi, M., T. Minagawa, and T. Kitamori. 2000. Integration of a microextraction system on a glass chip: Ion-pair solvent extraction of Fe(II) with 4,7-diphenyl-1, 10-phenanthrolinedisulfonic acid and tri-n-octylmethylammonium chloride. *Anal Chem* 72:1711.
27. Mary, P., V. Studer, and P. Tabeling. 2008. Microfluidic droplet-based liquid-liquid extraction. *Anal Chem* 80:2680.
28. Kutter, J. P., S. C. Jacobson, and J. M. Ramsey. 2000. Solid phase extraction on microfluidic devices. *J Microcolumn Sep* 12:93.
29. Oleschuk, R. D., et al. 2000. Trapping of bead-based reagents within microfluidic systems: On-chip solid-phase extraction and electrochromatography. *Anal Chem* 72:585.
30. Yu, C., et al. 2001. Monolithic porous polymer for on-chip solid-phase extraction and preconcentration prepared by photoinitiated in situ polymerization within a microfluidic device. *Anal Chem* 73:5088.

31. Jacobson, S. C., and C. T. Culbertson. 2006. Microfluidics: Some basics. In *Separation Methods in Microanalytical Systems*, 575. Boca Raton, FL: Taylor & Francis.
32. Woolley, A. T., and R. A. Mathies. 1994. Ultra-high-speed DNA fragment separations using microfabricated capillary array electrophoresis chips. *Proc Natl Acad Sci U S A* 91:11348.
33. Fu, L. M., et al. 2002. Electrokinetic injection techniques in microfluidic chips. *Anal Chem* 74:5084.
34. Jacobson, S. C., S. V. Ermakov, and J. M. Ramsey. 1999. Minimizing the number of voltage sources and fluid reservoirs for electrokinetic valving in microfluidic devices. *Anal Chem* 71:3273.
35. Kutter, J. P., and Y. Fintschenko. 2006. *Separation Methods in Microanalytical Systems*. Boca Raton, FL: Taylor & Francis.
36. Desmet, G., E. Chmela, and R. Tijssen. 2006. Pressure-driven separation methods on a chip. In *Separation Methods in Microanalytical Systems*, 575. Boca Raton, FL: Taylor & Francis.
37. Ocvirk, G., et al. 1995. High-performance liquid-chromatography partially integrated onto a silicon chip. *Anal Methods Instrum* 2:74.
38. Harris, C. M. 2003. Shrinking the LC landscape. *Anal Chem* 75:64A.
39. Shih, C. Y., et al. 2006. An integrated system for on-chip temperature gradient interaction chromatography. *Sens Actuators A Phys* 127:207.
40. Shih, C. Y., et al. 2006. On-chip temperature gradient interaction chromatography. *J Chromatogr A* 1111:272.
41. Chmela, E., et al. 2002. A chip system for size separation of macromolecules and particles by hydrodynamic chromatography. *Anal Chem* 74:3470.
42. Blom, M. T., et al. 2002. Design and fabrication of a hydrodynamic chromatography chip. *Sens Actuators B Chem* 82:111.
43. Malmstadt, N., et al. 2003. A smart microfluidic affinity chromatography matrix composed of poly(n-isopropylacrylamide)-coated beads. *Anal Chem* 75:2943.
44. Gale, B. K., K. D. Caldwell, and A. B. Frazier. 2001. Geometric scaling effects in electrical field flow fractionation. 1. Theoretical analysis. *Anal Chem* 73:2345.
45. Lacher, N. A., et al. 2001. Microchip capillary electrophoresis/electrochemistry. *Electrophoresis* 22:2526.
46. Verpoorte, E. 2002. Microfluidic chips for clinical and forensic analysis. *Electrophoresis* 23:677.
47. Paegel, B. M., et al. 2002. High throughput DNA sequencing with a microfabricated 96-lane capillary array electrophoresis bioprocessor. *Proc Natl Acad Sci U S A* 99:574.
48. Paegel, B. M., S. H. I. Yeung, and R. A. Mathies 2002. Microchip bioprocessor for integrated nanovolume sample purification and DNA sequencing. *Anal Chem* 74:5092.
49. Emrich, C. A., et al. 2002. Microfabricated 384-lane capillary array electrophoresis bioanalyzer for ultrahigh-throughput genetic analysis. *Anal Chem* 74:5076.
50. Paegel, B. M., et al. 2000. Turn geometry for minimizing band broadening in microfabricated capillary electrophoresis channels. *Anal Chem* 72:3030.
51. Herr, A. E., et al. 2003. On-chip coupling of isoelectric focusing and free solution electrophoresis for multidimensional separations. *Anal Chem* 75:1180.

6

Sensing and Detection Methods

6.1 Sensing and Detection

A *sensor*, like an actuator, is a type of *transducer*—a device that converts energy from one form to another for sensing or actuation. However, instead of converting electrical energy into a mechanical response, as in an actuator, a sensor detects a physical parameter in one form and reports on that detection event in another form. For instance, a pressure sensor receives a mechanical input and provides an electrical signal of the mechanical measurement. Regardless of the form of input, the most commonly used form of the measured signal is electrical (pneumatic, hydraulic, and optical are less common). This is in part attributed to the ease of manipulating the received measurement signal for display or permanent recording. The type of electrical signal is used to further distinguish transducers. For instance, a *primary* transducer converts energy to voltage, whereas a *secondary* transducer may convert energy to resistance or capacitance.

Sensors interface with a *measurand*—the physical parameter, process, property, or state in which a quantitative or qualitative measurement is desired—to acquire the desired measurement. The nature of this interface may be direct or indirect. In practice, many measurements are indirect and require manipulation of the acquired signal (e.g., by applying an equation) to extract the quantity of interest. The signal generated by the sensor may be *analog* or *digital*. Sensors also require power to operate. This may be sourced externally (as in *active* or *modulating* sensors) or internally by the input itself (as in *passive* or *self-generating* sensors).

A convenient way in which to understand and compare the different types of sensors is by applying a classification scheme. Although many classification schemes—ranging from simple to complex—have been devised, we will use the physical form of the sensed phenomena and the detection mode for classification in our discussions here. Tables 6.1 and 6.2 provide the framework for these classification schemes.

TABLE 6.1

Sensor Classification Based on the Sensed Phenomena

Sensed Phenomena	Measurands
Acoustic	Wave amplitude, phase, polarization, spectrum, velocity, and so on
Biological	Identity, concentration, state, and so on
Chemical	Humidity, pH, concentration, state, identity, and so on
Electrical	Charge, current, voltage, electric field (amplitude, phase, polarization, spectrum), resistance, conductance, dielectric permittivity, capacitance, inductance, and so on
Magnetic	Magnetic field (amplitude, phase, polarization, spectrum), flux, magnetic moment, magnetization, magnetic permeability, and so on
Mechanical	Position, displacement, velocity, acceleration, force, torque, moment, stress, pressure, strain, mass, density, flow (velocity and rate), shape, texture, stiffness, compliance, viscosity, structure, and so on
Optical	Amplitude/intensity, phase, polarization, spectrum, velocity, and so on
Radiational	Type (gamma, X-ray, ultraviolet, visible, infrared, microwave, radio, and so on), energy, intensity, and so on
Thermal	Temperature, heat, flux, entropy-specific heat, heat capacity, thermal conductivity, and so on

TABLE 6.2

Detection Methods of Sensors

Biological
Chemical
Electric, magnetic, or electromagnetic
Thermal (heat or temperature)
Mechanical/acoustic
Radiation

A sensor is not used in isolation but is typically part of a larger system. The term *instrumentation* includes sensors and refers to devices that make measurements. Typically, an instrument system consists of a measurand, sensor, electronic signal processor, and a display panel for the observer (Figure 6.1). The objective of these tools is to make the measurand clearly perceptible. Additional information on the design of instrumentation is found in several texts on the topic [3–6].

This chapter provides an overview of sensing and the major sensing mechanisms. The world of microsensing is vast, and the body of sensor literature is continuously expanding; for further study, the reader is referred to several

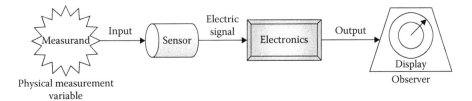

FIGURE 6.1
Simple model of a measurement instrument.

books [2,5,7,8] and reviews [9] dedicated to microsensors. Many other texts that focus on measurement, instrumentation, and sensors as a whole are also recommended for further study [3,4,6].

6.2 Sensor Characteristics

A wide variety of sensors exist in a number of fields. It is the task of the engineer to select the appropriate sensor for the measurement of interest. Sensors are distinguished from one another by their operating characteristics, which conveniently guide the engineer in the selection process. The key characteristics of sensors as related to performance are listed in Table 6.3. In addition to quantitative performance metrics, sensor selection is practically limited by considerations such as cost, size, and availability. Sensor performance characteristics are better understood by further comparing the desirable (ideal) and the undesirable sensor characteristics alongside each other, as in Table 6.4.

These lists of characteristics are not exhaustive, and many other performance specifications are typically provided in the manufacturer's product data sheet for commercial sensors. For example, a few device specifications about a medical-grade pressure sensor are listed in Table 6.5.

Another practical illustration of the limits of resolution in relation to the operating range for the selection of temperature sensors is presented in Figure 6.2. A similar comparison can be made between sensing frequency and operating range (Figure 6.3). Unfortunately, convenient collections of comparative information among similar sensors is not readily available, and the engineer must often collect, assemble, and evaluate a great deal of sensor data before appropriate sensor selection is possible.

The performance characteristics described thus far may be divided on the basis of *static* or *dynamic* responses. The static behavior of sensors is of interest when the measurand changes slowly over time. For example, sensitivity is a static sensor characteristic, whereas frequency response and response time are dynamic variables.

TABLE 6.3

Important Sensor Performance Characteristics

Performance Metric	Description
Operating range or dynamic range	Difference between the maximum and minimum measurable values
Resolution	Smallest discernible change in the measured value
Sensitivity	Ratio of the change in output signal for a given change in the measurand
Accuracy	Maximum error of the measurement as a percent of the full-scale measurement
Error	Deviation between the measured and actual values (can be random or systematic)
Precision	Variation in response to identical inputs due to random measurement errors
Operating frequency	Difference between the maximum and minimum frequencies at which a measurable signal is obtained
Reliability	Lifetime or number of cycles over which the sensor can be operated
Repeatability	Fluctuation in measured output when the measurand is the same
Selectivity	Sensitivity to the desired measurand divided by sum of sensitivities to all other signal sources
Operating-temperature range	Temperature range over which the sensor operates within the specified accuracy
Drift	Long-term stability of the measurement
Hysteresis	Measure of the inability to provide identical measurement signals, regardless of the direction of the measurand change
Linearity	Maximum deviation of the output-signal curve from a straight-line response
Offset	Measurement when the measurand is 0

TABLE 6.4

Comparison of Desirable and Undesirable Sensor Characteristics

Desirable Sensor Characteristics	Undesirable Sensor Characteristics
Linear and noise-free response	Nonlinear response with unwanted random noise signals
Zero baseline offset	Presence of baseline offset due to systematic error in output
No baseline drift	Baseline drift where output varies over time
Instantaneous response (zero response time)	Slow response where output takes time to reach the steady-state value
Infinite frequency bandwidth for instantaneous response	Limited or narrow frequency bandwidth for best response

TABLE 6.4 (*Continued*)

Desirable Sensor Characteristics	Undesirable Sensor Characteristics
Infinite operating range	Limited operating range
High sensitivity over operating range	Low sensitivity and response only for large input signals
Infinite resolution	Low resolution

TABLE 6.5

Selected Specifications for the MPX2300DT1 Medical Grade Compensated Piezoresistive Pressure Sensor Made by Freescale Semiconductor, Inc.

Sensor Characteristic	Description or Value
Supply voltage	6 Vdc
Pressure range	0–300 mmHg (0–40 kPa)
Pressure measurement mode	Gauge
Sensitivity	4.95–5.05 μV/V/mmHg
Maximum zero pressure offset	–0.75–0.75 mV
Maximum pressure	125 psi
Operating temperature range	15–40°C
Response time	1.0 ms
Biocompatibility	Class V approved

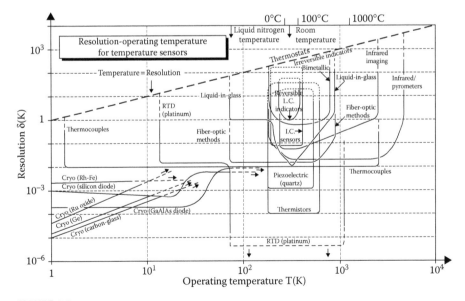

FIGURE 6.2

Comparison of resolution and sensor operating ranges for different types of displacement and proximity sensors. (From Shieh, J., et al. 2001. *Prog Mater Sci* 46:461. With permission.)

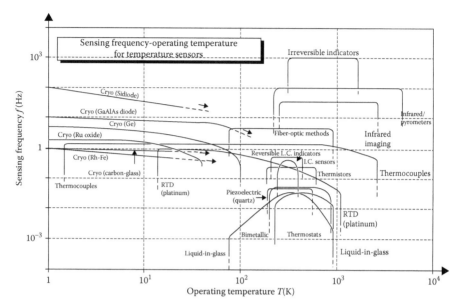

FIGURE 6.3
Comparison of sensing frequency and operating ranges for different types of displacement sensors. (From Shieh, J., et al. 2001. *Prog Mater Sci* 46:461. With permission.)

6.3 Principles of Physical Sensing

A convenient way to understand sensing mechanisms and their applications is to start with a discussion of measurement of the basic types of electronic sensor signals (resistance, capacitance, and inductance) for physical sensors. Then, we will delve into specific examples of sensing mechanisms that utilize these types of signals and their applications.

6.3.1 Resistive Sensors

Selection of appropriate sensor materials allows monitoring of the measurand by tracking the sensor's resistive change. The initial baseline resistance is expressed by

$$R_0 = \frac{\rho L}{A} \tag{6.1}$$

where ρ is the resistivity, L is the length, and A is the area. For a linear resistive sensor, the output is expressed as

$$R_x = R_0(1+x) \tag{6.2}$$

where the change in the output resistance R_x is due to a change in x, some factor related to the measurand that causes a resistive change. To detect this change, a voltage divider circuit is commonly used for detecting large changes in x (Figure 6.4) as follows:

$$V_{out} = \frac{R_x}{R_x + R_{ref}} V_{in} \qquad (6.3)$$

The sensitivity for this system is given by

$$K_{R,1} = \frac{dV_{out}}{dR_{sensor}} = \frac{V_{in} R_{ref}}{(R_x + R_{ref})^2} \qquad (6.4)$$

For small changes in x, a Wheatstone bridge configuration is used (Figure 6.4). A precision potentiometer (R_1) is used to operate the bridge in null balanced mode, in which R_1 is adjusted until V_{out} is zero initially. The circuit can also be operated in deflection or unbalanced mode. If the system is operated in balanced mode, we have

$$V_{out} = 0, \quad \text{and so,} \quad \frac{R_0}{R_3} = \frac{R_1}{R_2} = a \qquad (6.5)$$

The output of the bridge is then given by the expression

$$V_{out} = \left(\frac{R_x}{R_3 + R_x} - \frac{R_2}{R_1 + R_2} \right) V_{in} = \left(\frac{ax}{(a+1)(a+1+x)} \right) V_{in} \qquad (6.6)$$

The sensitivity of the Wheatstone bridge circuit is expressed as

$$K_{R,2} = \frac{dV_{out}}{d(xR_0)}\Big|_{x=0} = \frac{a}{R_0} \frac{1}{(a+1)^2} V_{in} \qquad (6.7)$$

Piezoresistive materials produce an electric potential when mechanically compressed and can be read out as a resistance change. Consider a wire that is placed longitudinally under tension. From Equation 6.1, we get

$$\frac{dR}{R} = \frac{d\rho}{\rho} + \frac{dL}{L} - \frac{dA}{A} \qquad (6.8)$$

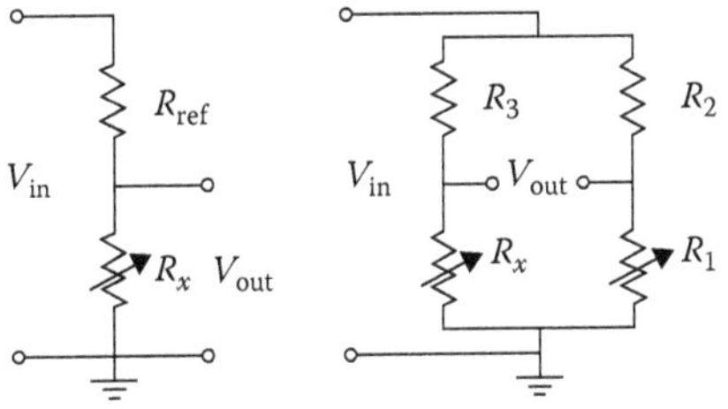

FIGURE 6.4
Left: Resistive voltage divider circuit. Right: Wheatstone bridge circuit.

The force required to produce this change in length is given by Hooke's law as follows:

$$\sigma = \frac{F}{A} = E\varepsilon = E\frac{dL}{L} \tag{6.9}$$

where σ is mechanical stress, F is force, A is area, E is Young's modulus, and ε is strain. For a wire of circular cross section, the longitudinal stress will also change the diameter (D), as follows:

$$A = \frac{\pi D^2}{4}; \quad \text{so,} \quad \frac{dA}{A} = \frac{2dD}{D} = -\frac{2vdL}{L} \tag{6.10}$$

where v $(0 < v < .5)$ is the Poisson ratio expressed as

$$v = -\frac{dt/t}{dL/L} \tag{6.11}$$

Here, t represents the transverse dimension. Equation 6.8 now becomes

$$\frac{dR}{R} = \frac{d\rho}{\rho} + \frac{dL}{L}(1+2v) \tag{6.12}$$

It is clear that the resistance change depends on both the dimension change and the resistivity change. The piezoresistive effect refers to the change in resistivity as a result of mechanical stress. From Equation 6.12, we can define the *gauge factor* (G) as follows:

$$G = \frac{\Delta R/R}{\Delta L/L} = \frac{\Delta R/R}{\varepsilon} = (1+2v) + \frac{\Delta\rho/\rho}{\Delta L/L} \tag{6.13}$$

The gauge factor is the ratio of the fractional change in resistance to the fractional change in length. This principle is used in strain gauges.

As discussed in Chapter 2, resistive materials also exhibit temperature sensitivity. Thus, a change in the ambient temperature may be measured as a change in the resistance that can be exploited for sensing purposes as follows:

$$R(T) = R(T_0)[1+\alpha(T-T_0)] \tag{6.14}$$

where T is the temperature, T_0 is the initial temperature, and α is the temperature coefficient of resistivity.

6.3.2 Capacitive Sensors

Capacitive sensors respond to measurands in multiple ways. For a parallel-plate capacitor, we have

$$C = \frac{\varepsilon A}{x} = \frac{\varepsilon_0 \varepsilon_r A}{x} \tag{6.15}$$

where ε_0 is the permittivity of vacuum, ε_r is the relative permittivity, A is the effective area of the plates, and x is the plate separation. The most common method for capacitive sensing relies on variations in plate separation. For small changes in separation, the sensitivity is given by

$$K_c = \frac{\Delta C}{\Delta x} = -\frac{\varepsilon_0 \varepsilon_r A}{x^2} \tag{6.16}$$

However, it is also possible to vary the capacitance by changing the permittivity or the area. A general expression for the variation in capacitance is

$$\partial C = \frac{dC}{d\varepsilon}\Big|_{A,x} \partial \varepsilon + \frac{dC}{dA}\Big|_{\varepsilon,x} \partial A + \frac{dC}{dx}\Big|_{\varepsilon,A} \partial x \tag{6.17}$$

Similar to resistive sensors, capacitive sensors can be measured using divider or bridge circuits.

6.3.3 Inductive Sensors

The self inductance of a coil of wire is expressed as

$$L = N\frac{\Phi}{i} \tag{6.18}$$

where N is the number of turns of the coil, Φ is the magnetic flux, and i is the current flowing in the wire. The magnetic flux may be written in terms of the magnetomotive force ($F_m = Ni$) and the reluctance ($\mathfrak{R}$) as

$$\Phi = \frac{Ni}{\mathfrak{R}} \tag{6.19}$$

Thus, the relationship between inductance and reluctance is given by

$$L = \frac{N^2}{\mathfrak{R}} \tag{6.20}$$

Magnetic reluctance is analogous to electrical resistance and is expressed as

$$\mathfrak{R} = \frac{L}{\mu_0 \mu_r A} + \frac{L_0}{\mu_0 A_0} \tag{6.21}$$

for a coil with cross section A and length L. The terms μ_0 and μ_r are the permeabilities of air and the magnetic core in the coil. A portion of the magnetic circuit path may cross through air with a path length of L_0 and cross section A_0. Inductive sensors rely primarily on the change in the reluctance of a magnetic circuit (changes in the material or coil geometry) but may also involve a change in the number of coil turns. In general, most self-inductance sensors are either of the variable gap (change in L_0) or moving core (change in μ) types (Figure 6.5). Mutual inductance sensors rely on the change in coupling between two separate coils due to their movement.

6.3.4 Resonant Sensors

A shift in the oscillation frequency caused by the measurand serves as the operational principle behind resonant sensors. Quartz crystal is a piezo-electric material that can be made to oscillate at its characteristic resonant frequency, f_0. This resonant frequency shifts as the crystal mass increases or decreases. For example, a quartz-crystal microbalance (QCM) can detect surface adsorption in the form of a resonant frequency shift. As the mass increases, the resonant frequency decreases. This relation is given by the Sauerbrey equation, as follows:

$$\Delta f = -\frac{1}{\rho_m k_f} f_0^2 \frac{\Delta m}{A}$$ (6.22)

where ρ_m is the density of the coating, k_f is the frequency constant, Δm is the change in mass, and A is the crystal surface area. This principle can be used in the detection of adsorbed water (or humidity sensing).

An alternate resonant-sensing mechanism uses surface acoustic waves (SAW), a type of elastic wave that propagates on the surface of a solid. An SAW is produced with a pair of interdigitated electrodes (transmitter). If each electrode element is separated by a gap measuring g and excited with a voltage signal of frequency f, the resulting surface wave has the velocity

$$v = 2fg$$ (6.23)

When material is adsorbed on the sensing area, the SAW detected downstream by an identical pair of interdigitated electrodes (receiver) registers

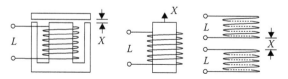

FIGURE 6.5
Inductive sensing of mechanical displacement.

a phase or frequency shift (Figure 6.6). Unlike QCMs, however, SAW sensors generally cannot be used with liquids due to significant wave attenuation. Thus, SAW sensors are used to measure gases and vapors. Both quartz resonators and SAW sensors may be applied for a variety of measurements, including temperature, force, and pressure.

6.3.5 Sensor Examples: Pressure Sensing

Pressure measurement has been historically significant in the context of microsystems because many early efforts focused on miniaturizing automotive sensors, which are now found in nearly all new cars. Physiological (Figure 6.7) and microfluidic pressures are also of interest, whether the

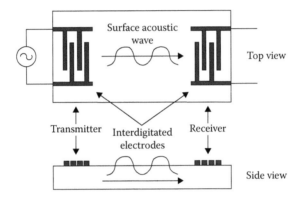

FIGURE 6.6
A simple surface acoustic wave sensor layout.

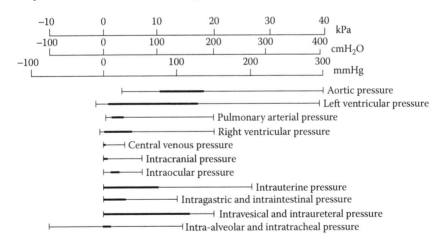

FIGURE 6.7
Body cavity pressures and their ranges for normal (thin lines) and abnormal (thick lines) conditions. (From Togawa, T., T. Tamura, and P. Å. Öberg. 1997. *Biomedical Transducers and Instruments*. Boca Raton, FL: CRC Press. With permission.)

purpose is for diagnostics and monitoring of patients or other devices. Clinical pressures are often expressed in terms of mmHg (1 mmHg = .133 kPa) or cmH_2O (1 cmH_2O = .098 kPa). In all these applications, pressure is usually measured with respect to atmospheric pressure (known as *gauge pressure*). Many methods, including piezoresistive, piezoelectric, capacitive, resonant, and optical, may be used for measurement of pressures.

An early example of a microelectromechanical systems pressure sensor for cardiovascular applications was described in reference [10]. The pressure sensor design was implemented in a 0.5-mm-diameter catheter for deployment in the coronary artery of the heart, used for tracking blood pressure variations (Figure 6.8). The pressure sensor operated on capacitive principles and contained a thin silicon diaphragm ($290 \times 550 \times 1.5$ μm^3) supported at the edges by a thicker frame (12 μm thick) overlying a metallized glass substrate. Deflection of the diaphragm in response to pressure altered the capacitance between the two plates. A thin but mechanically robust diaphragm was enabled using a special dissolved wafer process (Figure 6.9).

Similarly, in one study, a thin diaphragm attached to the end of an optical fiber (125-μm diameter) served as the pressure-sensitive element in an optical method for catheter-based pressure sensing [11]. Thin silicon wafers (50 μm thick) were used to create membranes (130×130 μm^2) suitable for integration with a 0.4-mm-diameter catheter. Deflection of the membrane in response to pressure resulted in modulation of the light intensity reflected from the gold coating on the back of the membrane. This optical signal could be processed using simple off-chip elements to extract pressure information. The catheter-mounted pressure sensors were used to track pressures and control inflation of polymer balloons in balloon catheters (1.5 mm in diameter).

Certain clinical conditions require indwelling pressure sensors. For example, endovascularly repaired abdominal aortic aneurysms require monitoring to ensure proper intra-aneurysm pressure following surgical repair and

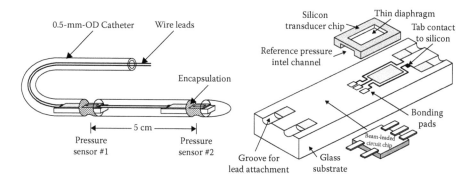

FIGURE 6.8
Left: Overall catheter design with two microelectromechanical systems pressure sensors. Right: A closeup schematic representation showing the construction of one pressure-sensing element. (From Chau, H. L., and K. D. Wise. 1988. *IEEE Trans Electron Devices* 35:2355. With permission.)

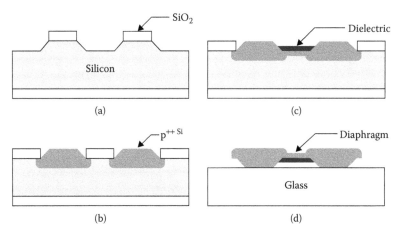

FIGURE 6.9
The dissolved wafer process for fabricating a thin silicon pressure-sensitive diaphragm: (a) anisotropic etching of silicon, (b) deep boron diffusion, (c) shallow boron diffusion and deposition of dielectric film, and (d) anodic bonding to glass substrate followed by wafer dissolution in anisotropic etchant.

to detect the presence of leaks or other complications. The surgical procedure strengthens the weakened aortic wall by using a stent graft. A wireless pressure sensor was introduced in other previous works [12–14] and was commercialized by CardioMEMS, Inc. In this implantable sensor (~5 mm wide and 30 mm long), pressure induces variations in the inductance and capacitance in a resonant LC (inductive-capacitive) circuit. This manifests as a shift in the resonant frequency, which is monitored wirelessly by magnetically coupling the circuit to an external coil. The earlier device consisted of a hermetically sealed cavity exposed to a pair of coils mounted on opposing walls constructed of thin fused silica membranes [13]. More recently, the substrate and construction were altered to enable two flexible sensors that can be rolled or folded for catheter-based delivery [14]. The difference between the two sensor designs was dictated by the intended duration of use: acute or chronic. Thus, either a polymer—liquid crystal polymer (LCP) or polytetrafluoroethylene (PTFE)—or a ceramic (sintered zirconia ceramic powder) was used in construction, respectively (Figure 6.10). The ceramic part allowed the hermetic sealing of the pressure in the reference cavity.

Elevated intraocular pressure is associated with glaucoma, a debilitating eye disease that damages the optic nerve and leads to progressive vision loss. Wireless pressure sensing approaches are sought to be able to continuously monitor the intraocular pressure to better understand the progression of the disease. In a previous study, a resonant LC circuit device was devised for implantable pressure monitoring in a small form factor device for the eye cavity [15]. Both variable capacitor and variable capacitor/inductor sensors were constructed (Figure 6.11), with reported pressure sensitivities of 7495 ppm/mmHg and 7595 ppm/mmHg, respectively.

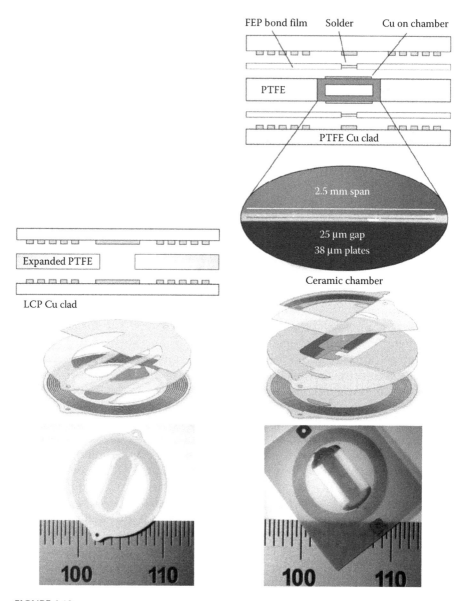

FIGURE 6.10
Left: Liquid crystal polymer sensor cross section, perspective view, and photograph. Right: Polytetrafluoroethylene/ceramic. (From Fonseca, M. A., et al. 2006. Flexible wireless passive pressure sensors for biomedical applications. In *Solid-state sensors, actuators, and microsystems workshop*. With permission.)

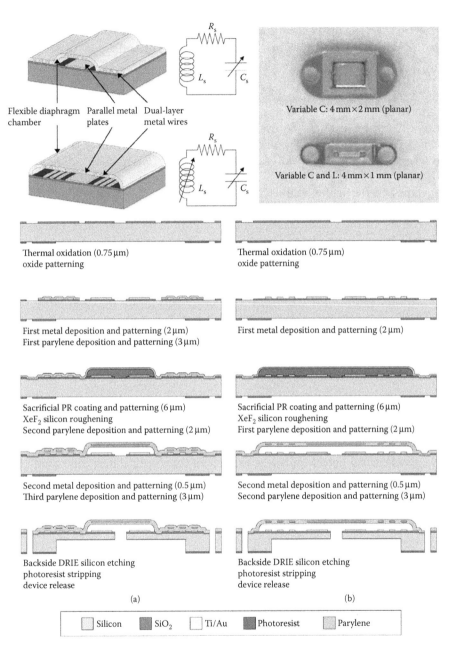

FIGURE 6.11
Top left: Two sensor designs in which the top device operates based on variable capacitance and the bottom on variable capacitance/inductance. Top right: Photograph of fabricated sensors showing the two suture holes flanking the central pressure-sensitive region. Bottom: Fabrication process for the intraocular pressure sensors: (a) variable capacitor and (b) variable capacitor/inductor. (From Chen, P. J., et al. 2008. *J Microelectromech Syst* 17:1342. With permission.)

6.4 Biological and Chemical Detection Methods

6.4.1 Biological Sensors

Biological sensors, or *biosensors*, use biological or molecular recognition systems in conjunction with a transducer to produce a measurable effect in the presence of an analyte (Figure 6.12). Examples of recognition systems, or *bioreceptors*, are biological molecules (e.g., antibodies, enzymes, proteins, and nucleic acids) and biological organisms (e.g., cells, organelles, cell receptors, tissue, and whole organisms). These bioreceptors are, by definition, highly specific to a particular analyte.

For biosensors to operate properly, the bioreceptor must be in close proximity to the signal transduction element. The biosensitive element is usually bound to the transducer by one of four mechanisms: (1) membrane entrapment, (2) physical adsorption, (3) matrix entrapment, or (4) covalent bonding. These mechanisms are summarized in Table 6.6 and are illustrated in Figure 6.13. The selection of the appropriate coupling mechanism is critical—the bioreceptor must not be inactivated by the immobilization process. Examples of biosensors include immunosensors, enzyme sensors, and cell-based sensors. These are discussed in more detail in Chapter 7.

6.4.2 Electrochemical Methods

Electrochemical detection involves conversion of a chemical signal into an electrical signal that can be detected by electrodes. This family of electroanalytical techniques can be miniaturized without any loss of sensitivity and have thus found an important place in microsystems. In addition, electrochemical detection is less expensive than optical detection methods. There are three major classes of electrochemical methods: amperometric,

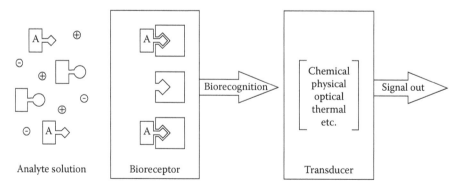

FIGURE 6.12
Principle of the biosensing process.

TABLE 6.6

Description of the Four Mechanisms that Couple a Bioreceptor to a Transducer in a Biosensor

Coupling Method	Description
Membrane entrapment	Semipermeable membrane traps bioreceptors against sensor surface
Physical adsorption	Bioreceptors are bound to sensor surface by a combination of van der Waals forces, hydrophobic forces, hydrogen bonds, and ionic forces
Matrix entrapment	Porous encapsulation matrix traps bioreceptors against sensor surface
Covalent bonding	Sensor surface contains reactive groups for covalent binding of bioreceptors

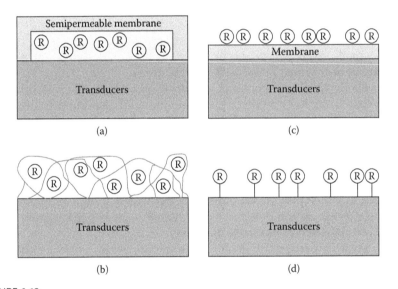

FIGURE 6.13
The four immobilization schemes where R stands for bioreceptor: (a) membrane entrapment, (b) matrix entrapment, (c) physical adsorption, and (d) covalent bonding.

potentiometric, and conductometric. These microelectrochemical detection methods were reviewed in references [17,18] and are briefly described here.

6.4.2.1 Amperometry

Amperometry, as the name would suggest, is a detection method centered on the sensing of current. Charge transfer arises from the oxidation or reduction of electroactive compounds at an energized electrode held at a constant potential. Although the method is only applicable for electro-active compounds, amperometric detection enjoys widespread use with microchip-based separations. The combination of capillary electrophoresis (CE) separation and amperometric detection, however, requires a careful decoupling of the high and low voltages used in each technique, respectively.

The most common layout for amperometry is a three-electrode electrochemical cell, which includes reference, auxiliary, and working electrodes. Two-electrode detection with only working and counter electrodes is also possible. The working electrode is energized and charge transfer occurs at its surface. The current i that is recorded is proportional to the moles of analyte oxidized or reduced. This is captured by Faraday's law and expressed as follows:

$$i = \frac{dQ}{dt} = nF \frac{dN}{dt} \tag{6.24}$$

where Q is the charge at the electrode surface, t is the time, n is the number of moles of electrons per mole of analyte, F is the Faraday constant, and N is the number of oxidized or reduced analyte moles.

The selectivity of amperometric sensors is tunable by the selection of an appropriate detection potential. Electrode material also impacts selectivity, with carbon electrodes having the most widespread use due to their ability to analyze a wide variety of samples including catechols, phenols, neurotransmitters, and peptides [19]. However, carbon is not easily applied by conventional microfabrication techniques. A clever technique to obtain carbon films for electrochemical detection based on lithography was devised by Kim et al. [20], in which photoresist films were pyrolyzed at high temperatures to produce various carbon patterns. Other metals such as gold, platinum, and copper are also used as electrode materials.

The use of amperometric detection in microsystems is reviewed in references [17,21–24]. The major motivations for on-chip integrated amperometric detection include minimization of dead volume, reduction of response time, and the integration of miniaturized electrodes. In reference [25], CE was integrated with an amperometric electrochemical detector. A sheath flow was applied to the detector to decouple the influence of the separation voltages on detection. The system was constructed of two glass wafers: the top containing etched channels and the bottom substrate containing detection electrodes (gold working and silver/silver chloride [Ag/AgCl] reference electrodes; Figure 6.14). Two sheath flow channels joined the separation channel (15 μm deep and 60 μm wide) near its end, and after the intersection, the separation channel widened into a detection reservoir (1000×1500 μm²). The sheath flow channels actively transported and focused the analyte stream onto the working electrode and prevented band broadening due to diffusion, enabling the use of larger separation voltages and larger channels. A dramatic improvement in the detected

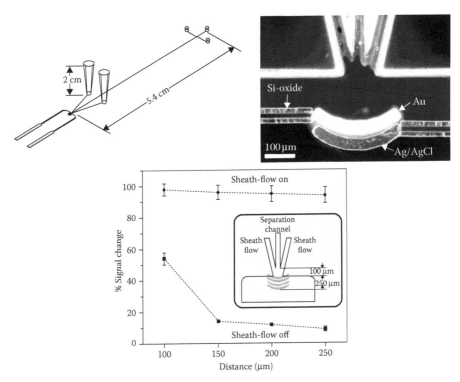

FIGURE 6.14
Top left: Layout of a capillary electrophoresis–electrochemical detection device with pipette tips inserted at the sheath flow channel inlets to provide gravity-driven flow. Top right: Magnified view of the sheath flow/separation channel intersection, detection electrodes, and the detection region. Bottom: Impact of sheath flow and electrode placement (100, 150, 200, and 250 μm from the separation channel exit) on the catechol detection signal. (From Ertl, P., et al. 2004. *Anal Chem* 76:3749. With permission.)

signal was obtained for the detection of catechol separations (limit of detection: 4.1 μM, for electrodes placed 250 μm away from the separation channel exit).

6.4.2.2 *Voltammetry*

Voltammetric detection is related to amperometry, but in this type of detection, the current–voltage relationship contains the information about the analyte. Here, voltage is applied to electrodes, and the resulting current is measured. In one example, a sinusoidal excitation voltage was used to obtain voltammetric signatures [26]. This system integrated a detector with a CE separation channel for the detection of neurotransmitters such as dopamine. A key feature of this system was a pyrolyzed photoresist carbon film electrode (external Ag/AgCl reference electrode). The photoresist pattern, supported on a thermally resistant quartz substrate, was heated to 1000°C for 60 minutes to produce a smooth carbon pattern. This was then integrated with a replica molded silicone rubber substrate containing 7-μm-deep and 20-μm-wide microchannels. The silicone rubber–quartz chip achieved a 100-nM detection limit for electrophoretically separated dopamine by sinusoidal voltammetry.

6.4.2.3 Potentiometry

Selectivity in electrochemical detection may also be achieved by combining an ion-selective membrane with an electrode. Potentiometry requires two electrodes: an ion-selective electrode and a reference electrode. The voltage measured across the ion-selective membrane then reflects the sample composition by selectively measuring signals related to the ions of interest. The signal arises as a result of charge separation induced by the membrane. This measurable Nernst potential (E) is related to the ionic concentration as follows:

$$E = E^0 + \frac{RT}{F} \ln\left(\sum K_i c_{i,z}\right) \tag{6.25}$$

where E^0 is a constant, R is the gas constant, T is the temperature, K_i is the selectivity coefficient for species i, and $c_{i,z}$ is the concentration of the species i of charge z. Potentiometric detection methods, although straightforward to implement on the microscale, have not received as much attention as other electrochemical methods. Microscale potentiometric approaches are reviewed in reference [27].

An example of a potentiometric sensor is the ion-selective field effect transistor developed by Bergveld in the 1970s. These devices are modified metal oxide semiconductor field effect transistors (MOSFETs) that expose the gate insulator directly to an ionic solution. This modification was spurned by the recognition that the sensitivity of MOSFETs to surface contamination could be used for sensing the presence of ions. The presence of ionic "contamination" in the gate insulator results in a detectable change in the drain current.

In a study by Liao et al. [28], a potentiometric sensor was fabricated using a platinum microelectrode modified with a silicone rubber membrane with potassium (valinomycin) and calcium (ETH 1001) ionophores for detection of K^+ and Ca^{2+}, respectively. An integrated silicone rubber pneumatic peristaltic pump transported fluids to the detection sites. The 10^{-6}-M detection limit was comparable to that of conventional electrodes. A similar device was used for pH sensing in a previous study [29]. Platinum electrodes with a sputtered SiO_2-LiO_2-BaO-TiO_2-La_2O_3 layer and a reference electrode (Ag/AgCl) were used for pH sensing. The electrodes were housed in a silicone rubber microchannel with multiple pneumatic peristaltic pumps to increase delivery (20 μL/min) of the sample to the detection zone (Figure 6.15). Hydrogen ions in the sample interacted with the sputtered coating, resulting in a detectable potential change (with a sensitivity of –55 mV/pH).

6.4.2.4 Conductometry

Conductivity detection monitors the change in solution resistance between the bulk and the analyte. Unlike other electrochemical detection methods, conductometric detection is widely applicable as the species need not be electroactive. The measured conductivity (G) is related to the concentration of analyte as follows:

$$G = \frac{A}{L}\sum \lambda_i c_i \qquad (6.26)$$

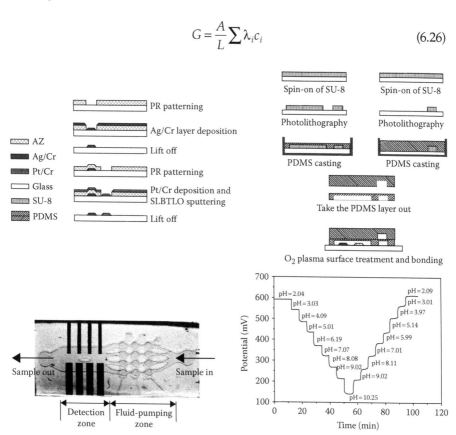

FIGURE 6.15

Top left: Fabrication process for electrodes. Top right: Fabrication process for silicone rubber microfluidics and assembly of system. Bottom left: Photograph of the detection/pumping system layout. The thinner electrodes in the upper part of the channel are the working electrodes and the bottom electrodes are reference electrodes. Bottom right: Recorded potential resulting from known pH changes. (From Lin, C. F., et al. 2006. *Biosens Bioelectron* 21:1468. With permission.)

where A is the electrode area, L is the distance between the two electrodes, λ_i is the ion's molar conductivity, and c_i is the ion concentration. Measurements are made using two electrodes and are made under an applied alternating current (AC) potential to avoid Faradaic electron transfer reactions. Microscale conductometric detection has been reviewed in references [17,22,24,27].

Galvanic contact between the electrodes and the solution were used for monitoring the progress of mixing in another previous work [30]. A silicone rubber–glass CE microchip was used for electroosmotic flow (EOF)–driven transport of analytes to a conductivity sensor (Figure 6.16). Both detection of EOF and gradient mixing (solutions with different concentrations of N-tris(hydroxymethyl) methyl-2-aminoethanesulfonic acid [TES]) was achieved using evaporated titanium/gold electrodes.

Solution contact may result in electrode fouling or solution electrolysis. Thus, a contactless alternative was developed, in which the signal is instead obtained from capacitive coupling with the electrolyte in a channel with a sinusoidal voltage signal applied at the sensing electrode. For a single-capacitor system, the capacitive reactance is measured instead of the conductivity, as follows:

$$Z_c = \frac{x}{\omega \varepsilon_0 \varepsilon_r A} \tag{6.27}$$

where ω is the angular frequency (in rad/s). A review dedicated solely to contactless conductometric detection can be found in [31].

A contactless conductometric sensor integrated with a separation channel is shown in Figure 6.17 [32]. Separation between the platinum electrodes and the detection region was achieved with a thin glass wall (10–15 μm thick).

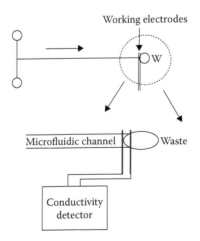

FIGURE 6.16
Layout of a conductivity detector. (From Liu, Y., D. O. Wipf, and C. S. Henry. 2001. *Analyst* 126:1248. With permission.)

To maintain a flat surface for glass-to-glass fusion bonding, electrodes were deposited in etched channels (12 μm deep and 50 μm wide). Detection of K⁺ at 18 μM was reported; however, lower detection limits were possible in other microfabricated sensors.

6.4.3 Optical Detection Methods

6.4.3.1 Fluorescence

A molecule absorbs a photon or photons, bringing an electron to an excited singlet state. Subsequently, the molecule emits detectable light as these excited electrons return to the ground state. This phenomenon is known as *fluorescence*. Usually, the photon is absorbed in the ultraviolet (UV) spectrum and the emitted light is in the visible spectrum (this phenomenon is called the *Stokes shift*). The emitting molecules possess a fluorescence efficiency that is quantified by the quantum yield or ratio of the number of emitted photons to the number absorbed. Some molecules possess native fluorescence, but most are derivatized with fluorescent species, or *fluorophores*. In conventional systems, the excitation source is commonly a laser, a light-emitting diode, or

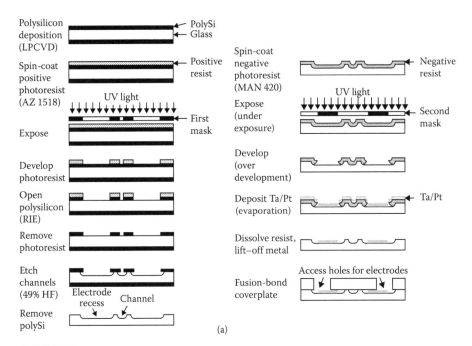

FIGURE 6.17
(a) Fabrication process of a conductometric detector, starting with channel etching, followed by electrode patterning and packaging of the whole system. (b) Device photograph and layout, showing the capillary electrophoresis separator and integrated contactless conductometric detector. (c) Conductometric detection of capillary electrophoresis separated K⁺, Na⁺, and Li⁺ cations. (From Lichtenberg, J., N. F. de Rooij, and E. Verpoorte. 2002. *Electrophoresis* 23:3769. With permission.)

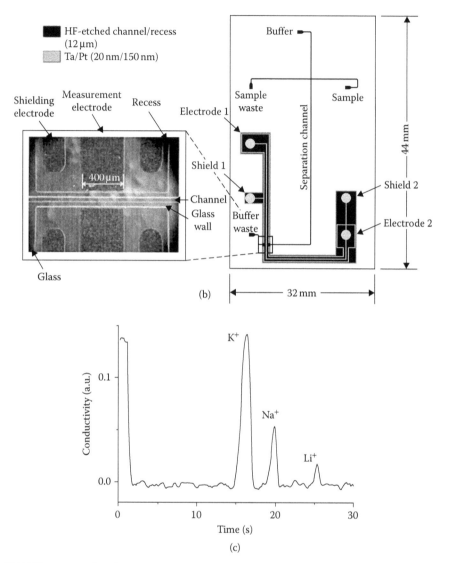

FIGURE 6.17 (*Continued*)

a lamp (e.g., a xenon arc lamp). Laser-induced fluorescence (LIF) is popularly used for detection in microfabricated systems because lasers possess narrow wavelengths and small areas of focus to match the small detection areas defined on lab-on-a-chip (LOC) systems. The most common wavelength used is 488 nm, which corresponds to the excitation of fluorescein isothiocyanate (520-nm emission) and Alexa Fluor 488 (519-nm emission). Many other common fluorophores covering wide range of wavelengths are used in LIF and other forms of fluorescence detection. Examples of these fluorophores and their corresponding analytes are given in reference [19].

The detected fluorescence is correlated with the concentration of the target analyte. The *Beer–Lambert law* provides a useful relationship between the optical absorbance and the concentration of the species of interest, and is represented as follows:

$$A = \varepsilon L c \qquad (6.28)$$

where A is the absorbed radiation, ε is the molar absorptivity, L is the path length through the medium, and c is the concentration of the absorbing species. High levels of both molar absorptivity and quantum yield are desirable fluorophore features. In practical fluorescence applications, interference from background fluorescence or light scattering from the materials of the device is present. Thus, careful selection of the optical properties of the materials and their processing are key factors in system design. In addition, fluorophore selection is critical due to variations in their limited photochemical lifetime. Photochemical damage may occur following excitation, resulting in fading or *photobleaching*.

The detection of emitted photons is usually performed with a charge-coupled device (CCD), a photomultiplier tube (PMT), or an avalanche photodiode. CCDs are typically used for detection in conjunction with a microscopy system. This allows the use of the existing optics to magnify the detection region. A PMT detects photons using a photoemissive cathode, a series of dynodes, and an anode. Photons collide with the cathode and eject electrons, which are then accelerated by an electric field to the series of dynodes. This starts an amplification cascade in which electrons collide with dynode after dynode, releasing more electrons with each successive collision. Eventually, the electrons reach the anode, where they are recorded as a current pulse that corresponds to the photons originally collected at the cathode. High-sensitivity detection of photons is possible with avalanche photodiodes, which also produce current amplification, but do so by impact ionization. Some detection schemes may include the use of optical fibers or waveguides to assist in the collection of emissions for transport to the detection device. Some common configurations for LIF detection in microsystems are shown in Figure 6.18. Optical elements focus the laser excitation onto the detection region of a microchannel. In a continuous flow system, fluorophores passing this laser region will be excited and then detected.

A clever method for the automated detection of multiple parallel CE-based separations is shown in Figure 6.19. In this system, a four-color rotary confocal fluorescence scanning system was implemented for the detection of signals from 384 radially oriented CE lanes on a single chip. An Ar^+ laser (488 nm) was used as the excitation source. Light was focused onto individual separation lanes using a microscope objective attached to a motorized optical system. The same objective collected the fluorescence and directed the signal to a four-color confocal detector (520, 550, 580, and 605 nm) [33].

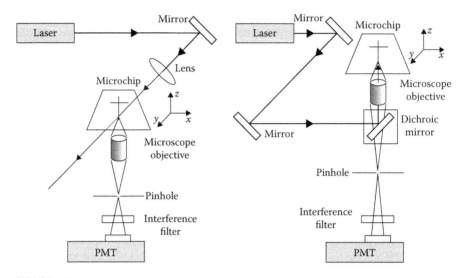

FIGURE 6.18
Left: Laser-induced fluorescence detection scheme. Right: Laser-induced fluorescence detection scheme with confocal fluorescence.

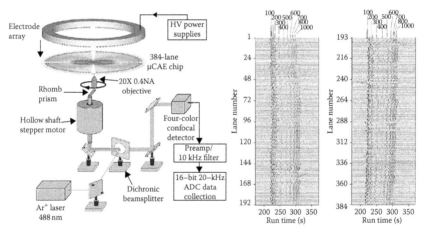

FIGURE 6.19
Left: Four-color confocal fluorescence scanner for detecting electrophoretic separations (excitation at 488 nm). Stepper motor rotates at 5 rev/s and allows scanning of electrophoresis lanes from the backside of chip. Right: Electropherograms collected from different lanes on the chip. (From Emrich, C.A., et al. 2002. *Anal Chem* 74:5076. With permission.)

6.4.3.2 Absorbance

Some molecules have a region known as a *chromophore*, in which absorbed energy causes a transition between the ground state and the excited state corresponding to the UV and visible ranges, respectively (UV/Vis absorbance). Detection of this absorbance is strongly dependent on the optical path length.

Thus, one strategy to improve absorbance detection in microsystems is to increase the effective path length. This may be accomplished by using internal reflection within flow cells, by transverse detection across the width of a microchannel ($w \gg h$), or by creating a U-shaped bend in the channel in which the detection occurs across the bottom of the U structure (Figure 6.20). Integrated waveguides also assist in collecting emissions for detection [34].

6.4.3.3 Chemiluminescence

Chemical reactions may also produce a detectable optical emission that correlates with the analyte concentration. This phenomenon, known as *chemiluminescence* (CL), does not require an external light source. Other benefits include wide dynamic range, high selectivity, and improved sensitivity as a result of low background interference. Optical detection of the presence of the analyte concentration may occur due to native luminescence following a chemical reaction or due to the modification of the luminescence of other participating compounds. Although an extra preparation step is necessary, nonluminescent analytes may be labeled to allow direct optical detection.

CL was first demonstrated in an LOC system using a horseradish peroxidase (HRP)-catalyzed reaction of luminol with peroxide to detect CE separations [35]. HRP is an enzyme commonly used for labeling immunoreagents (e.g., goat antimouse immunoglobulin [IgG]) and luminol in

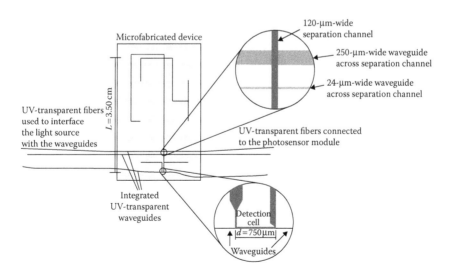

FIGURE 6.20
Layout of a capillary electrophoresis chip with waveguides and an U-channel for ultraviolet absorbance detection. The detection region is 750 μm long (compared to the channel that is only 12 μm high. (From Petersen, N. J., K. B. Mogensen, and J. P. Kutter. 2002. *Electrophoresis* 23:3528. With permission.)

immunoassay detections (e.g., that of mouse IgG with goat antimouse IgG). Thus, immunoassay detection was performed in a glass CE system with an integrated injector, a separator, and a postseparation reactor in 10- or 40-μm-deep channels. A metallic mirror was sputtered below the detection channel and the collection efficiency was enhanced to enable a detection limit of 35 nM (a 1-nL plug in 10-μm-deep channels) or 7 nM (an 8-nL plug in 40-μm-deep channels) using a calibration sample (fluorescein-conjugated HRP).

Another version of CL involves oxidation or reduction at an electrode. This is referred to as *electrochemiluminescence* (ECL) and requires the application of a potential using an electrode to control the chemical reaction. A standard ECL detection scheme starts with electrochemical detection at the cathode, followed by optical reporting of the event at the anode in a single microfluidic stream (Figure 6.21). However, this simple scheme allows detection of only those targets that can be reduced. An advanced approach, demonstrated by Zhan et al. [36], utilizes a three-channel, two-electrode

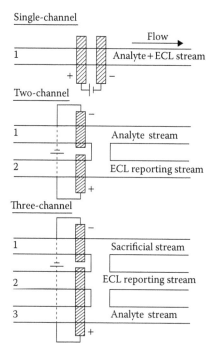

FIGURE 6.21
Three different electrochemiluminescence detection schemes. The single-channel device involves detection at the cathode and optical reporting at the anode. The two-channel device works similarly but chemically separates the detection and reporting into different streams. The single- and two-channel devices allow detection of only targets that can be reduced. The three-channel scheme allows detection of analytes that undergo anodic oxidation and reports their presence.

system. The contents of each channel are distinct. Here, the sacrificial electroactive molecule ($Ru(NH_3)_6^{3+}$) was in channel 1, the reporting molecules (tripropylamine [TPA] and $Ru(bpy)_3^{2+}$) were in channel 2, and the oxidizable analyte (dopamine) was in channel 3. The latter two channels shared a common anode. This allowed chemical separation of the detection and reporting reactions, but correlated the electrochemical oxidation events. First, an applied potential reduced $Ru(NH_3)_6^{3+}$ to $Ru(NH_3)_6^{2+}$, and also oxidized TPA and $Ru(bpy)_3^{2+}$. The analyte competed with the contents of the central channel during the reduction process in channel 1. Thus, the presence of the target analyte was detected by a reduction of light emission in channel 2. This microfluidic system was constructed from silicone rubber, and transparent indium–tin oxide microelectrodes were used to facilitate optical observation.

6.4.3.4 Other Optical Detection Methods

Several other optical detection methods are used, but to a lesser extent than the methods previously described. Optical detection of compounds having poor absorbance and fluorescence is achieved by a *refractive index* (RI) technique, in which the ratio of the speed of light through vacuum and through the solution is determined. In *Raman spectroscopy*, the analyte's structure can be determined for identification of functional groups and molecular fingerprints. This information is obtained by applying monochromatic light to molecules and measuring the scattered light possessing upward or downward wavelength shifts resulting from molecular vibrations. Light scattered by surface plasmons (a type of surface electromagnetic wave) carries information concerning surface-bound chemical species. This technique, known as *surface plasmon resonance* (SPR), is performed at the boundary between a metal (substrate or particle) and a fluid. Surface plasmons are very sensitive to adsorbed species and can be used, for instance, to detect binding events in immunoassays.

6.4.4 Mass Spectrometry

Identification of the chemical composition of a sample is a fundamental necessity in many chemical and biological applications. For example, the analysis of protein samples with high sensitivity and specificity is critical in proteomics. *Mass spectrometry* (MS) allows the measurement of the mass-to-charge (m/z) ratio of ionized compounds. The mass spectrum obtained is a plot of the m/z ratio versus the relative or absolute abundance or intensity (Figure 6.22).

Mass spectrometry detection is the result of many steps including sample handling, ionization, mass analysis, and ion detection. For example, the sample is first separated and then transferred for analysis using the following sample ionization methods: (1) *electrospray ionization* (ESI) or (2) *matrix-assisted laser desorption ionization* (MALDI). The sample transfer and ionization

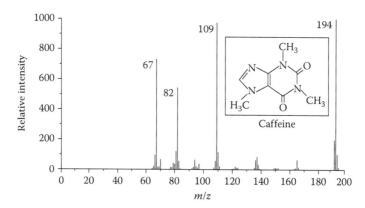

FIGURE 6.22
Mass spectrum of caffeine obtained through mass spectrometry, with major peaks identified. (Redrawn from http://www.massbank.jp/jsp/Dispatcher.jsp?type=disp&id=PR010011&site=1. With permission.)

steps are of great importance in the overall detection performance. Both are amenable to microfabricated solutions, which are reviewed in references [37–39]. Both single- and multiple-channel systems exist, and some systems even integrate sample pretreatment. Multiple mass analysis and ion detection methods have been devised, including single quadrupole, triple quadrupole, ion-trap, tandem mass spectrometry (MS/MS), and Fourier transform ion cyclotron resonance (FTICR). However, the detection portion of the MS system is not currently amenable to miniaturization and, therefore, is not discussed further here. Instead, we will focus on ESI and MALDI.

6.4.4.1 Electrospray Ionization

ESI for MS was introduced by Fenn et al. in 1984 [40] and involves the use of an electric field to create a fine spray of charged droplets from a liquid sample through a small opening (typically a fine nozzle). The electric field induces electrostatic dispersion of the liquid into charged droplets and then the droplets are further transformed into free ions in a gas phase through desolvation and ionization (Figure 6.23). This gaseous output can then be analyzed. Conventional macroscale ESI devices use nozzles with large internal diameter (~100 μm) with flow rates in the μL/min range. However, the efficiency of MS is greatly improved by scaling down the ESI system (smaller nozzle and flow rate) without losing sensitivity. The smaller nozzle size allows smaller droplet sizes to facilitate sample ionization and concentrates the electric field so that lower voltages can be used.

In microfabricated devices, three ESI configurations in which spray emanates from (1) the edge of a chip [41–43], (2) an inserted capillary [44–46], or (3) an integrated capillary [47,48] have been implemented (Figure 6.24). These are reviewed in references [37,49]. The electric field used in ESI is typically established between the nozzle and the MS inlet, as shown in Figure 6.25.

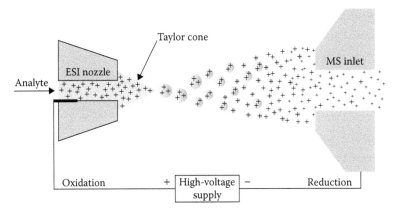

FIGURE 6.23
Schematic representation of an electrospray ionization interface for mass spectrometry showing the characteristic Taylor cone formed at the electrospray ionization nozzle.

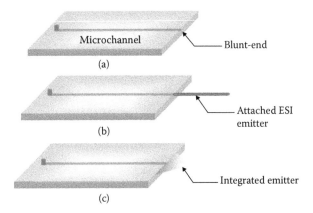

FIGURE 6.24
Three electrospray ionization configurations for mass spectrometry using the spray from (a) the edge of a chip, (b) an inserted capillary emitter, or (c) an integrated capillary emitter. (From Lee, J., S. A. Soper, and K. K. Murray. 2009. *J Mass Spectrom* 44:579. With permission.)

The first microfabricated approach to ESI simply used small openings formed from an exposed microchannel at the edge of a chip [41–43]. Both single and multiple ESI channel chips were developed at nearly the same time [41,42]. In an early single-channel device (Figure 6.26), a simple glass microchannel network (10 μm deep and ~60 μm wide) with integrated CE separation was used [41]. The electrospray opening was revealed by manually scoring and breaking the glass chip. The opening was placed 3–5 mm from the MS inlet orifice. External platinum wires were used to apply the electric field for electroosmotic pumping, electrophoretic separation, and electrospray of the fluid sample (tetrabutylammonium iodide in 60% water/40% methanol).

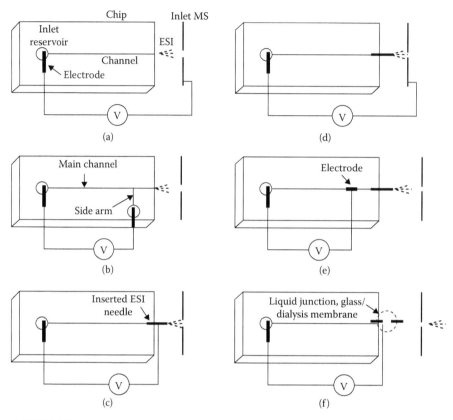

FIGURE 6.25
Configurations common to microfabricated electrospray ionization systems. (a–b) Spray from the edge of a chip with different electrical configurations, (c–d) spray from an inserted fused silica capillary with different electrical configurations, (e) an integrated electrode system, (f) spray using an integrated liquid junction or dialysis membrane. (From Koster, S., and E. Verpoorte. 2007. *Lab Chip* 7:1394. With permission.)

A similar multiple-channel approach is shown in Figure 6.27. The glass ESI chip contained nine separate channels (25 μm deep and 60 μm wide). Flow was generated using an external syringe pump. Although ESI-MS was demonstrated using these simple interfaces, the hydrophilic samples used tended to spread along the device edge and shift the location of spray generation. Surface coatings can be used to minimize this; however, other more elegant methods have been devised to avoid such issues.

The interface between CE separations and ESI for MS was simplified by attaching a conventional fused-silica capillary to a microchip [44]. A small-diameter drill bit (200-μm outer diameter) was used to make a hole to receive and interface the capillary (185-μm outer diameter and 50-μm inner diameter) with etched glass microchannels (13 μm deep and 40 μm wide; Figure 6.28). By controlling the type of drill bit used, the dead volume was

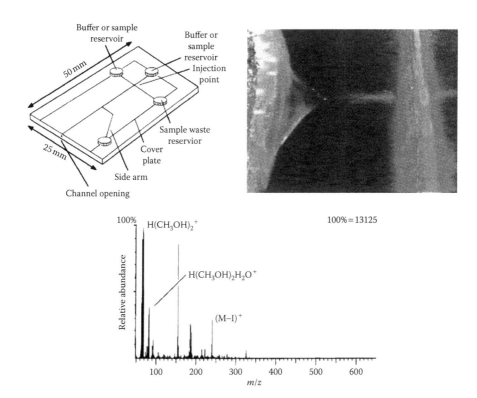

FIGURE 6.26

Top left: Schematic representation of an electroosmotically pumped electrospray ionization chip. Top right: Taylor cone and electrospray generated by the chip using a 60% water/40% methanol solution. Bottom: Mass spectrum from a 10-μM solution of tetrabutylammonium iodide in 60% water/40% methanol. (From Ramsey, R. S., and J. M. Ramsey. 1997. *Anal Chem* 69:1174. With permission.)

minimized. In addition, the tip of the capillary was shaped using a glass puller to further reduce the tip diameter (60-μm outer diameter and 15-μm inner diameter). The mass spectra results obtained for the detection of peptide mixtures using a microchip with a capillary emitter attached in such a manner were compared to conventional CE-ESI-MS interfaces using a sheath flow arrangement [45]. Although the conventional system outperformed the microchip, this preliminary effort demonstrated the feasibility of such an approach and indicated that further investigation was required to improve the performance.

Although manual coupling of capillaries to ESI chips is labor-intensive and susceptible to dead volume at the interface, a multiple-channel system approach has been previously demonstrated [46]. A cast-polymer array of 96 wells (5 mm in diameter and 5 mm deep) was constructed, and each well was connected to an independent fused-silica electrospray capillary (140-μm

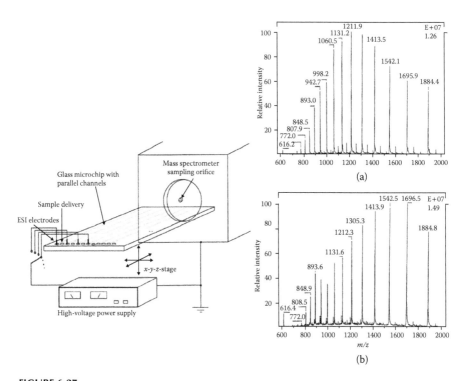

FIGURE 6.27
Left: Schematic representation of the 9-channel electrospray ionization chip. Right: A comparison of the mass spectra of a 6-μM myoglobin solution in 75% methanol/0.1% acetic acid from
(a) a conventional capillary (50-μm inner diameter) and (b) a microchip. (From Xue, Q. F., et al.
1997. *Anal Chem* 69:426. With permission.)

outer diameter and 26-μm inner diameter; Figure 6.29). A microchannel with
an integrated electrode was present at the interface. The sample was loaded
by pressure-driven flow using a pneumatic source (N_2 gas). Successful demonstration of the multiple-channel format ESI microchip pointed to the pos-
sibility of high throughput, automated analysis, and polymer construction
for disposable chips.

Batch fabricated and integrated ESI needles offer distinct advantages over
the previously described approaches [47,48]. In an earlier work, Parylene
electrospray emitters having tapered tips with 5- by 10-μm^2 rectangular openings were fabricated using surface micromachining techniques (Figure 6.30)
[47]. The tips extended from microchannels (5 μm high and 100 μm^2 wide)
supported by the silicon substrate. Emitter tips were sputtered with 60%
platinum/40% palladium to allow the application of voltage. Microfluidic
connections to the ESI emitters were formed by etching through the substrate. An array of pillars was fabricated around each port prior to connecting with the microchannels; these integrated filter structures removed debris.
The sample was delivered by pressure-driven flow, and the mass spectra of

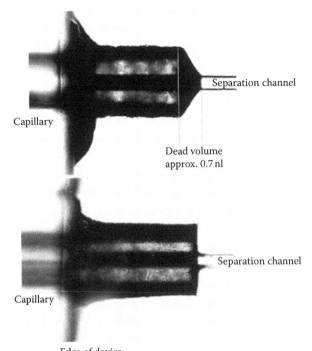

FIGURE 6.28
Photographs showing the interface between a conventional fused-silica capillary (185-μm outer diameter and 50-μm inner diameter) and a microchannel (13 μm deep and 40 μm wide) using two different drill bit types. In the bottom panel, the dead volume at the capillary-channel interface is minimized by using a flat-tipped drill to make the capillary receptacle. (From Bings, N. H., et al. 1999. *Anal Chem* 71:3292. With permission.)

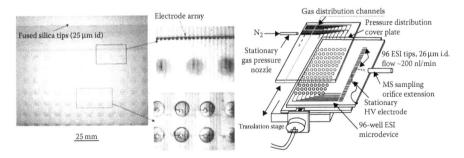

FIGURE 6.29
Left: An arrayed electrospray ionization microchip with magnified insets showing the microchannels connected to each sample well (top) and the electrodes embedded in the microchannels (bottom). Right: Schematic representation of a 96-channel electrospray ionization system with a pneumatic distribution system positioned over a mass spectrometry inlet orifice. A motorized translation stage allows positioning of each electrospray ionization channel over the mass spectrometry inlet. (From Liu, H. H., et al. 2000. *Anal Chem* 72:3303. With permission.)

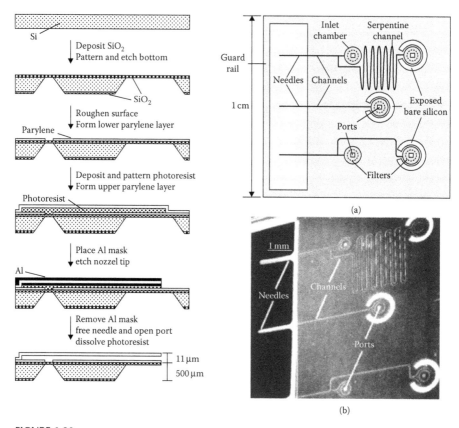

FIGURE 6.30
Left: Fabrication process for a Parylene electrospray ionization emitter. Right: (a) Schematic representation showing the device layout and (b) a scanning electron micrograph of fabricated electrospray ionization emitters. (From Liu, H. H., et al. 2000. *Anal Chem* 72:3303. With permission.)

peptide mixtures with both blunt- and tapered-tip emitters were obtained (Figure 6.31).

The next generation of this device showed further improvement involving the integration of multiple functions into a single microchip [48]. Three electrolysis-based pumps for driving the sample, a sample injector, a static mixer, a packed-bead separation column (reverse-phase column), and an ESI emitter were all integrated on a silicon substrate for performing a liquid chromatography and tandem mass spectrometry (LC–MS/MS) procedure to analyze the peptide mixtures (Figure 6.32). System integration reduces the number of fluidic interconnects and reduces the overall dead volume. The reverse-phase column was packed with 3-µm-diameter C18-A silica beads and retained in place by a constriction in the microchannel. Two solvent reservoirs (A and B) flanked the sample reservoir and were loaded with 95% water/5% methanol/0.1% formic acid, 40% water/60% methanol/0.1%

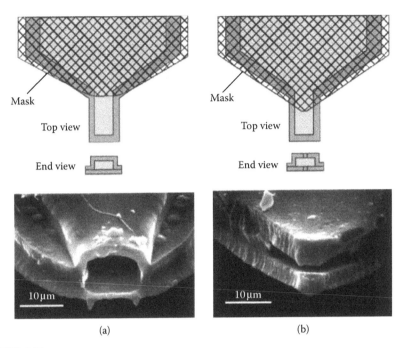

FIGURE 6.31
Two types of Parylene electrospray ionization emitters: (a) blunt and (b) tapered tip. The top view shows the masks used to generate the corresponding tips shown in the scanning electron microscopy images below them. (From Licklider, L., et al. 2000. *Anal Chem* 72:367. With permission.)

formic acid, and 5 µL of 1 pmol/µL of trypsin-digested bovine serum albumin. First, LC separation of the peptide mixture was performed by controlling the current provided to each pumping chamber (Figure 6.33). Briefly, a column wash, column equilibration, sample injection, sample wash, and gradient elution of the separated components were performed (total of 65 minutes with ~600 nL of injected sample). Gradient elution was achieved by varying the current to the solvent pumps. Following elution, ESI was performed (20×5-μm^2 opening), and the resulting mass spectrum was obtained.

6.4.4.2 Matrix-Assisted Laser Desorption Ionization

MALDI has received less attention than ESI; nevertheless, it is amenable to miniaturization. Compared to ESI–MS, MALDI–MS is preferred due to its excellent sensitivity for high molecular weight species such as proteins and polymers with masses above 100 kDa. However, MALDI underperforms for low molecular weight species with masses below 500 Da as a result of interference from the matrix [50]. Microfluidic MALDI techniques are reviewed in reference [50].

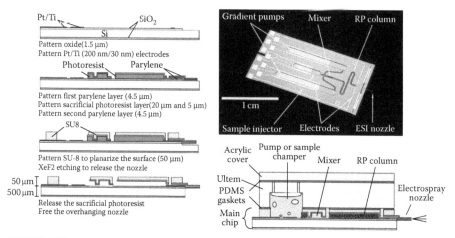

FIGURE 6.32

Left: Fabrication process for an integrated Parylene microfluidic platform for liquid chromatography and tandem mass spectrometry that includes an electrospray ionization emitter. Top right: Photograph of a fabricated microfluidic platform base. Bottom right: Schematic representation of a completely packaged liquid chromatography and tandem mass spectrometry microfluidic platform. (From Xie, J., et al. 2005. *Anal Chem* 77:6947. With permission.)

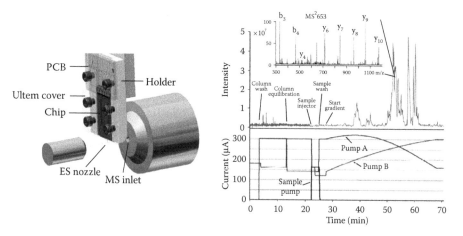

FIGURE 6.33

Left: Three-dimensional schematic representation showing the orientation of the electrospray ionization emitter in relation to the mass spectrometry inlet and the chip housing. Right: Upper panel shows the chromatogram of a peptide mixture and the lower panel shows the pump control settings for each chamber. The inset is the mass spectrum obtained for the ion (m/z 653.3) for the peptide with the sequence HLVDEPQNLIK. (From Xie, J., et al. 2005. *Anal Chem* 77:6947. With permission.)

MALDI was first introduced by Karas et al. [51] in 1985, and uses an alternate approach to achieve ionization of the sample. First, the sample is mixed with low molecular weight aromatic acids, such as nicotinic acid, benzoic acid derivatives, and cinnamic acid derivatives (energy-absorbing matrix). The mixture is evaporated onto a metallic plate and then ionized with a laser beam. The ionized vapors are usually analyzed with a time-of-flight (TOF) MS instrument (i.e., MALDI–TOF–MS).Microfabricated approaches to MALDI include (1) target containment plates, (2) off-chip target preparation, and (3) microfluidic integration of on-chip targets with target preparation.

Well-defined MALDI targets were created by liquid-phase deposition on microfluidic devices. Target confinement is desired to prevent sample spreading in the liquid phase, which results in a lowered local analyte concentration and ultimately in a reduction in MALDI-MS sensitivity. To confine the targets, two approaches were devised. Sample spreading was controlled by targeting the deposition onto patterned hydrophilic regions surrounded by hydrophobic barriers. Target samples were also physically confined in lithographically patterned and etched vial arrays (Figure 6.34).

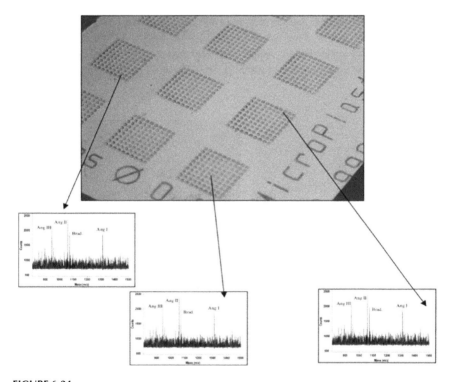

FIGURE 6.34
Matrix-assisted laser desorption ionization–mass spectrometry results from etched nanovial arrays. (From Marko-Varga, G. A., J. Nilsson, and T. Laurell. 2004. *Electrophoresis* 25:3479. With permission.)

Microfluidics can be used to deposit MALDI targets on a standard off-chip target plate. This avoids the sample confinement issues associated with direct mechanical spotting of targets. Piezoelectric dispensers can spot MALDI targets at high speed (Figure 6.35). Multiple spots deposited on a single target allow on-spot enrichment of the sample [52]. Target samples can also be deposited by electrospray onto a target plate [53]. In one example, polycarbonate channels (30 μm deep, 12 μm wide, and 4.5 cm long) were fabricated by hot embossing with a bulk micromachined silicon master mold. The electrospray channel was exposed by milling the chip, and a porous PTFE membrane (50 μm thick) was bonded to the exposed microchannel to provide a hydrophobic surface (Figure 6.36). Angiotensin samples, followed by the matrix solution (α-cyano-4-hydroxycinnamic acid [CHCA] and 50% acetonitrile/40% water/10% acetic acid) were spotted on a MALDI target plate using both the electrospray chip and a conventional mechanical spotter. Using the electrospray deposition method, MALDI targets with small crystal sizes and uniform films were produced.

MALDI targets can also be crystallized on-chip. An interesting approach to cocrystallize the sample and matrix involves the use of the electrowetting on dielectric (EWOD) technique for manipulation, mixing, and deposition of sample-matrix droplets on a surface. In one previous study, EWOD–MALDI target preparation and analysis were performed on Teflon-AF™-coated surfaces [54]. The EWOD system was constructed from a quartz base chip supporting individually addressable, doped polysilicon electrodes and a top glass chip with

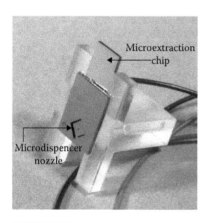 

FIGURE 6.35
Left: A piezoelectric dispensing system for matrix-assisted laser desorption ionization, including a bead-based extraction chip. The single dispenser nozzle at the bottom of the dispensing unit is shown. Right: A matrix-assisted laser desorption ionization array spotting system. (From Marko-Varga, G. A., J. Nilsson, and T. Laurell. 2004. *Electrophoresis* 25:3479. With permission.)

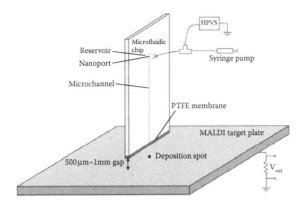

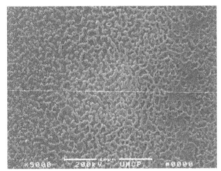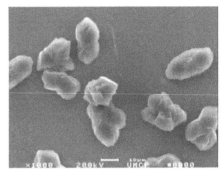

FIGURE 6.36
Top: Electrospray spotting of matrix-assisted laser desorption ionization targets. Bottom: Scanning electron micrographs of 10.5-pmol angiotensin spotted by electrospray deposition (left) and mechanical spotting (right). (From Wang, Y. X., et al. 2005. *Electrophoresis* 26:3631. With permission.)

a single transparent indium tin oxide electrode (Figure 6.37). Droplets of 0.5 μL were dried for ~1–2 minutes to form MALDI targets. In the study, insulin was first transported and dried, followed by the matrix droplet. Then, the EWOD chips were placed on a conventional MALDI target to interface with a TOF-MS.

6.5 Problems

1. Pick a full-length article from the current year from either *Biomedical Microdevices, Lab on a Chip, Biosensors and Bioelectronics, Sensors and Actuators B: Chemical*, or *Analytical Chemistry*. The article you choose must describe a microdevice in which a sensing or detection operation is performed. The focus of the article may be on the sensor or the detector itself. Using your choice of presentation software,

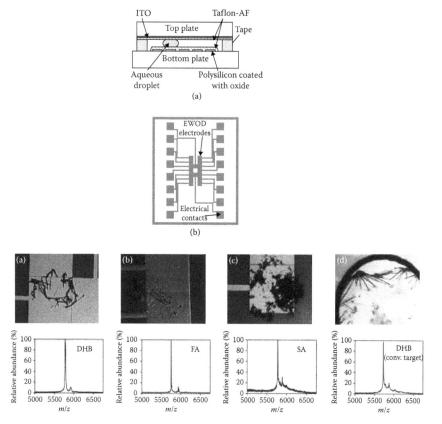

FIGURE 6.37
Top: (a) side and (b) top views of an electrowetting on dielectric device for matrix-assisted laser desorption ionization target preparation. Bottom: Matrix-assisted laser desorption ionization–mass spectrometry results for insulin (5733 Da) cocrystallized with different matrices (DHB = 2,5-dihydroxybenzoic acid, FA = ferulic acid, and SA = sinapinic acid). (a–b) Electrowetting on dielectric technique for both insulin and matrix deposition (insulin first, followed by the matrix) was used. (c) Combined manual deposition of the matrix, followed by insulin deposition by the electrowetting on dielectric technique. The results in (d) used manual deposition of both matrix and insulin on a standard stainless steel target. (From Wheeler, A. R., et al. 2004. *Anal Chem* 76:4833. With permission.)

create a presentation of 1–2 slides containing all the following information:

a. Briefly describe the device and its specific applications.

b. Indicate the sensing or detection method used. Identify the key criteria used by the authors to benchmark the performance of the method against previously reported devices.

c. Include at least one image of the device or a figure summarizing the major findings of the article.

 d. Write a brief summary capturing the significant contribution to the field made by the article in your own words.

 e. Cite the article using the Institute of Electrical and Electronics Engineers (IEEE) reference format.

2. A popular sensor format is to embed piezoelectric strain gauges on microfabricated cantilevers. In static mode, changes in strain on the cantilever may be induced by the absorption or bonding of molecules of interest. Cantilevers may also be operated in resonant mode. In this case, the resonant frequency of the beam is obtained from the following expression:

$$f_0 = \frac{1}{2\pi}\sqrt{\frac{k}{m}}$$

where k is the stiffness constant and m is the mass. For a silicon cantilever with dimensions of $100 \times 0.3 \times 10$ μm^3 and a stiffness constant of 0.013 N/m, calculate the unloaded resonant frequency. The cantilever is anchored at one end. Assume the density of silicon is 2330 kg/m^3. What is the unloaded resonant frequency of the cantilever? If a cell weighing 10 pg attaches to the beam, what is the resonant frequency shift?

3. What advantages are gained by using a pair of interdigitated combs to form the two conducting plates in a capacitive sensor over a simple parallel plate format? Justify your answer and be specific. What potential disadvantages exist in this method?

4. Find an example of a commercially available microfabricated sensor. What sensing principle is used? Find the data sheet and list the most relevant sensor specifications (at least three). How much does this sensor cost? Provide citations of all data sheets and other references used to answer this problem.

5. Biorecognition of the target analyte is often confounded by *nonspecific binding*. Define nonspecific binding in your own words. Find a recent article that addresses nonspecific binding of a molecule of your choice and discuss the methods employed to reduce nonspecific binding. Provide any references used to answer this question in the IEEE format.

6. A biological sensing application requires single molecule detection and you have a cantilever-based sensor capable of doing so. The cantilever is immersed in a 1-μL solution in a 1-mm^3 cubic sample well containing the single molecule of interest. Assume that no interfering species are present and that detection is carried out at room

temperature. If the molecule travels by diffusion alone ($D = 10^{-11}$ m^2/s) and starts at the maximum possible distance away from the sensor, how long will it take the sensor to detect the molecule? What implications does this detection time have on practical single molecule detection?

7. Select a major disease of your choice. Research the potential biomarkers that will allow disease diagnosis. List the biomarkers (at least three) and provide a brief description of how each is used to determine the disease state. For example, the level of a biomarker or its presence/absence may provide valuable information to healthcare providers. Cite any references you use in the IEEE format.

8. Choose an electrochemical or optical detection method and write a short summary on the current state-of-the-art (about 1 page). The report should include the key figures of merit used to compare the performances of different systems and the detection limits. Also include a discussion of the issues that need to be addressed to further improve the detection limits. Provide any references used in the IEEE format.

References

1. White, R. M. 1987. A sensor classification scheme. *IEEE Trans Ultrason Ferroelectr Freq Control* 34:124.
2. Gardner, J. W. 1994. *Microsensors: Principles and Applications*. New York: Wiley, Chichester.
3. Togawa, T., T. Tamura, and P. Å. Öberg. 1997. *Biomedical Transducers and Instruments*. Boca Raton, FL: CRC Press.
4. Webster, J. G. 1999. *The Measurement, Instrumentation, and Sensors Handbook*. Boca Raton, FL: CRC Press published in cooperation with IEEE Press.
5. Pallás-Areny, R., and J. G. Webster. 2001. *Sensors and Signal Conditioning*, 2nd ed., Vol., New York: Wiley.
6. Webster, J. G., and J. W. Clark. 2009. *Medical Instrumentation: Application and Design*, 4th ed., Vol., Hoboken, NJ: John Wiley.
7. Ristic, L. 1994. *Sensor Technology and Devices*. Boston: Artech House.
8. Sze, S. M. 1994. *Semiconductor Sensors*. New York: Wiley.
9. Shieh, J., et al. 2001. The selection of sensors. *Prog Mater Sci* 46:461.
10. Chau, H. L., and K. D. Wise. 1988. An ultraminiature solid-state pressure sensor for a cardiovascular catheter. *IEEE Trans Electron Devices* 35:2355.
11. Tohyama, O., et al. 1998. A fiber-optic pressure microsensor for biomedical applications. *Sens Actuators A Phys* 66:150.
12. Ohki, T., et al. 2004. Wireless pressure sensing of aneurysms. *Endovasc Today* 4:47.

13. Allen, M. G. 2005. Micromachined endovascularly-implantable wireless aneurysm pressure sensors: From concept to clinic, in *Transducers 2005*, eds.,. Piscataway, NJ, USA: IEEE.
14. Fonseca, M. A., et al. 2006. Flexible wireless passive pressure sensors for bio-medical applications. In *Solid-state sensors, actuators, and microsystems workshop*.
15. Chen, P. J., et al. 2008. Microfabricated implantable parylene-based wireless pas-sive intraocular pressure sensors. *J Microelectromech Syst* 17:1342.
16. Dewa, A. S., and W. H. Ko. 1994. Biosensors. In *Semiconductor sensors*, eds., New York: Wiley.
17. Nyholm, L. 2005. Electrochemical techniques for lab-on-a-chip applications. *Analyst* 130:599.
18. Pumera, M., A. Merkoci, and S. Alegret. 2006. New materials for electrochemical sensing vii. Microfluidic chip platforms. *TRAC - Trends Analyt Chem* 25:219.
19. Pasas, S., et al. 2006. Detection on microchips: Principles, challenges, hyphen-ation, and integration. In *Separation methods in microanalytical systems*, eds., Boca Raton, FL: Taylor & Francis.
20. Kim, J., et al. 1998. Electrochemical studies of carbon films from pyrolyzed pho-toresist. *J Electrochem Soc* 145:2314.
21. Lacher, N.A., et al. 2001. Microchip capillary electrophoresis/electrochemistry. *Electrophoresis* 22:2526.
22. Schwarz, M. A., and P. C. Hauser. 2001. Recent developments in detection meth-ods for microfabricated analytical devices. *Lab Chip* 1:1.
23. Vandaveer, W. R., et al. 2002. Recent developments in amperometric detection for microchip capillary electrophoresis. *Electrophoresis* 23:3667.
24. Matysik, F. M. 2008. Advances in amperometric and conductometric detection in capillary and chip-based electrophoresis. *Microchim Acta* 160:1.
25. Ertl, P., et al. 2004. Capillary electrophoresis chips with a sheath flow supported electrochemical detection system. *Anal Chem* 76:3749.
26. Hebert, N. E., et al. 2003. Performance of pyrolyzed photoresist carbon films in a microchip capillary electrophoresis device with sinusoidal voltammetric detec-tion. *Anal Chem* 75:4265.
27. Tanyanyiwa, J., S. Leuthardt, and P. C. Hauser. 2002. Conductimetric and poten-tiometric detection in conventional and microchip capillary electrophoresis. *Electrophoresis* 23:3659.
28. Liao, W. Y., et al. 2006. Development and characterization of an all-solid-state potentiometric biosensor array microfluidic device for multiple ion analysis. *Lab Chip* 6:1362.
29. Lin, C. F., et al. 2006. Microfluidic pH-sensing chips integrated with pneumatic fluid-control devices. *Biosens Bioelectron* 21:1468.
30. Liu, Y., D. O. Wipf, and C. S. Henry. 2001. Conductivity detection for monitoring mixing reactions in microfluidic devices. *Analyst* 126:1248.
31. Pumera, M. 2007. Contactless conductivity detection for microfluidics: Designs and applications. *Talanta* 74:358.
32. Lichtenberg, J., N. F. de Rooij, and E. Verpoorte. 2002. A microchip electropho-resis system with integrated in-plane electrodes for contactless conductivity detection, *Electrophoresis* 23:3769.
33. Emrich, C.A., et al. 2002. Microfabricated 384-lane capillary array electrophore-sis bioanalyzer for ultrahigh-throughput genetic analysis. *Anal Chem* 74:5076.

34. Petersen, N. J., K. B. Mogensen, and J. P. Kutter. 2002. Performance of an in-plane detection cell with integrated waveguides for uv/vis absorbance measurements on microfluidic separation devices. *Electrophoresis* 23:3528.
35. Mangru, S. D., and D. J. Harrison. 1998. Chemiluminescence detection in integrated post-separation reactors for microchip-based capillary electrophoresis and affinity electrophoresis. *Electrophoresis* 19:2301.
36. Zhan, W., et al. 2003. A multichannel microfluidic sensor that detects anodic redox reactions indirectly using anodic electrogenerated chemiluminescence. *Anal Chem* 75:1233.
37. Sung, W. C., H. Makamba, and S. H. Chen. 2005. Chip-based microfluidic devices coupled with electrospray ionization-mass spectrometry. *Electrophoresis* 26:1783.
38. Lazar, I. M., J. Grym, and F. Foret. 2006. Microfabricated devices: A new sample introduction approach to mass spectrometry. *Mass Spectrom Rev* 25:573.
39. Lee, J., S. A. Soper, and K. K. Murray. 2009. Microfluidic chips for mass spectrometry-based proteomics. *J Mass Spectrom* 44:579.
40. Yamashita, M., and J. B. Fenn. 1984. Electrospray ion-source - another variation on the free-jet theme. *J Phys Chem* 88:4451.
41. Ramsey, R. S., and J. M. Ramsey. 1997. Generating electrospray from microchip devices using electroosmotic pumping. *Anal Chem* 69:1174.
42. Xue, Q. F., et al. 1997. Multichannel microchip electrospray mass spectrometry. *Anal Chem* 69:426.
43. Rohner, T. C., J. S. Rossier, and H. H. Girault. 2001. Polymer microspray with an integrated thick-film microelectrode. *Anal Chem* 73:5353.
44. Bings, N. H., et al. 1999. Microfluidic devises connected to fused-silica capillaries with minimal dead volume. *Anal Chem* 71:3292.
45. Li, J. J., et al. 1999. Integration of microfabricated devices to capillary electrophoresis-electrospray mass spectrometry using a low dead volume connection: Application to rapid analyses of proteolytic digests. *Anal Chem* 71:3036.
46. Liu, H. H., et al. 2000. Development of multichannel devices with an array of electrospray tips far high-throughput mass spectrometry. *Anal Chem* 72:3303.
47. Licklider, L., et al. 2000. A micromachined chip-based electrospray source for mass spectrometry. *Anal Chem* 72:367.
48. Xie, J., et al. 2005. Microfluidic platform for liquid chromatography-tandem mass spectrometry analyses of complex peptide mixtures, *Anal Chem* 77:6947.
49. Koster, S., and E. Verpoorte. 2007. A decade of microfluidic analysis coupled with electrospray mass spectrometry: An overview. *Lab Chip* 7:1394.
50. Devoe, D. L., and C. S. Lee. 2006. Microfluidic technologies for MALDI-MS in proteomics. *Electrophoresis* 27:3559.
51. Karas, M., D. Bachmann, and F. Hillenkamp. 1985. Influence of the wavelength in high-irradiance ultraviolet-laser desorption mass-spectrometry of organic-molecules. *Anal Chem* 57:2935.
52. Marko-Varga, G. A., J. Nilsson, and T. Laurell. 2004. New directions of miniaturization within the biomarker research area. *Electrophoresis* 25:3479.
53. Wang, Y. X., et al. 2005. Electrospray interfacing of polymer microfluidics to MALDI-MS. *Electrophoresis* 26: 3631.
54. Wheeler, A. R., et al. 2004. Electrowetting-based microfluidics for analysis of peptides and proteins by matrix-assisted laser desorption/ionization mass spectrometry. *Anal Chem* 76:4833.

7

Applications to Cells, Nucleic Acids, and Proteins

7.1 Cells

Chips for cells may perform many functions in the fields of biological science, drug discovery, diagnostics, and clinical testing.

7.1.1 Cell Culture Reactors

In conventional laboratory practice, cell culture is performed in sterile glass or polymer dishes that have often been pretreated such that the surfaces are hydrophilic. There are many advantages to miniaturizing cell cultures; however, the transfer of conventional techniques to microscale formats is not always an easy process. It has been shown that culture conditions associated with miniaturization and the use of new materials can have a profound impact on the resulting cellular behavior [1]. Even so, many researchers have demonstrated successful cultures of prokaryotic, insect, and mammalian cells in microsystems. These results contribute to the continued development of microfluidic techniques in cell culture.

The typical steps in mammalian cell culture are sterilization, surface treatment (to promote cell adhesion and thus survival), cell plating, cell feeding, and culture maintenance. These steps may differ for other cell types. These are time-consuming and labor-intensive regardless of the scale. At the microscale, cells may need to be confined when further study or use of cells in sensing applications is required.

Sterilization was discussed in Section 2.3.4. In addition to the aforementioned techniques, ethanol immersion and bleach treatment are also popular sterilization methods used before performing cell culture on a chip. Surface treatments are discussed in Section 7.1.2. The focus of this section is on cell-culture methods in microsystems.

Huang and Lee have demonstrated an automated culture system that mimicked the static nature of conventional cell culture processes, as opposed to continuous perfusion cell culture microsystems [2]. Microfluidic delivery of

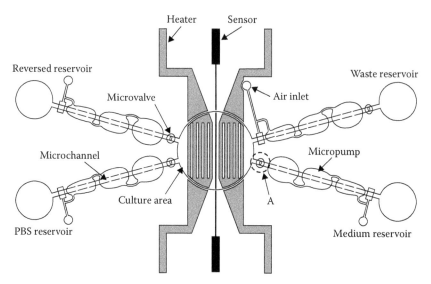

FIGURE 7.1
Layout of the automated cell culture system, including pumps, valves, reservoirs, channels, heaters, and a sensor. (From Huang, C. W., and G. B. Lee. 2007. *J Micromech Microeng* 17:1266. With permission.)

nutrients and removal of metabolites were accomplished by pneumatic micro-pumps and valves to automate these processes that are normally performed manually (Figure 7.1). This portion of the cell culture chip was cast in silicone rubber using standard soft lithography processes. The thermal conditions were monitored, and constant temperature was maintained using integrated heaters (indium tin oxide [ITO]) and a thermal sensor (platinum [Pt]) patterned on a glass substrate. The two halves were joined together to form the cell culture system. Human lung carcinoma cells (A549) were cultured on a chip (1 cm diameter) and in a Petri dish (10 cm diameter) for comparison. The culture medium was exchanged with the integrated pumps in the device and manually in the Petri dish. Nearly identical pH conditions and cell proliferation were observed in the chip compared to that in the Petri dish culture.

Conventional cell culture bioreactors permit high density and large population cell culture. However, efficient delivery of nutrients and oxygen along with waste removal are difficult at this scale. In one study, microbioreactors with large surface areas were achieved by stacking multiple layers of structured silicone rubber to create multiple culture chambers (Figure 7.2) [3]. The microstructures provided a scaffold for cell attachment. Microfluidic culture chambers were designed to allow homogeneous distribution of culture medium. Adequate oxygen supply to the cultures was easily achieved by taking advantage of the high gas permeability of the silicone rubber.

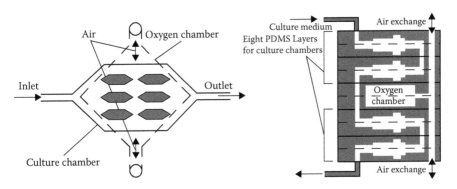

FIGURE 7.2
Single culture chamber design (left) and stacked culture chamber concept (right). (From Leclerc, E., Y. Sakai, and T. Fujii. 2004. *Biotechnol Prog* 20:750. With permission.)

Cultures were maintained for up to 12 days, and cell densities comparable to macroscale bioreactors (~10^7 cells/cm^3) were achieved.

Multiple cell types combined in a cleverly designed microfluidic cell culture reactor allow an on-chip replica of animal systems. Microscale Cell Culture Analogs (µCCAs) [4] provide an *in vitro* model that mimics the body's dynamic response without requiring animals or humans to test the response to toxins [5], drugs [6], and other chemicals. Microfluidic networks were etched in silicon and enclosed in Plexiglas sheets, which contained multiple interconnected cell culture chambers that were appropriately proportioned to represent different organs or tissues (e.g., lung, liver, or fat). The circulatory system was implemented by recirculating the culture media (a blood surrogate; Figure 7.3).

7.1.2 Cell Adhesion

Some cell types must adhere to surfaces or form aggregates for viability. Thus, a control of cell adhesion to surfaces is necessary. In addition, selective control of cell adhesion to the surfaces allows precisely defined cocultures of different cell types. The scale of cell adhesion control ranges from single focal adhesion sites to the patterning of groups of cells. Control of cell patterning was reviewed in reference [7]. This technique enables the investigation of many interesting and fundamental questions in biology (such as cell alignment, migration, metabolism, and behavior).

A variant of soft lithography uses replica molded silicone rubber structures as microstructured stamps for patterning selective cell adhesion regions on a surface [8,9]. This approach is sometimes called microcontact printing, and it avoids photolithography on the patterned surfaces. Although printing on standard silicon substrates is possible, the optically opaque surface is not

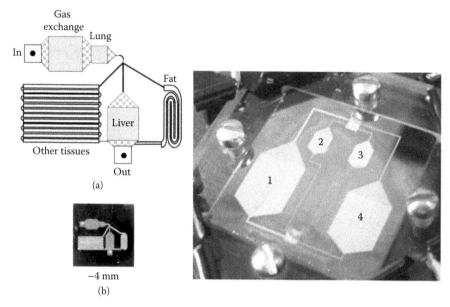

FIGURE 7.3

Left: (a) Schematic and (b) photograph of a Microscale Cell Culture Analogs (μCCA) for evaluating toxins (from Viravaidya, K., A. Sin, and M. L. Shuler. 2004. *Biotechnol Prog* 20:316. With permission.) Right: μCCA for evaluating drugs for combating drug resistance cancers. The cell types in the numbered chambers are (1) hepatocytes (C3A), (2) multidrug resistant cancer (DX5), (3) wild type cancer (MESSA), and (4) megakaryocytes (MEG01) (from Tatosian, D. A., and M. L. Shuler. 2009. *Biotechnol Bioeng* 103:187. With permission.)

compatible with conventional inverted microscopy. By combining microcontact printing and micromolding in capillaries [10], cell patterning on transparent and contoured surfaces was performed [8]. Thin gold (Au) films (10–12 nm for optical transparency with 1.5-nm titanium [Ti] adhesion layer) were deposited on the contoured polyurethane films produced by the micromolding in capillaries technique. The raised plateau areas were stamped with one self-assembled monolayer (SAM) and then immersed in a second SAM to coat the grooves (both SAMs were alkanethiols). The different SAMs controlled both protein adsorption and cell adhesion; bovine capillary endothelial cells attached preferentially to either the grooves or the plateaus depending on the properties of the SAM present in those regions (Figure 7.4).

Controlled cell patterning on surfaces was also used to control the formation of cultured neural networks on surfaces [11,12]. This was an extension of the earlier technique that allowed patterning of neural networks directly on microelectrode arrays, which used alkanethiol SAM chemistry on thin gold films (~5–8 nm) to facilitate long-term electrophysiological recordings [12]. Polylysine was linked to the SAM by microcontact printing to allow

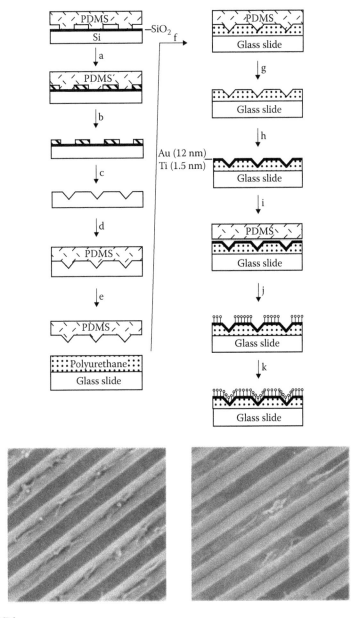

FIGURE 7.4
Top: Fabrication process for producing contoured and functionalized surfaces for controlling cell adhesion. Bottom: Scanning electron microscope images showing selective patterning of bovine capillary endothelial cells on grooves (left) and plateaus (right). (From Mrksich, M., et al. 1996. *Proc Natl Acad Sci USA* 93:10775. With permission.)

adhesion of hippocampal neurons from rat embryos in features as small as 5 µm lines. The surrounding regions were treated with polyethylene glycol (PEG) to prevent cell adhesion [13]. PEG reduces nonspecific protein adsorption and cell adhesion (Figure 7.5).

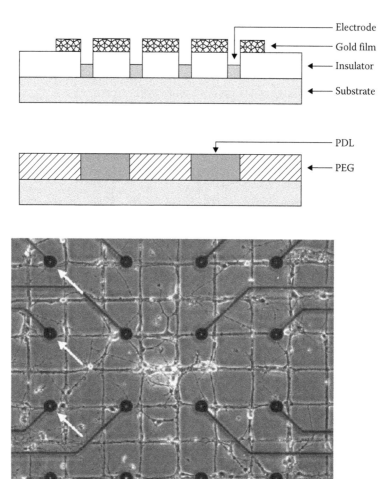

FIGURE 7.5
Top: Cross section of a multielectrode array with a thin gold film coating for self-assembled monolayer chemistry. Middle: Cross section of a representative final surface pattern with alternating poly-D-lysine (PDL) and polyethylene glycol (PEG) regions. Bottom: Microelectrode-array substrate with neurons arranged on polylysine patterns in a regular grid pattern. The PEG was patterned between the polylysine grid lines to prevent neuron adhesion and growth in these regions. Scale bar = 200 µm. (From Nam, Y., et al. 2004. *IEEE Trans Biomed Eng* 51:158. With permission.)

7.1.3 Retention: Filters, Weirs, and Polymer Matrix

Filtering was previously discussed in Chapter 5. In this section, the retention of cells for further interrogation is addressed in more detail. Microorganisms can be trapped for further study. Two different microbes were trapped using weirs etched into silicon. *Cryptosporidium parvum* measured 2–6 μm in diameter and *Giardia lamblia* were elongated and measured 8–13 μm in length and 7–10 μm in width. The gap in the weir structures measured 1–2 μm or 3–4 μm at the interface between the silicon substrate and Pyrex top cover. Weirs flanked reaction chambers (50 μm deep) intended for trapping and concentrating the microbes for fluorescent detection [14]. As discussed in Chapter 5, this technique is suitable for groups of cells.

Retention of single cells facilitates single cell analysis. For example, long-term extracellular recordings are possible on isolated electrogenic cells without requiring careful positioning of a saline-filled glass micropipette electrode on the cell membrane. In addition, selective delivery of drugs to this single cell may also be performed. In 1976, Erwin Neher and Bert Sakmann developed a technique called patch clamping, which was used to study electrogenic cells. Their invention was recognized with a Nobel Prize in 1991. This method allows the recording of currents (~pA) associated with the opening and closing of single ion channels in a cell membrane by using a glass pipette and suction to establish a GΩ resistive seal. Glass pipettes were processed by pulling and fire polishing to obtain a tip diameter on the order of 1 μm suitable for probing a portion of the cell membrane. The suction forces a small patch of the cell membrane into the bore of the pipette to establish an electrical circuit with ion channels in the membrane and prevents leakage (Figure 7.6). In practice, glass pipette patch clamping is a difficult technique to master and requires intensive effort to obtain adequate seals. Many microfabricated tools made from a range of materials (silicon, glass, and polymers) have been devised to address the need for automation for high throughput. Some commercial systems that feature microfabricated patch clamp interfaces to cells have been introduced. All these are reviewed in reference [15].

An out-of-plane patch clamp device in glass was developed that avoided the limitations of lithography on the dimensions of patch clamp aperture. Ion track etching with a gold beam enabled controlled fabrication of single pores (50 to <1 μm in diameter) in thinned glass membranes (~20–80-μm thick in 300–500-μm diameter areas on a 200-μm thick substrate). Only a single ion was needed to produce one pore (Figure 7.7) [17]. To capture cells, a Chinese hamster ovary or mouse neuroblastoma (N1E-115) cell suspension was applied as a drop over the pore and the current across the pore was monitored with silver/silver chloride electrodes connected on either side. A single cell was drawn to the pore by suction and sealed against the pore, triggering a current response indicating a GΩ seal. The same chips were also used to record from the lipid bilayers formed across the pore [16,18].

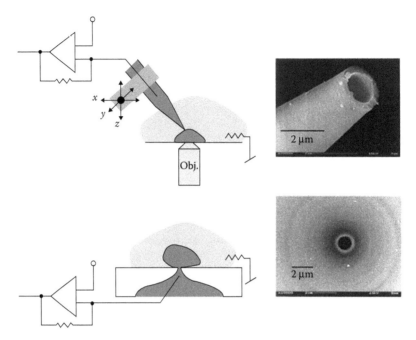

FIGURE 7.6
Comparison of a traditional glass pipette patch clamp (top) to a microfabricated system
(bottom). (From Fertig, N., R. H. Blick, and J. C. Behrends. 2002. *Biophys J* 82:3056. With
permission.)

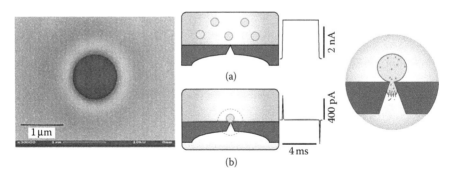

FIGURE 7.7
Left: Scanning electron microscope image showing an ion track etched pore (from Fertig,
N., et al. 2001. *Phys Rev E* 64. With permission.) Right: Procedure for trapping one cell
from a cell suspension with the glass patch clamp chip. The recorded current corrrespond-
ing to application of the (a) cell suspension and (b) trapping of a cell are shown to the
right of the schematic depictions. (from Fertig, N., et al. 2002. *Appl Phys Lett* 81:4865. With
permission.)

Single cells may also be trapped in-plane. Single adult murine cardiac myocytes were trapped in a 100-pL microchamber (125 μm long, 25 μm wide, and 30 μm high) by a 5-μm constriction constructed on a silicone rubber–glass chip (Figure 7.8) [19]. The trap isolated a single cell at a time from a cell suspension by using a pressure gradient that was established using a suction channel. Integrated recording electrodes allowed measurement of extracellular potential. Localized drug delivery was performed through small channels connected to the side of the trap.

Other special structures may be used for trapping single cells. U-shaped traps were patterned in T-intersections in silicone rubber microfluidic chips using soft lithography techniques [20]. Integrated pumps and valves allowed nL delivery of reagents to trapped cells. Cell channels were 20 μm deep, and drain channels were 5 μm deep. Two types of cells were investigated. Both Jurkat T and U937 cells were trapped at the flow stagnation point formed at the center of a T-intersection indicated in Figure 7.9. Only a single Jurkat T cell (15 μm diameter) was captured in a cell trap; however, multiple smaller

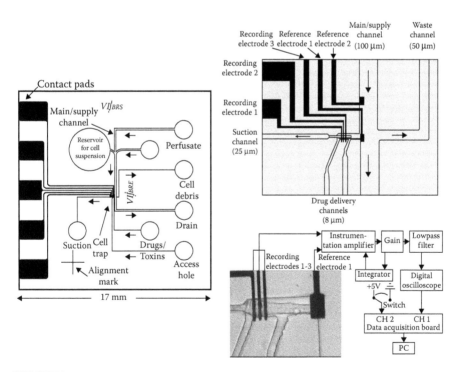

FIGURE 7.8
Left: Cardiac myocyte chip layout. Top right: Close-up photograph of a cell trap showing the electrode and drug delivery channel configuration. Bottom right: Experimental setup for extracellular recordings from a trapped cardiac myocyte. (From Werdich, A. A., et al. 2004. *Lab Chip* 4:357. With permission.)

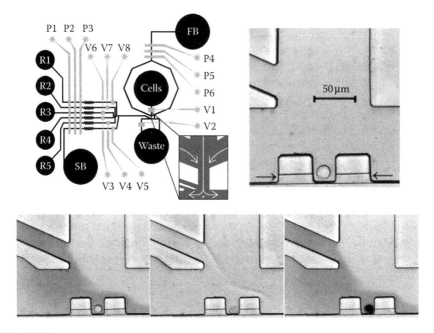

FIGURE 7.9

Top left: Layout of a single cell analysis device in silicone rubber with microfluidic (dark) and pneumatic control (light) channels (R1–5 = reaction inlets, SB = shield buffer inlet, FB = focusing buffer inlet, V1–8 = control lines for valves 1–8, and P1–6 = control lines for peristaltic pumps). Inset shows the direction of flow for trapping a single cell at the stagnation point indicated by *. Top right: Close-up of a trapped cell. Bottom: Selective perfusion of a Jurkat T cell in successive images with (from left to right) trypan blue dye, methanol, and trypan blue dye. (From Wheeler, A. R., et al. 2003. *Anal Chem* 75:3581. With permission.)

U937 cells (10 µm in diameter) could occupy a single trap. This indicated the relationship between the geometrical design of the flow stagnation trap and the cell size. Trapped cells were also selectively perfused using additional channels located near the U-trap.

Single neurons were also trapped and cultivated in neural networks using 4 × 4 arrayed Parylene cage structures [21]. The cages consisted of a Parylene chimney with tunnels radiating from the base anchored to a silicon substrate (Figure 7.10). The anchor structures prevented the delicate "neurocages" from delaminating during cell culture, which required immersion in culture media. The chimney (30 µm diameter, 4 µm high, and 15 µm hole) included a slight overhang to accept top loading of embryonic neurons but prevent their escape after growth. Tunnels (5 or 10 µm wide, 4 or 40 µm long, and 1.5 µm high) permitted the growth of neurites, allowing the formation of synaptic connections with neighboring trapped neurons.

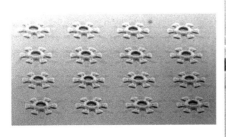
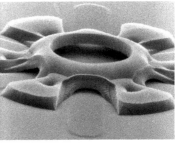

Cross section through anchors Cross section through tunnels

Pattern and etch oxide for anchors

Pattern photoresist for chimneys (4 μm) and tunnels (1.5 μm)

Etch silicon for anchors using DRIE

Deposit parylene (4 μm)

Pattern and etch parylene and release chimneys and tunnels

Silicon Oxide Photoresist Parylene

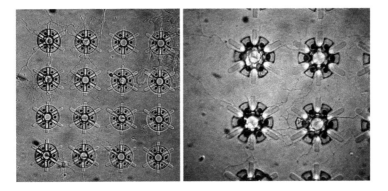

FIGURE 7.10

Top: Scanning electron microscope images of a neurocage array and close-up of a single neurocage. Middle: Fabrication process for neurocages with cross-sections drawn through the tunnels or the anchors. Bottom: Optical images showing neurons with neurites growing in individual cages. (From Tooker, A., et al. 2005. *IEEE Eng Med Biol Mag* 24:30. With permission.)

Instead of trapping cells in artificial two-dimensional (2D) substrates, polymer matrix entrapment techniques were introduced to provide cells with an environment that was similar to their natural three-dimensional (3D) surroundings (i.e., the extracellular matrix [ECM]). Microbes and single or multiple mammalian phenotypes were entrapped with the potential to be used as biorecognition elements in biosensing applications [22,23]. One such polymer matrix is PEG hydrogel. A mixture of PEG hydrogel precursor, a photoinitiator, and cells was loaded into silicone rubber-glass microchannels and selectively cross-linked following UV exposure [23]. Since PEG is nonadhesive, hydrogels were modified with peptide sequence Arg-Gly-Asp (RGD). The addition of RGD, a cell-adhesion peptide, improved cell adhesion and spreading; both are factors that improve cell viability (Figure 7.11). Murine fibroblasts, hepatocytes, and macrophage cell lines encapsulated in this manner were viable over one week in PEG hydrogels.

Another method to provide a 3D matrix around entrapped cells involves a layer-by-layer approach, which further mimics the heterogeneous 3D structure found in tissues. By patterning different polymer matrices and cells, artificial biomimetic engineered tissue constructs can be created. A microfluidic approach to layer-by-layer cell matrix patterning was

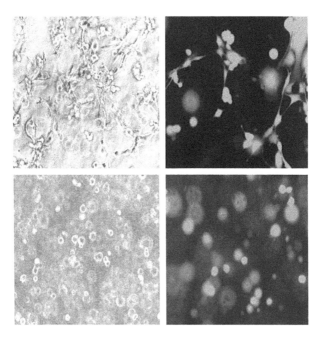

FIGURE 7.11
Encapsulated cells in optical (left column) and fluorescent (right column) images with (top row) and without (bottom row) RGD (Arg-Gly-Asp) modification of the hydrogel matrix. (From Koh, W. G., L. J. Itle, and M. V. Pishko. 2003. *Anal Chem* 75:5783. With permission.)

performed using silicone rubber–silicon microchannels produced by soft lithography. Different biopolymer matrices (collagen, collagen–chitosan, matrigel, and fibrin) were mixed with cell suspensions (human lung fibroblasts, umbilical vein smooth muscle cells, and umbilical vein endothelial cells) and introduced at 0°C into the microchannels (350–500 μm wide and 300 μm high) using a syringe pump (Figure 7.12). Prior to depositing the first layer, the bottom silicon surface of the microchannel was silanized (with 3-aminopropyltriethoxysilane [APTES] and glutaraldehyde [GA]), to improve the retention of the first deposited cell-laden prepolymer layer. Each deposited layer was polymerized at 37°C. By flowing successive biopolymer mixtures with distinct cells, heterogeneous tissue-mimicking matrices were formed [24].

7.1.4 Cell Manipulation

Cells may also be positioned or manipulated without physical constraints or contact using magnetic, acoustic, ultrasound, optical, or dielectrophoretic forces [25]. We will now take a closer look at two of these methods: optical tweezers and dielectrophoretic (DEP) trapping. Following cell capture, these techniques may also be used either independently or simultaneously

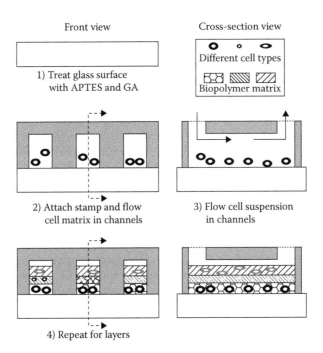

FIGURE 7.12
Microfluidic layer-by-layer approach to three-dimensional polymer-entrapped cell matrices.

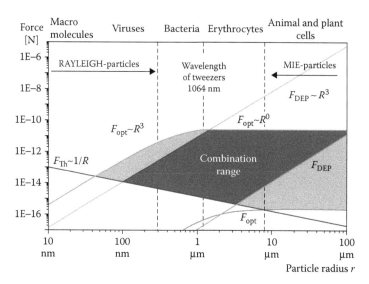

FIGURE 7.13
Force scaling for manipulation of particles and cells ranging from 10 nm to 100 μm with optical, thermal, and dielectrophoretic forces. (From Mietchen, D., et al. 2002. *J Phys D Appl Phys* 35:1258. With permission.)

to manipulate cells. The force scaling of optical (F_{opt}) and DEP (F_{DEP}) forces, as well as thermal (Brownian; F_{Th}) forces, are shown in Figure 7.13.

Optical tweezers generate radiation pressure from a laser beam that can trap particles or cells as a function of the radius and wavelength used. Combinations of light scattering and gradient forces enable both trapping and movements of microscopic objects. Since this is an optical technique, a transparent chip is required. In another study, silicone rubber chips with optical tweezers were used to pick and place yeast and human red blood cells into regular square arrays [27]. Selection and removal of single *E. coli* cells was conducted on etched glass chips to study the growth and division variations in generations stemming from the same initial cell [28].

DEP trapping and manipulation of cells or other microscopic objects require electrodes to produce a nonuniform electric field to generate a net force acting on a dipole. Cell dipoles are induced by the electric field and are less polarizable than the surrounding media. Thus, cells experience negative DEP and are drawn toward the field minimum (Figure 7.14). In one study, high frequency AC fields (1–20 MHz), and groups of four electroplated electrodes were used to trap and sort cells to avoid electrophoretic drift of cells and unwanted electrochemical reactions [29]. The experiments were performed on glass substrates with titanium/gold electrodes contained within an SU-8 flow chamber (2 mm wide, 8 mm long, and 150 μm high). Quadrupole DEP traps consisting of four electrode posts (50 mm high and ~19 μm in diameter)

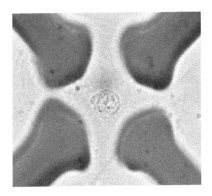

FIGURE 7.14
Jurkat T lymphoma cell trapped in an octopole dielectrophoretic cage (4 titanium/platinum electrodes on top and bottom substrate separated by 40 µm) formed at the center of electrodes (5 MHz and 8 V_{pp}). (From Reichle, C., et al. 1999. *J Phys D Appl Phys* 32:2128. With permission.)

arranged in a 2×2 array captured human leukemia (HL-60) cells (9.3–14.4 µm in diameter) in the trap center, which corresponded to the field minimum (Figure 7.15). Following optical observation, selected cells were released by turning off their respective traps.

7.1.5 Electroporation

Electroporation is a means to gain access to the interior of a cell by using pulsed electric fields to form pores in the cell membrane. This capability is used for the release of cell contents or for the delivery of foreign molecules (i.e., proteins, DNA, and drugs) into a cell. This phenomenon is not completely understood even though the study of the interaction between electric fields and cell membranes can be traced to the 1960s [31].

In electroporation, a pulsed electric field (~kV/cm and ~µs) applied to a pair of electrodes, acts on the cell. This induces a transmembrane potential (TMP) that is a function of both the field strength and the cell size. For spherical cells in a DC field, the potential is

$$\Delta\Phi_m = 1.5rE_0 \cos\theta - \Delta\Phi_{m0} \qquad (7.1)$$

where $\Delta\Phi_m$ is the transmembrane potential, r is the cell radius, E_0 is the electric field, θ is the angle between the applied electric field and the position on the cell membrane where potential is measured, and $\Delta\Phi_{m0}$ is the resting membrane potential [31].

Phospholipids present in the cell membrane become mobile and rearrange themselves to allow the formation of hydrophobic pores (nanometers in size)

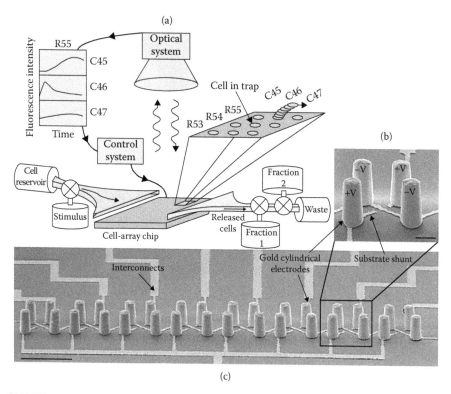

FIGURE 7.15
(a) Dielectrophoretic (DEP) cell trapping and sorting-device operation. (b) Magnified scanning electron microscope (SEM) image of a single quadrupole DEP trap. (c) SEM image of a 1 × 8 quadrupole DEP trap array. (From Voldman, J., et al. 2002. *Anal Chem* 74:3984. With permission.)

in the membrane. Once the threshold TMP is reached (~0.3–1 V), the pores become hydrophilic, allowing the transport of material across the pore. For sufficiently low TMPs, pore formation is reversible allowing the cell membrane to reseal in the mammalian cells. This is thought to occur in bacteria as well. At even larger TMPs, irreversible electroporation or cell lysis occurs. Cell lysis is discussed in Section 7.1.6.

Electroporation microdevices are reviewed in references [25,32]. The key advantages of miniaturizing electroporation are that field strengths are reduced due to the small spacing between the electrodes, and cell handling is facilitated by microfluidics. Flow through devices for poration of both multiple cell and single cell devices have been developed.

Electroporation devices require electrode pairs and cell chambers. A simple electroporation system for plant protoplasts that featured radio frequency (RF) pulsing has been developed [33]. This method improved cell viability and permeability during electroporation as compared to DC or exponential decay pulsing. A cell-chamber array (20 × 24) was patterned using SU-8 on top and titanium/gold electrodes (separated by a 30 μm gap) deposited on a glass substrate (Figure 7.16). The cubic chamber dimensions (30 μm on a side) were selected to match plant protoplast diameter (~30 μm). Rapeseed, cabbage, and spinach protoplants were obtained after enzymatic digestion, and loaded into the cell chambers. Electroporation was verified by using dyes that fluoresced once bound to nucleic acids within the cell. The individual addressability of each cell chamber allowed electroporation experiments to be performed in parallel. This enabled the development of electroporation "phase diagrams," which provide guidance on pulse duration and field strength for achieving electroporation, cell lysis, and no electroporation.

A flow through system was used for electroporation of cells flowing in a single file within a microchannel [34]. Individual cells were captured at a microhole by suction and were electroporated. Permeabilized cells were

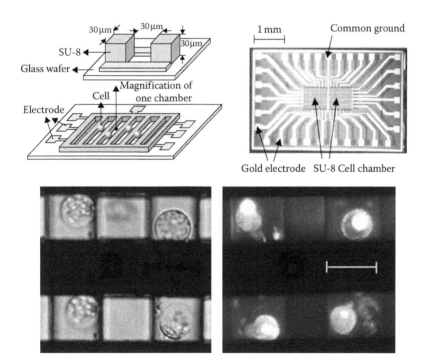

FIGURE 7.16
Top left: Schematic showing key features of electroporation chip. Top right: Photograph of the electroporation chip. Bottom left: Optical image of electroporated cabbage cells. Bottom right: Fluorescence image of electroporated cabbage cells. Scale bar is 30 μm. (From He, H. Q., D. C. Chang, and Y. K. Lee. 2006. *Bioelectrochemistry* 68:89. With permission.)

transfected with foreign genetic material and then released to allow the next cell to be processed. This method was used to deliver fluorescent dyes and enhanced green fluorescent protein (EGFP) genes into single prostate adenocarcinoma (ND-1) cells. The overall device was constructed from a stack of three substrates (Figure 7.17). A glass cover slip with two openings formed the top piece and sealed the flow channel. The flow channel, microhole, and electroporation electrodes were included in the middle piece of the stack. A silicon substrate was anisotropically etched to reveal a nitride membrane (1 mm wide, 1 mm long, and 1 µm thick), perforated with a lithographically defined hole for pneumatic cell capture (2–6 µm diameter). Three pairs of chromium/platinum electrodes were included:

1. A pair of electrodes around the microhole was used for electroporation.
2. A second pair of electrodes was used for impedance detection of cells.
3. The final pair of electrodes was located at the inlet and was used to lyse any cells blocking the channel.

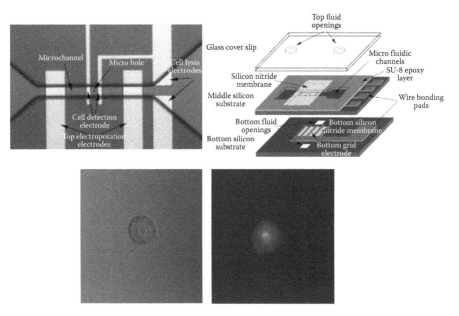

FIGURE 7.17
Top left: Top view of the flow through electroporation chip showing the arrangement of the electrodes and the flow chamber. Top right: Illustration showing an exploded view of the electroporation device components. Bottom left: ND-1 cell trapped over a 3 µm hole before gene transfection. Bottom right: Fluorescent image of ND-1 cell after enhanced green fluorescent protein (EGFP) was transfected. (From Huang, Y., and B. Rubinsky. 2003. *Sens Actuators A Phys* 104:205. With permission.)

An SU-8 flow channel (30 μm wide, 200 μm long, and 25–30 μm deep near the hole region) with a width 1.5 times that of the cell diameter was patterned on the top. The bottom layer of the stack was used to form a bottom electroporation chamber and contains a platinum electrode, fluidic openings, and wire bonding pads. The stack of three chips was assembled using both UV and heat curable adhesives.

Irreversible electroporation is a pulsed electric field (PEF) process that may be used for pasteurization in the food industry to inactivate bacteria and other unwanted cells. A high throughput flow through electroporation device is required for this application. By focusing the electric field within a constriction in a microchannel, continuous electroporation of cells passing through this region is possible [35]. Such a device is shown in Figure 7.18. Borofloat glass wafers were etched using HF and bonded to form the flow through system. The top half contained channels (1 mm wide and 50 μm deep) separated by a 30 μm gap. A 10 μm deep channel in the bottom substrate was aligned with the top to form a 30 μm long constriction in the flow path. Platinum electrodes were placed at either end of the constriction. Carboxyfluroescein (CF) filled vesicles (152 nm in diameter) were introduced with a syringe pump and continuously electroporated with efficiencies up to 51%.

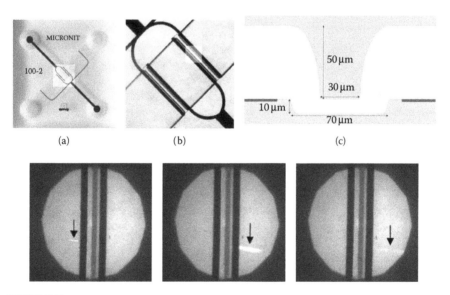

(a)   (b)   (c)

FIGURE 7.18
Top: (a and b) Photo and close-up of the pulsed electric field reactor. (c) Cross-sectional illustration of the constriction region. The dark lines are the electrodes. Bottom: Sequence of images showing the break up of a vesicle following electroporation, with flow moving from left to right. The two darker lines are the electrodes and the constriction is the gray line in between. The vesicle is tracked by the arrows. (From Fox, M., et al. 2005. *Lab Chip* 5:943. With permission.)

7.1.6 Cell Lysis

Analysis of the contents of cells such as nucleic acids, proteins, and enzymes requires collection of these molecules by disrupting or *lysing* the cell membrane. Many methods, including optical, mechanical, acoustic, electrical, and chemical methods may be used to attain access to the cell contents. These are reviewed in references [25,36].

A simple mechanical cell membrane disruption device was devised using scallop structures formed during the silicon deep reactive ion etching (DRIE) process [37]. Scallops were enhanced by increasing the duration of the isotropic etch cycle. The scallops formed nanostructured barbs along the wall of the filter pillars that sliced membranes as cells were forced through the gaps between the pillars (Figure 7.19). HL-60 and whole blood were lysed in this manner.

Increasing both the electrical field strength and the duration used in electroporation will result in cell lysis. The precise field strength is a function of cell size and shape, and membrane composition. Electrical cell lysis has been performed on cells at rest and in motion [39]. Cells (yeast [*Saccharomyces cerevisiae*], Chinese cabbage protoplasts, radish protoplasts, and *E. coli*) loaded into

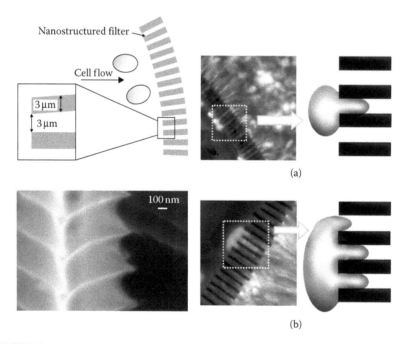

FIGURE 7.19
Top left: Layout of nanobarb-laden filter pillars. Bottom left: Scanning electron microscope image showing nanobarbs. Right: Cell lysis by single (a) and multiple (b) protrusions into the filter with flow from left to right. (From Di Carlo, D., K. H. Jeong, and L. P. Lee. 2003. *Lab Chip* 3:287. With permission.)

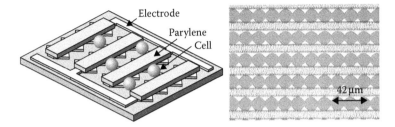

FIGURE 7.20
Left: Illustration of a cell lysis device. Right: Photograph of cell lysis electrodes on an oxidized silicon substrate. (From Lee, S. W., and Y. C. Tai. 1999. *Sens Actuators A Phys* 73:74. With permission.)

a microfluidic chamber (30 μm high) with chromium/gold electrode pairs were lysed under a field from 1–10 kV/cm (Figure 7.20) [38]. Moving Jurkat cells in a microfluidic system were lysed by first hydrodynamically focusing the cells and then applying an electric field. Etched glass channels were first coated with silicone rubber and then pluronic F-127 to minimize cell adhesion. Hydrodynamic focusing was performed at the intersection between the cell, buffer, waste, and the separation channel intersection. Electrochemically generated hydroxide ions were used for local cell lysis [40]. Palladium electrodes were used to hydrolyze water to create hydroxide ions that porate the cell membrane and lead to lysis. Two devices were created to perform lysis on cells moving in a flow chamber and on trapped cells. The flow through chamber (1.5 mm wide and 5 mm long) included a pair of palladium electrodes separated by 600 μm (Figure 7.21). Lysis chambers (3 mm wide and 40 μm high) were used in the second device that included an integrated filter (2 μm wide and 2 μm high) between the anode and the cathode to keep the cells intact prior to lysis. The devices were constructed using molded silicone rubber microfluidic channels bonded to glass substrates. Three cell types were successfully lysed using this method (red blood cells, tumor [HeLa] cells, and Chinese hamster ovary cells).

7.1.7 Cell-Based Sensors

A single cell is the smallest self-sustaining biological entity, and exquisite natural biorecognition mechanisms occur within this structure. Cells respond to extracellular stimuli. For example, a T cell is triggered by a single antigenic peptide. In the central nervous system, olfactory and retinal neurons respond to single odorant molecules or photons, respectively. Compared to other biological recognition elements such as proteins, cells possess machinery that maintains detection receptors and may include natural amplification cascades that further improve detection sensitivity. Cells may be transfected to express the desired receptors. All these characteristics make them attractive biorecognition elements in biosensors. While this discussion is focused

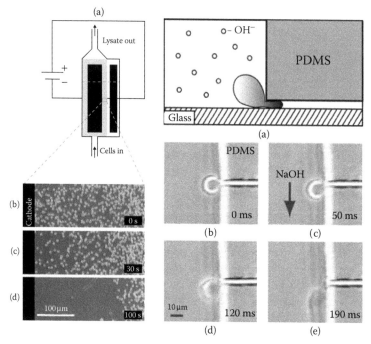

FIGURE 7.21
Left: (a) Flow through lysis device layout. (b-d) Close-up photographs showing red blood cell density following increasing exposure duration to the hydroxide ions. Right: (a) Illustration of a single cell trapped by applying suction. (b-e) Photographic sequence showing lysis of a trapped cell following exposure to sodium hydroxide ions over time. (From Di Carlo, D., et al. 2005. *Lab Chip* 5:171. With permission.)

on whole cells, the cell may be substituted with cell layers, cell networks, tissues, and whole animals or plants.

A key difference between cell-based sensors and other sensing forms is the type of information that is obtained. Typically, most sensors provide analytical information. For example, an electrochemical sensor may provide the concentration of an analyte. Cellular sensors, however, provide functional information because they detect extracellular stimuli and elicit a physiological response. In other words, these sensors provide the effect of an analyte on biological functions. Both the presence of an analyte and its induced response are detected.

Despite all these advantageous elements, cellular sensors face many challenges in their practical implementation. Detailed knowledge of microbiology and cell culture is required to understand cellular responses to stimuli. Cells have a limited lifetime and may be difficult to handle. Cell-based sensors are reviewed in references [41–45].

Cell-based sensors may utilize metabolic products as indicators of functional information following a sensing event. Microphysiometers introduced in the 1990s combined cells with silicon-based, light-addressable potentiometric sensors (LAPS) for the measurement of acidic energy metabolism products [46]. A microphysiometer with LAPS is shown in Figure 7.22. The LAPS detects pH changes due to the relationship between the applied potential to generate a photocurrent and the surface potential at the electrolyte–insulator interface. The photocurrent is only present in the area above the light emitting diode (LED), and thus only local sensing of pH in this defined region is obtained. Microphysiometers were used in another example to detect the response of Chinese hamster ovary –K1 cells that were maintained on glucose following the substitution of glucose with pyruvate. Another application of this sensor is the determination of the efficacious concentration of pharmaceutical compounds by obtaining dose–response curves.

Other indicators used in cell-based sensing include cytotoxic, genomic, and electric responses. Application areas include drug discovery, toxicology, pharmacology, pathogen screening, and environmental monitoring. Zinc toxicity was investigated using murine neuronal networks cultured on microelectrode arrays [47]. Transparent ITO electrodes were deposited on glass plates in 64-electrode arrays (Figure 7.23). Electrophysiological recordings showed a concentration dependency on cell excitation followed by significant electrical activity loss with increasing concentration of zinc acetate.

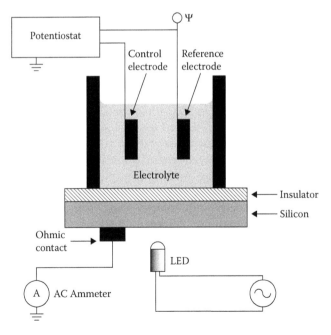

FIGURE 7.22
Schematic of microphysiometer.

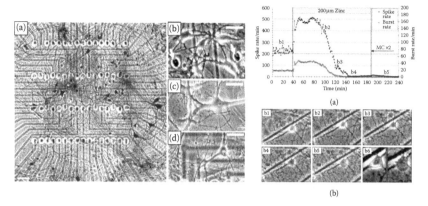

FIGURE 7.23
Left: (a) Cultured neurons on indium tin oxide (ITO) microelectrode arrays and (b-d) close-up photographs of neurons. Arrows indicate electrode sites and the scale bars are all 50 mm. Right: (a) The concentration dependency of electrical activity. (b) The morphological progression of a single neuron following a administration of 200 mM zinc. (From Parviz, M., and G. W. Gross. 2007. *Neurotoxicology* 28:520. With permission.)

Interestingly, changes in electrical activity preceded morphological changes (cell swelling and lysing).

7.2 Nucleic Acids

The popularity and growth of microfabricated devices for nucleic acid analysis is partially attributed to the intense interest in the Human Genome Project, which was completed in 2003. Over the 13-year span of the effort, all genes in human DNA were identified and the sequence of their base pairs was determined. These achievements were in part attributed to our ability to isolate, manipulate, and detect genetic information. In general, performing genetic analysis tasks at the microscale level has many distinct benefits, one of which is the scarcity of source material. Many researchers have contributed to microscale nucleic acid analysis tools, from devices that perform one step operations to integrated devices that start from a sample preparation and end in a detection result. Devices for DNA analysis are reviewed in reference [48], and a more general review on nucleic acids (including ribonucleic acid [RNA]) is found in reference [49].

Nucleic acid analysis starts with obtaining materials from the cells. Cells may be trapped and sorted, followed by lysis. Purification allows isolation of the desired genetic material. Cell trapping, manipulation, and lysis were introduced in Section 7.1. Cell sorting is covered in Chapter 8. In DNA analysis, the next step in the process may be to amplify the target DNA to obtain an adequate number of copies, followed by separation and finally detection. DNA amplification is discussed here. Separation (usually by CE) and detection were discussed in Chapters 5 and 6, respectively.

7.2.1 Purification

Following cell lysis, nucleic acids must be purified prior to subsequent steps such as amplification and hybridization. Microfluidic flow through devices utilize the high surface area-to-volume ratio to increase the efficiency of DNA and RNA retention. Filter structures such as pillar arrays, which were introduced in Chapter 5, can be used for concentrating nucleic acids. One example is a silicon substrate that is etched using DRIE to create pillar arrays in a microchamber measuring 200 μm high (10 or 18 μm in diameter spaced 18 or 34 μm apart, respectively) with corresponding surface areas of about 36 mm^2 [50]. In Figure 7.24, the layout and electron microscope images of a pillar array fabricated in a serpentine channel are shown [51]. The square-shaped pillars were coated with silica in order to purify bacteriophage lambda DNA and bacterial chromosomal DNA. DNA was bound in the presence of the chaotropic salt guanidinium isothiocyanate, washed with ethanol, and finally eluted from the system with a low-ionic strength buffer.The surfaces were oxidized to facilitate binding of DNA and RNA in the presence of a chaotropic salt solution. Other examples have demonstrated nucleic acid extraction

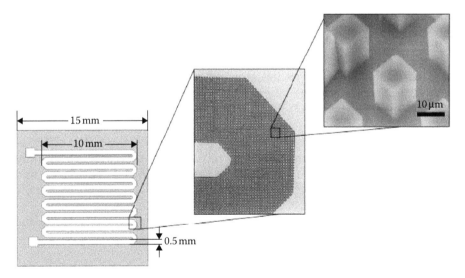

FIGURE 7.24
A serpentine channel lined with an array of silica pillars (10 mm width and 10 mm spacing). The height of the pillars was adjusted from 20 to 50 mm. The first inset is an optical micrograph showing a close-up of one area of the microchannel. The second inset is a scanning electron micrograph showing a close-up of the pillars. (From Cady, N. C., S. Stelick, and C. A. Batt. 2003. *Biosens Bioelectron* 19:59–66. With permission.)

with beads or particles [52–55]. In one example, purification of nucleic acids with derivatized beads (2.8 µm diameter) was combined with other operations such as cell isolation and lysis in integrated microfluidic chips [55]. Two chips were described: one for purifying mRNA and the other for DNA. The DNA chip contained three identical parallel processing channels for performing cell isolation, chemical cell lysis, and bead-based purification. Chips were made using multilayer soft lithography that incorporated pneumatic controls lines for operating the valves and pumps for fluid manipulation in the system. The DNA-purification process was performed using only nanoliter (0.4–1.6 nL) sample volumes by the steps illustrated in Figure 7.25. Using this methodology, *E. coli* DNA was purified.

7.2.2 Amplification: Polymerase Chain Reaction

Polymerase chain reaction (PCR) is a thermal and chemical process by which DNA or RNA (*reverse transcriptase-PCR*; RT-PCR) of interest can be copied and replicated for further processing or for facilitating detection. Since its development in 1984 by Kary Mullis, PCR has become a common technique in genetic analysis. PCR is a cyclical process comprising three separate thermal processes called *denaturing, annealing,* and *extension* (Figure 7.26).

DNA is usually present in a double-stranded form in which the complementary strands are joined in a double helix structure. To replicate this original DNA copy, first, the strands must be thermally separated by denaturing the DNA at a high temperature (~90–95°C). Next, the complementary strands hybridize, or join, with primers, that is, single-strand oligonucleotides (short nucleotide polymer) present in the PCR cocktail at lower temperatures (~50–60°C). This results in partially formed double-stranded DNA and starts the replication process. If a single DNA copy was present initially, two partial copies are present here. The copies are completed or extended with the assistance of the DNA polymerase enzyme at elevated temperatures (~70°C). DNA polymerase assembles the remaining DNA strand from the nucleotides (deoxynucleoside triphosphates or dNTPs) present in the PCR cocktail. This process is repeated from the denaturing step and is typically conducted for 20–40 cycles. For RNA replication by RT-PCR, the RNA is first converted to DNA using reverse transcriptase and then amplified using the same procedure as before.

One would expect 2^n copies, where n is the number of cycles. However, the amplification factor is modified by an efficiency factor as follows:

$$\Gamma = \left[1 + E(n)\right] \qquad (7.2)$$

where Γ is the amplification factor and E is the efficiency factor, which is a function of n. Efficiency tends to drop with higher cycle numbers ($n > 20$).

Since PCR is inherently a thermal process, many performance improvements are gained by miniaturization. A lower thermal mass leads to faster

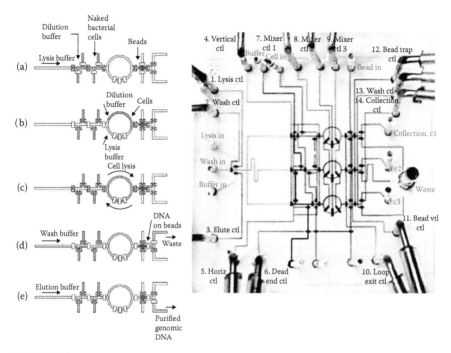

FIGURE 7.25
Left: Illustration of the process sequence leading to DNA purification from *E. coli* cells. The rectangles represent valves; an "x" denotes a closed valve and an empty rectangle denotes an open valve. (a) Bacteria, dilution buffer, and lysis buffer are introduced. (b) Individual sample plugs are loaded in order into the rotary mixer. (c) The peristaltic pump mixes the three solutions resulting in cell lysis. (d) The mixer contents are flushed over a packed bead column. (e) DNA capture in the bead column is recovered by elution. Right: DNA purification chip with dye-filled channels to facilitate visualization. (From Hong, J. W., et al. 2004. *Nat Biotechnol* 22:435. With permission.)

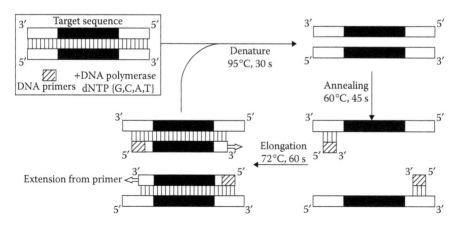

FIGURE 7.26
Illustration of the polymerase chain reaction process.

cycle speeds. In particular, the thermal ramping rate between temperature transitions is faster (up to ~80°C/s for heating and ~40°C/s for cooling [55]). The high surface area-to-volume ratio improves reaction efficiency compared to conventional PCR equipment. Many researchers in the field, who have pioneered and continued to develop chip-based PCR devices, have recognized these advantages. Their work is reviewed in references [55–58]. Two major approaches to PCR can be identified: microreactor (or spatial domain PCR) and flow through (or temporal domain PCR). Heating is provided either by off-chip devices, integrated resistive heaters, or even noncontact thermal sources such as infrared heating. Some devices include on-chip temperature sensing.

The first miniaturized PCR device was a simple microreactor etched in silicon with integrated polysilicon heaters separated by a nitride membrane [59,60]. In microreactor-based PCR devices, the PCR cocktail is stationary, and the temperature is cycled throughout the entire volume of the reaction chamber. Therefore, these devices are sometimes referred to as spatial-domain PCR devices. The structure of an early PCR reactor is shown in Figure 7.27. The total volume of the reactor is either 25 or 50 µL. In this device, amplification of human immunodeficiency virus (HIV) and cystic fibrosis (CF) from human DNA as well as other DNA was demonstrated and verified through off-chip detection. Many other studies have followed this example and have contributed to further performance enhancements.

An interesting variation on micro-PCR reactors was developed in which a droplet served as a single PCR reactor [61]. The PCR reaction was tailored for high throughput DNA methylation detection (methylation specific PCR or MSP); DNA methylation is a process by which DNA is chemically modified and linked to tumorigenesis. Multiple reactor clusters, each consisting of 12 open (uncovered) reactors were arrayed in a snowflake pattern to allow 108 simultaneous MSP reactions. The reactor clusters consisted of a central input port (0.64 mm diameter) feeding 12 open reactors (4 mm diameter) through closed connecting channels (500 µm wide). Silicone rubber reactor arrays were cast from SU-8/silicon masters and then bonded to glass slides. The ports and reactors were created using punches. Primers for the PCR reaction were predeposited into each reaction chamber and

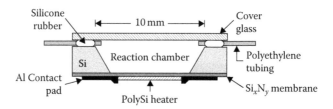

FIGURE 7.27
Early polymerase chain reaction reactor.

allowed to evaporate. The channel and reactor surfaces were first immersed in bovine serum albumin (BSA), which prevented enzyme adsorption to the walls and thus increased the PCR efficiency. MSP was performed on human DNA obtained from leukocytes isolated from whole blood. The droplet-based PCR reactions were carried out using the method shown in Figure 7.28. PCR cocktail was introduced via the central port (either 50 or 10 µL) followed by an injection of mineral oil. The oil forced the sample into the channels, and in each chamber, where sample droplets (4 or 0.8 µL, respectively) surrounded by the oil formed. Thus, the introduction of oil metered and compartmentalized the sample in a single step. The oil also prevented evaporation of the sample contained in the open reaction chamber. Then, PCR was carried out by cycling the array temperature on an external Peltier heating plate.

Reactor-based PCR devices can also be integrated with other functions such as on-chip detection. In one such study, PCR followed by DNA hybridization and washing were integrated into a carbon dioxide laser machined polycarbonate chip [62]. Channels (150 µm wide) were limited in size by the resolution of the laser beam (150 µm) and carved into an absorbing black polycarbonate sheet (250 µm thick). Structured polycarbonate sheets were thermally bonded in a hydraulic press (139°C and 2 tons of pressure for

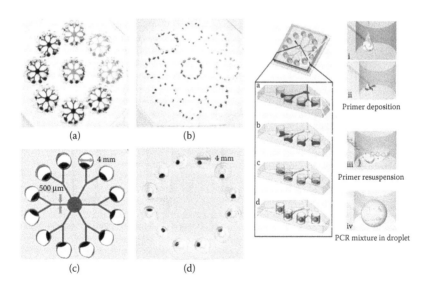

FIGURE 7.28
Left: Images showing (a) the overall layout of the droplet-based polymerase chain reaction (PCR) device and initial loading of sample into each array; (b) 108 PCR droplet reactions after droplet isolation with mineral oil; (c) layout of a single reactor array unit with a central loading port and 12 connected reactors; (d) a close-up of the droplets in each reaction chamber. Right: Sequence of images depicting the loading of the PCR mixture into each reaction chamber. (From Zhang, Y., et al. 2009. *Lab Chip* 9:1059. With permission.)

45 minutes). This simple polymer construction allowed for low cost and disposable devices. Valving was integrated within each channel by taking advantage of the phase change properties of Pluronic gel. The PCR compatible gel is solid at room temperature and liquefies upon cooling to ~5°C. These are single-use valves (because of the disposable nature of the chip) that are normally-closed and permanently open after activation. The valves were required to isolate the PCR reaction and withstand pressures generated from expansion at elevated temperatures (~8 psi). Additional valves separated the hybridization region from the wash and waste ones. Pumping was performed with off-chip syringe pumps. Both heating and cooling were achieved with a Peltier electrothermal device in the serpentine PCR reactor (38 μL; Figure 7.29). Hybridization-based detection was performed on-chip in an adjacent channel (7 μL) following amplification. Both *E. coli* and *Enterococcus faecalis* genes were amplified and detected.

Shortly after the first microreactor-based PCR devices were introduced, a new concept called spatial domain PCR was developed [63,64]. In this concept, instead of cycling the temperatures throughout a single reactor, distinct constant temperature zones are defined on a chip. Switching between denaturing, annealing, and extension is achieved by flowing the sample through the channels that are passing through each temperature region. The sample

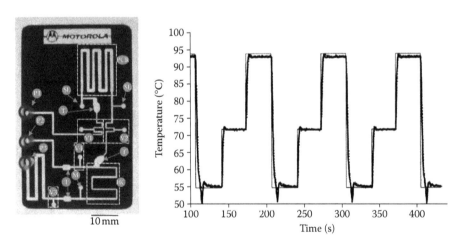

FIGURE 7.29
Left: Integrated polycarbonate DNA analysis device including a continuous flow polymerase chain reaction channel (PCR), hybridization channel (HC), Pluronics valves (V1–V4), Pluronics traps (T), hydrophobic gas permeable membrane (M), sample loading ports (SL), syringe pump inputs (P1 and P3), and syringe pump output (P2). The overall device dimensions are 5.4 cm × 8.6 cm × 0.75 mm. Right: Temperature profile produced by a Peltier heating and cooling system showing the Peltier system temperature (thick line), device temperature (dotted line), and temperature set point (thin line). (From Liu, Y. J., et al. 2002. *Anal Chem* 74:3063. With permission.)

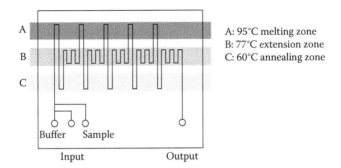

A: 95°C melting zone
B: 77°C extension zone
C: 60°C annealing zone

FIGURE 7.30
Layout of the first continuous flow polymerase chain reaction device.

flow rate and the channel length in each zone determine the step time for each cycle. Since the temperatures are fixed, faster PCR cycle times are possible. The first continuous flow PCR device pushed sample plugs through a serpentine channel that meandered through three temperature zones, one for each step in the PCR process (Figure 7.30). Completion of 20 cycles was achieved in just 1.5 minutes. However, one drawback was that the number of cycles was fixed and determined by the device layout. Spatial domain PCR is also possible in loop channels with fixed temperature zones [64]. These rotary PCR reactors combined attributes from both spatial and temporal domain PCR in a multilayer soft lithography silicone rubber device with integrated heaters (tungsten with aluminum electrical leads; Figure 7.31). The loop channel width was asymmetric (120 μm on the left half and 70 μm on the right; 8.5 μm high and 2.5 mm in radius), which resulted in differing fluid velocities in these regions. Sample volumes as small as 12 nL were handled in this rotary loop reactor. Both human and λ-phage DNA were amplified, although cycle times were longer than in the serpentine channel device.

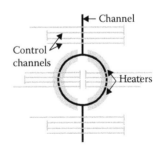

7.2.3 Hybridization: Microchannel and Microarray

Hybridization is a step in the PCR that is used for the detection of the amplified products. In this application, single-stranded DNA (ssDNA) oligonucleotide probes are used as biorecognition elements and perform a biosensing function. In this context, biorecognition is essentially

FIGURE 7.31
Layout of rotary Polymerase chain reaction (PCR) device (top view) showing the PCR loop, pneumatic control channels, and integrated heaters. (From Liu, J., M. Enzelberger, and S. Quake. 2002. *Electrophoresis* 23:1531. With permission.)

a biochemical reaction with kinetics that depend on temperature, concentration, and other factors. Both microfluidic in-channel and microarray hybridization biochips have been developed and used in conjunction with a suitable detection method (e.g., those discussed in Chapter 6).

Hybridization reactions are conveniently performed using probes immobilized on surfaces in microsystems. Examples include channel walls [62,65] and microbeads [66]. Streptavidin-coated microbeads were further derivatized by binding to ssDNA probes labeled with biotin (via streptavidin–biotin binding). These beads could then be used to capture complementary DNA target strands. To do this, the beads were loaded into serial chambers connected by microchannels in silicone rubber replicas molded from photoresist-glass masters (Figure 7.32). The beads were captured by a combination of weirs and UV cross-linked hydrogel plugs fabricated directly in the channels between the bead chambers. The hydrogel plugs prevented the passage of DNA under a pump-induced pressure gradient, but allowed passage under an electrokinetic flow. The fluorescein-labeled DNA target strands were introduced by electrophoresis and detected optically. Beads could be reused for subsequent detection reactions by releasing and recovering them with a 0.1 N sodium hydroxide wash.

Arrayed hybridization reactions were carried out in a single polymer channel [67]. Injection molded polycarbonate enabled a single use, disposable hybridization microarray with sufficient optical transparency for fluorescence detection through the chip (Figure 7.33). The 10 μL hybridization channel could accommodate up to 1000 separate hybridization reactions, but only 512 spots were utilized by printing the probes in a 4 × 128 array with a spot pitch of 250 μm using conventional biochip spotter. Following the deposition of the capture probes (with varying concentrations of 0.5–40 μM), they were covalently linked to the substrate by a UV-irradiation process. Hybridization was performed using 12–20-mer capture probes and 30-mer targets and was detected by optical fluorescence.

7.2.4 Sequencing

Sequencing is a technique that allows the determination of the order of bases in a target strand of DNA. Its development in recent years has been driven by applications requiring high speed sequencing, such as the Human Genome Project. However, while great advances have been made, sequencing, which is usually performed using capillary array electrophoresis (CAE), is still expensive. For example, a complete human genome consists of 6 billion base pairs (bp) when counting both sets of chromosomes (about 60,000 genes) and costs approximately $20 million to sequence, according to a recent report [68]. Thus, we are still far from the new goal set by the National Human Genome Research Institute to make an individual's genome available for under $1000 [69]. This new goal may further fuel the push to microfabricated chip

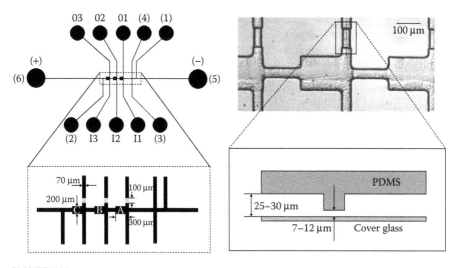

FIGURE 7.32
Left: Layout and dimensions of the key features on the hybridization chip. Microbeads were introduced though I1–I3. Right: Photograph of the three bead chambers and a cross-section of the weir used to capture the beads (15.5 μm diameter). (From Seong, G. H., W. Zhan, and R. M. Crooks. 2002. *Anal Chem* 74:3372. With permission.)

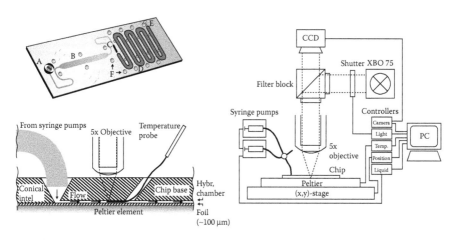

FIGURE 7.33
Top left: Hybridization chip layout including the hybridization channel (B), hydrophobic stop valve (C), and waste storage channel (D). Bottom left: Cross-sectional view of the chip in the hybridization area showing the experimental setup. Right: Schematic showing the complete experimental setup. (From Noerholm, M., et al. 2004. *Lab Chip* 4:28. With permission.)

sequencing technologies that may prove more economical than the slow and costly CAE technology used today. For example, CAE requires ~1–2 hours to read out a 600–800-bp sequence of DNA compared to only ~25 minutes using a microfabricated chip with a 20 cm long separation channel [70,71]. Recently, it was reported that only 6.5 minutes were required to sequence 600 bp [72].

Sequencing is predominantly achieved using the Sanger dideoxy chain termination reaction developed in 1975 (for which Frederick Sanger and Walter Gilbert won the Nobel Prize in chemistry in 1980). The process is a combination of PCR and electrophoretic separation (Figure 7.34). Enzymatic replication of the target ssDNA using DNA polymerase is modified to include dideoxynucleoside triphosphates (ddNTPs) that terminate DNA elongation by blocking the addition of the next nucleotide. Four separate reactions are performed using one ddNTP (ddTTP, ddCTP, ddATP, and ddGTP) for each of the four types of bases (T, C, A, and G) to produce complementary ssDNA fragments terminated at all possible locations for each base. Therefore, matching strands differ in length by only one base. Electrophoretic separation is

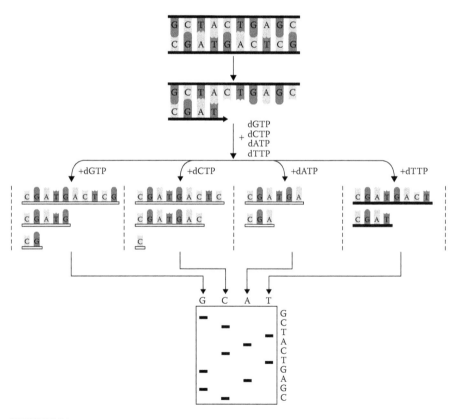

FIGURE 7.34
Sanger sequencing process. (From Burns, M. A., et al. 1998. *Science* 282:484. With permission.)

performed with single base resolution in a sieving matrix (cross-linked gel or polymer solution). This is necessary to achieve an adequate resolution and to compensate for the otherwise size-independent electrophoretic mobility of DNA in free solution. Each nucleotide is labeled with a unique identifier, such as a radioactive or fluorescent marker, which is read out following separation. The results are then superimposed to obtain the fragment's sequence information.

Current sequencing technology provides error-free reads up to 800 bases per separation channel. However, human genes run ~100 kb in length. Sequencing a single human gene requires the gene to be chopped up into shorter overlapping segments of ~1 kb that are then sequenced individually and reassembled to obtain the sequence for the whole gene. It is clear that there is room for improvement.

Historically, sequencing was performed using slab-gel electrophoresis, which has now been replaced by CAE. The use of higher electrical fields resulted in a faster transition from the larger gel to the capillary format, which in turn provided faster separations. This is possible in smaller separations formats that minimize the impact of Joule heating. Long separation channels typically must accommodate both longer segments for high-resolution separation with single base resolution. In microfabricated chips, long channels are obtained using a serpentine channel. However, the turns induce dispersion and loss of resolution. High resolution DNA separation in microfabricated sequencing systems was made possible by careful optimization of turn geometry that had a small radius of curvature (~250 μm) with a short tapering length (~55 mm) but a large tapering ratio (4 separation channels:1 turn channel) [73]. With these modifications, parallel sequencing in up to 384 serpentine lanes was achieved as discussed in Chapter 5. A review of the microfabricated systems for sequencing is found in reference [74].

Recently, faster separations for increased sequencing throughput were achieved in a microchip system [72]. A commercially available borosilicate glass chip with a 7.5 cm long separation channel was used in conjunction with a four-color laser induced fluorescence (LIF) detection system to produce a sequencing read out in only 6.5 minutes for 600 bp. The introduction of a more efficient separation matrix along with a coating on the channel wall enabled this improvement in speed. The channel wall coating (poly(N-hydroxyethylacrylamide) [pHEA]) minimized interaction of the sequencing fragments with the channel wall. In standard DNA separations, linear polyacrylamide gels are used which results in a process time of 15–35 minutes. By switching the gel to a new polymer (poly(N,N-dimethylacrylamide) [pDMA]) with a higher molar mass (> 1×10^6 Da), DNA separation was promoted.

7.2.5 Integrated Genetic Analysis Systems

Many processing steps are required in genetic analysis. Many microfabricated devices only focus on a single step, but some have demonstrated the potential of microfabricated systems in removing human intervention altogether. An elegant example of a complete DNA analysis device fabricated in silicon and glass is shown in Figure 7.35 [75]. An nL sample injector, mixer, sample positioner, thermal reaction chamber, electrophoretic separation column, and fluorescence detectors were included in this chip that occupied a volume of 47 mm × 5 mm × 1 mm. The only external components required to

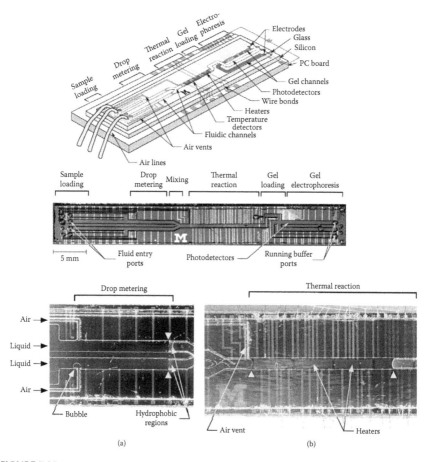

FIGURE 7.35
Top: Integrated nL DNA-analysis device that includes sample preparation (loading, metering, and mixing), polymerase chain reaction reactor, gel electrophoresis, and detection. Bottom: Close-up images of the (a) drop metering and (b) thermal reaction portions of the device. (From Liu, R. H., et al. 2004. *Anal Chem* 76:1824. With permission.)

operate the system were an optical excitation source, a pressure source, and an electronic control circuitry.

The silicon–glass chip was mounted on a printed circuit board (PCB) to make external electrical connections. The silicon base included the photo-diodes for fluorescence detection, heaters, temperature detectors, electrodes, and hydrophobic patches for the stop valves. Microfluidic channels (500 μm wide and 50 μm high) were etched into the glass substrate and the two pieces were bonded prior to assembly on the PCB.

The sample was first introduced into the device with external tubing. Nanoliter sample droplets were metered using a hydrophobic stop valve in conjunction with air pressure pulses from a vent line. The volume was deter-mined by the distance between the valve and the vent (120 nL droplet, shown below in Figure 7.35). Droplets in separate lanes merged and mixed prior to positioning in a thermal reaction chamber, where PCR was performed. The DNA products were moved to the separation column by pressure and then electrokinetically loaded onto the polyacrylamide gel. Integrated photodiodes below the channel detected fluorescing DNA products. In this manner, a 106-bp DNA fragment was amplified and detected in an automated fashion.

Another fully integrated genetic analysis system has been constructed from polymers [76]. This system, however, required no external fluidic com-ponents. In fact, the device provided "sample-to-answer" genetic analysis of DNA by automating sample preparation (cell capture, pre-concentration, purification, lysis), PCR, hybridization, and electrochemical detection. To achieve all these functions, the device was assembled from a polymer chip ($60 \times 100 \times 2$ mm^3), PCB, and a Motorola eSensor microarray chip that con-tained all the following components: mixers, valves, pumps, channels, cham-bers, heaters, and sensors (Figure 7.36).

The polymer chip consisted of polycarbonate sheets that were convention-ally machined and joined together. The channels and chambers (1–5 mm wide, 300 μm–1.2 mm high) present in the chip were used for mixing, precon-centration, purification, lysis, PCR, and hybridization. Valves were formed by loading molten paraffin into the channel regions. Cooling solidified the wax and created a normally closed valve. Thermal activation of the valve removed the paraffin and allowed electrochemically and thermopneumati-cally pumped fluids to pass. Electrochemical detection of hybridized PCR products was performed on the eSensor chip, consisting of a 4×4 array of gold electrodes functionalized using self-assembled thiol-terminated DNA oligonucleotides.

First, the samples were loaded and the device was plugged into an electri-cal control unit. In the sample storage chamber, blood was mixed with immu-nomagnetic beads to bind the target cells. The cell-bead conjugates were pumped into the PCR chamber and preconcentrated by magnetic capture. Purified cells were thermally lysed, and the exposed DNA was amplified by PCR. The amplified products were pumped into the hybridization/sensing chamber. The redox reaction of the ferrocene-labeled signaling probes was

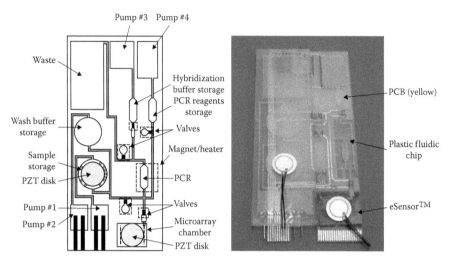

FIGURE 7.36
Left: Schematic of the overall device layout. Pumps 1–3 are electrochemical and pump 4 is thermopneumatic. Right: Photograph of the integrated device. (From Hatch, A., et al. 2001. *Nat Biotechnol* 19:461. With permission.)

detected electrochemically with the eSensor chip. This device detected the presence of pathogenic bacteria from whole blood and performed an analysis of single-nucleotide polymorphism from dilute blood.

7.3 Proteins

Completion of the Human Genome Project has turned the attention of many to addressing the much more challenging area of *proteomics*, in which the entire *proteome* (a set of proteins encoded by one gene) is the goal. Proteins, however, are far more complex than nucleic acids. They possess a greater number of fundamental building blocks (20 amino acids compared to four bases) and intricate 3D structures. Unfortunately, no convenient means of chemical amplification presently exists for replicating large numbers of protein copies. Separation methods in microchannels are used to extract proteins (including diffusion-based methods); however, proteins adsorb well to many surfaces (glass, silicon, and polymers). Thus, surface treatments, reviewed in reference [77], are required to either enhance or prevent adsorption. Detection is also challenging, as large proteins must be digested prior to MS detection. Protein samples must also be concentrated or desalted (purification by removal of low-molecular-weight compounds) prior to detection.

These techniques have been discussed in Chapter 6. Here, our discussion on proteins focuses on their use as biorecognition elements in biosensors.

7.3.1 Immunoassays

The specific nature of the *antigen–antibody* binding interaction is a biorecognition event that may be used in biosensing (*immunoassay* or *immunosensing*). This was recognized as early as 1959, when Rosalyn Yalow and Salomon Berson used a radioactive marker to track binding events (called *radioimmunoassay*). They received the Nobel Prize in medicine for this achievement in 1977. Antibodies and antigens are complex biomolecules composed of hundreds of individual amino acid sequences that have a specific, high affinity "lock-and-key" relationship. In the body, the immune system (B-lymphocytes) produces proteins (~10 nm) in the immunoglobulin family in response to the presence of foreign antigenic substances. Specific binding sites are coded in the protein's structure; antibodies have a Y-shape with specific bonding sites on the arms (Figure 7.37).

There are many types and formats of immunoassays, which conveniently fall into two major classifications: *homogeneous* and *heterogeneous*. Homogeneous assays are performed in a free solution, whereas heterogeneous assays require immobilization of the antibody on a solid support. Since homogeneous assays do not require an extra immobilization step, they are faster. Heterogeneous assays, however, benefit from the high surface-area-to-volume ratio.

Two additional classifications for immunoassays are *competitive* or *noncompetitive*. Competitive assays involve at least two analytes with different affinities that compete for limited antibody binding sites. An unlabeled analyte competes with a labeled analyte, resulting in a decrease in the signal with an increase in the unlabeled analyte concentration (which displaces the labeled analyte). The opposite may also occur: the signal increases if

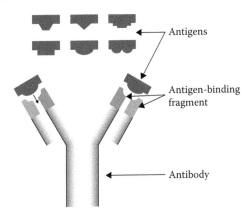

FIGURE 7.37
Simplified structure of an antibody indicating the antigen-specific binding sites. (From Choi, J. W., et al. 2002. *Lab Chip* 2:27. With permission.)

the binding of the labeled analyte increases. For a noncompetitive assay, the signal increase is proportional to the target analyte concentration. Noncompetitive *sandwich* (heterogeneous) reactions are perhaps the most commonly used. The antibody is immobilized to a surface and a binding reaction captures an antigen and produces a detectable response (i.e., by fluorescent tag). However, this requires that the target be labeled, which can prove to be a complicated process. A competitive assay does not require target labeling. Instead, the major challenge is to select an appropriate analyte as an analog to the intended target of the assay. These techniques have been implemented in microfabricated systems either as static arrays [78,79] or as microfluidic assays [80,81].

A simple T-channel structure was used to perform rapid diffusion-based competitive homogeneous immunoassays [82]. Nanomolar concentrations of a fluorescently labeled antigen (phenytoin, an antiepileptic drug) were detected in <1 minute. The diffusion immunoassay was performed at the interface between two laminar fluid streams (antigen and antibody) in a device called the T-sensor (Figure 7.38). The chip was constructed from a laser machined mylar film (main channel dimensions of 1200 µm wide and

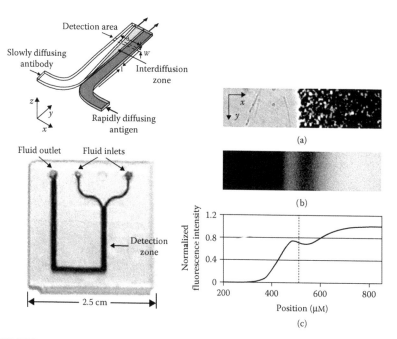

FIGURE 7.38
Top left: T-sensor immunoassay schematic. Bottom left: Photograph of T-sensor with dye-filled channels. Right: Phenytoin diffusion immunoassay results: (a) bright-field photograph of blood-antigen stream (right) and antibody stream (left), (b) fluorescence photograph, and (c) intensity plot showing the diffusion of the labeled antigen into the antibody stream [82] (Reprinted by permission from Macmillan Publishers Ltd: *Nat Biotechnol* 19:461–5; copyright 2001.).

100 µm high) sandwiched between two glass coverlips. The antibody and antigen were introduced intially as two separate streams at the same flow rate. Once in the main channel, the two solutions mixed only by diffusion. In this case, the antigen was contained in diluted blood, and a key advantage of this approach over conventional methods was that the blood cells did not have to be removed prior to conducting the immunoassay. Detection was achieved using the molecular mass-dependent differences in diffusion rate (~10×) between small antigens (relative molecular mass <10,000) and large antibodies (relative molecular mass ~1,50,000) to drive the antigens into the antibody stream, where they binded to antibodies and fluoresced.

An electrochemical detection method was used to detect a heterogeneous noncompetitive sandwich assay performed on the surface of magnetic beads [83]. First, derivatized magnetic beads (biotinylated sheep anti-mouse immunoglobulin G antibodies) were introduced into a channel and immobilized next to interdigitated detection electrodes by applying a magnetic field (Figure 7.39). Next, antigens were introduced and immobilized on the beads. A second enzyme-labeled antibody was introduced and immobilized onto

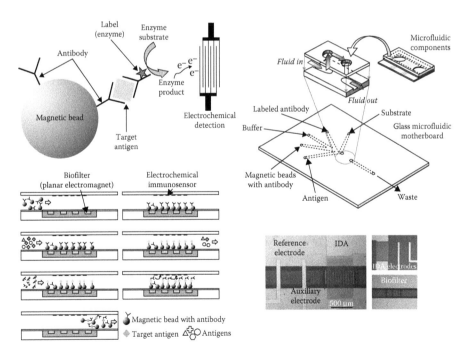

FIGURE 7.39
Top left: Schematic of the immunosensing concept. Bottom left: Step-by-step illustration of the immunoassay procedure. Top right: Major components of the bead-based immunosensing device. Bottom right: Photographs of the immunoassay chamber with details of the interdigitated electrodes and planar electromagnet biofilter [83] (Reproduced by permission of the Royal Society of Chemistry.).

the captured antigens. Finally, electrochemical detection of the bound species was performed in the presence of an enzyme substrate injected into the detection chamber. The total chamber volume was 750 nL and the microfluidic channels were 400 μm wide and 100 μm deep.

7.3.2 Enzymatic Assays

Enzymes catalyze many useful reactions. For example, polymerase enables the PCR amplification reaction. In the final example described in Section 7.3.1, an enzymatic reaction was used to produce an electrochemically detectable product. The latter type of reaction is particularly useful in converting non-detectable analytes (e.g., glucose, glutamate, alcohol, and so on) into detectable species (e.g., hydrogen peroxide). As in immunoassays, enzymatic assays may be homogeneous or heterogeneous. Similar concerns exist here as well; namely, the immobilization of enzymes without rendering them inactive is required for heterogeneous assays. To avoid the immobilization procedure, many have turned to using soluble enzymes in homogeneous assays. These, however, require greater enzyme volumes. Reviews of enzymatic assay systems are provided in references [84–86].

7.4 Problems

1. Some cells demonstrate chemotaxis; they are able to move in response to the presence of particular chemicals in their environment. Suppose you want to construct a simple microfludic device that will allow you to position a cell in a long microchannel across which a chemical gradient can be selectively introduced. Assume that the cell is a 5 μm sphere. The introduction of a chemical signal at the opposite end of the system would either trigger the cell to move in response or not. Design a system that will allow adequate support of the cell and have the ability to test chemotactic responses to different chemical species. Describe your approach and draw the layout of the device, its cross section, and the supporting equipment required to run the experiments. Describe how cell movement would be measured. Search the literature and find a cell that exhibits chemotaxis and the chemicals to which it responds. Provide any references used in IEEE format.

2. Many cell retention devices are demonstrated using microspheres. What are the issues that are encountered when moving from inanimate spheres to real cells or organelles?

3. Find a recent article describing a cell-based sensor. In the particular application described, why was it necessary to use a cell-based sensor over another sensing method? Provide your references in IEEE format.

4. Compare and contrast the relative advantages and disadvantages of temporal and spatial PCR devices.

5. Integrated genetic analysis systems represent a holy grail of sorts for many working on microfabricated devices. Locate a recent article describing an integrated genetic analysis device. What are the functions integrated in the device? How does the level of integration compare to the devices described in this chapter? What technological challenges are cited in the article? Provide your references in IEEE format.

6. It is better to have an excess of analyte or antibody in a noncompetitive immunoassay? Justify your answer.

7. Draw the typical calibration curve (signal versus analyte concentration) for a competitive and noncompetitive immunoassay, assuming that the readout is optical (e.g., fluorescence). Also indicate the nonspecific background signal in the both plots.

References

1. Paguirigan, A. L., and D. J. Beebe. 2009. From the cellular perspective: Exploring differences in the cellular baseline in macroscale and microfluidic cultures. *Integr Biol* 1:182.

2. Huang, C. W., and G. B. Lee. 2007. A microfluidic system for automatic cell culture. *J Micromech Microeng* 17:1266.

3. Leclerc, E., Y. Sakai, and T. Fujii. 2004. Microfluidic PDMS (polydimethylsiloxane) bioreactor for large-scale culture of hepatocytes. *Biotechnol Prog* 20:750.

4. Sin, A., et al. 2004. The design and fabrication of three-chamber microscale cell culture analog devices with integrated dissolved oxygen sensors. *Biotechnol Prog* 20:338.

5. Viravaidya, K., A. Sin, and M. L. Shuler. 2004. Development of a microscale cell culture analog to probe naphthalene toxicity. *Biotechnol Prog* 20:316.

6. Tatosian, D. A., and M. L. Shuler. 2009. A novel system for evaluation of drug mixtures for potential efficacy in treating multidrug resistant cancers. *Biotechnol Bioeng* 103:187.

7. Fink, J., et al. 2007. Comparative study and improvement of current cell micropatterning techniques. *Lab Chip* 7:672.

8. Mrksich, M., et al. 1996. Controlling cell attachment on contoured surfaces with self-assembled monolayers of alkanethiolates on gold. *Proc Natl Acad Sci U S A* 93:10775.

9. Chen, C. S., et al. 1998. Micropatterned surfaces for control of cell shape, position, and function. *Biotechnol Prog* 14:356.

10. Kim, E., Y. N. Xia, and G. M. Whitesides. 1995. Polymer microstructures formed by molding in capillaries. *Nature* 376:581.
11. Martinoia, S., et al. 1999. A simple microfluidic system for patterning populations of neurons on silicon micromachined substrates. *J Neurosci Methods* 87:35.
12. Nam, Y., et al. 2004. Gold-coated microelectrode array with thiol linked self-assembled monolayers for engineering neuronal cultures. *IEEE Trans Biomed Eng* 51:158.
13. Sharma, S., R. W. Johnson, and T. A. Desai. 2004. Evaluation of the stability of nonfouling ultrathin poly(ethylene glycol) films for silicon-based microdevices. *Langmuir* 20:348.
14. Zhu, L., et al. 2004. Filter-based microfluidic device as a platform for immunofluorescent assay of microbial cells. *Lab Chip* 4:337.
15. Lehnert, T., and M. A. M. Gijs. 2004. Patch clamp microsystems. In *Lab-on-chips for Cellomics: Micro and Nanotechnologies for Life Science*. Dordrecht, The Netherlands: Kluwer Academic Publishers.
16. Fertig, N., R. H. Blick, and J. C. Behrends. 2002. Whole cell patch clamp recording performed on a planar glass chip. *Biophys J* 82:3056.
17. Fertig, N., et al. 2001. Microstructured glass chip for ion-channel electrophysiology. *Phys Rev E* 64.
18. Fertig, N., et al. 2002. Activity of single ion channel proteins detected with a planar microstructure. *Appl Phys Lett* 81:4865.
19. Werdich, A. A., et al. 2004. A microfluidic device to confine a single cardiac myocyte in a sub-nanoliter volume on planar microelectrodes for extracellular potential recordings. *Lab Chip* 4:357.
20. Wheeler, A. R., et al. 2003. Microfluidic device for single-cell analysis. *Anal Chem* 75:3581.
21. Tooker, A., et al. 2005. Biocompatible parylene neurocages. *IEEE Eng Med Biol Mag* 24:30.
22. Heo, J., et al. 2003. A microfluidic bioreactor based on hydrogel-entrapped e. Coli: Cell viability, lysis, and intracellular enzyme reactions. *Anal Chem* 75:22.
23. Koh, W. G., L. J. Itle, and M. V. Pishko. 2003. Molding of hydrogel multiphenotype cell microstructures to create microarrays. *Anal Chem* 75:5783.
24. Tan, W., and T. A. Desai. 2004. Layer-by-layer microfluidics for biomimetic three-dimensional structures. *Biomaterials* 25:1355.
25. Yi, C. Q., et al. 2006. Microfluidics technology for manipulation and analysis of biological cells. *Anal Chim Acta* 560:1.
26. Mietchen, D., et al. 2002. Automated dielectric single cell spectroscopy—Temperature dependence of electrorotation. *J Phys D Appl Phys* 35:1258.
27. Ozkan, M., et al. 2003. Optical manipulation of objects and biological cells in microfluidic devices. *Biomed Microdevices* 5:61.
28. Wakamoto, Y., et al. 2003. Development of non-destructive, non-contact single-cell based differential cell assay using on-chip microcultivation and optical tweezers. *Sens Actuators B Chem* 96:693.
29. Voldman, J., et al. 2002. A microfabrication-based dynamic array cytometer. *Anal Chem* 74:3984.
30. Reichle, C., et al. 1999. Electro-rotation in octopole micro cages. *J Phys D Appl Phys* 32:2128.

31. Rubinsky, B. 2004. Micro-electroporation in cellomics. In *Lab-on-chips for Cellomics: Micro and Nanotechnologies for Life Science*. Dordrecht, The Netherlands: Kluwer Academic Publishers.
32. Fox, M. B., et al. 2006. Electroporation of cells in microfluidic devices: A review. *Anal Bioanal Chem* 385:474.
33. He, H. Q., D. C. Chang, and Y. K. Lee. 2006. Micro pulsed radio-frequency electroporation chips. *Bioelectrochemistry* 68:89.
34. Huang, Y., and B. Rubinsky. 2003. Flow-through micro-electroporation chip for high efficiency single-cell genetic manipulation. *Sens Actuators A Phys* 104:205.
35. Fox, M., et al. 2005. A new pulsed electric field microreactor: Comparison between the laboratory and microtechnology scale. *Lab Chip* 5:943.
36. Brown, R. B., and J. Audet. 2008. Current techniques for single-cell lysis. *J R Soc Interface* 5(Suppl 2):S131.
37. Di Carlo, D., K. H. Jeong, and L. P. Lee. 2003. Reagentless mechanical cell lysis by nanoscale barbs in microchannels for sample preparation. *Lab Chip* 3:287.
38. Lee, S. W., and Y. C. Tai. 1999. A micro cell lysis device. *Sens Actuators A Phys* 73:74.
39. McClain, M. A., et al. 2003. Microfluidic devices for the high-throughput chemical analysis of cells. *Anal Chem* 75:5646.
40. Di Carlo, D., et al. 2005. On-chip cell lysis by local hydroxide generation. *Lab Chip* 5:171.
41. Ziegler, C. 2000. Cell-based biosensors. *Fresenius J Anal Chem* 366:552.
42. Kovacs, G. T. A. 2003. Electronic sensors with living cellular components. *Proc IEEE* 91:915.
43. Wang, P., et al. 2005. Cell-based biosensors and its application in biomedicine. *Sens Actuators B Chem* 108:576.
44. Haruyama, T. 2006. Cellular biosensing: Chemical and genetic approaches. *Anal Chim Acta* 568:211.
45. Banerjee, P., and A. K. Bhunia. 2009. Mammalian cell-based biosensors for pathogens and toxins. *Trends Biotechnol* 27:179.
46. McConnell, H. M., et al. 1992. The cytosensor microphysiometer—Biological applications of silicon technology. *Science* 257:1906.
47. Parviz, M., and G. W. Gross. 2007. Quantification of zinc toxicity using neuronal networks on microelectrode arrays. *Neurotoxicology* 28:520.
48. Sun, Y., and Y. C. Kwok. 2006. Polymeric microfluidic system for DNA analysis. *Anal Chim Acta* 556:80.
49. Chen, L., A. Manz, and P. J. R. Day. 2007. Total nucleic acid analysis integrated on microfluidic devices. *Lab Chip* 7:1413.
50. Christel, L.A., et al. 1999. Rapid, automated nucleic acid probe assays using silicon microstructures for nucleic acid concentration. *J Biomech Eng Trans ASME* 121:22.
51. Oleschuk, R. D., et al. 2000. Trapping of bead-based reagents within microfluidic systems: On-chip solid-phase extraction and electrochromatography. *Anal Chem* 72:585.
52. Wolfe, K. A., et al. 2002. Toward a microchip-based solid-phase extraction method for isolation of nucleic acids. *Electrophoresis* 23:727.
53. Chung, Y. C., et al. 2004. Microfluidic chip for high efficiency DNA extraction. *Lab Chip* 4:141.

54. Hong, J. W., et al. 2004. A nanoliter-scale nucleic acid processor with parallel architecture. *Nat Biotechnol* 22:435.
55. Kricka, L. J., and P. Wilding. 2003. Microchip PCR. *Anal Bioanal Chem* 377:820.
56. Andersson, H., and A. van den Berg, eds. 2004. *Lab-on-chips for Cellomics: Micro and Nanotechnologies for Life Science*. Dordrecht, The Netherlands: Kluwer Academic Publishers.
57. Auroux, P. A., et al. 2004. Miniaturised nucleic acid analysis. *Lab Chip* 4:534.
58. Zhang, Y., and P. Ozdemir. 2009. Microfluidic DNA amplification—A review. *Anal Chim Acta* 638:115.
59. Northrup, M. A., et al. 1993. DNA amplification with a microfabricated reaction chamber. In *Transducers '93. Yokohama, Japan*, 924-26.
60. Northrup, M. A., et al. 1995. A MEMS-based miniature DNA analysis system. In *Transducers '95, Stockholm, Sweden* 764-67.
61. Zhang, Y., et al. 2009. DNA methylation analysis on a droplet-in-oil PCR array. *Lab Chip* 9:1059.
62. Liu, Y. J., et al. 2002. DNA amplification and hybridization assays in integrated plastic monolithic devices. *Anal Chem* 74:3063.
63. Kopp, M. U., A. J. de Mello, and A. Manz. 1998. Chemical amplification: Continuous-flow PCR on a chip. *Science* 280:1046.
64. Liu, J., M. Enzelberger, and S. Quake. 2002. A nanoliter rotary device for polymerase chain reaction. *Electrophoresis* 23:1531.
65. Lee, T. M. H., M. C. Carles, and I. M. Hsing. 2003. Microfabricated PCR-electrochemical device for simultaneous DNA amplification and detection. *Lab Chip* 3:100.
66. Seong, G. H., W. Zhan, and R. M. Crooks. 2002. Fabrication of microchambers defined by photopolymerized hydrogels and weirs within microfluidic systems: Application to DNA hybridization. *Anal Chem* 74:3372.
67. Noerholm, M., et al. 2004. Polymer microfluidic chip for online monitoring of microarray hybridizations. *Lab Chip* 4:28.
68. National Human Genome Research Institute National Institutes of Health. *Nhgri Seeks Next Generation of Sequencing Technologies*. 2004 [cited January 1, 2010]; Available from: http://genome.gov/12513210.
69. National Human Genome Research Institute National Institutes of Health. 2010 [cited January 1, 2010]; Available from: http://www.genome.gov/.
70. Woolley, A. T., and R. A. Mathies. 1995. Ultra-high-speed DNA-sequencing using capillary electrophoresis chips. *Anal Chem* 67:3676.
71. Schmalzing, D., et al. 1998. DNA sequencing on microfabricated electrophoretic devices. *Anal Chem* 70:2303.
72. Fredlake, C. P., et al. 2008. Ultrafast DNA sequencing on a microchip by a hybrid separation mechanism that gives 600 bases in 6.5 minutes. *Proc Natl Acad Sci USA* 105:476.
73. Paegel, B. M., et al. 2000. Turn geometry for minimizing band broadening in microfabricated capillary electrophoresis channels. *Anal Chem* 72:3030.
74. Kan, C. W., et al. 2004. DNA sequencing and genotyping in miniaturized electrophoresis systems. *Electrophoresis* 25:3564.
75. Burns, M. A., et al. 1998. An integrated nanoliter DNA analysis device. *Science* 282:484.

76. Liu, R. H., et al. 2004. Self-contained, fully integrated biochip for sample preparation, polymerase chain reaction amplification, and DNA microarray detection. *Anal Chem* 76:1824.
77. Bohringer, K. F. 2003. Surface modification and modulation in microstructures: Controlling protein adsorption, monolayer desorption and micro-self-assembly. *J Micromech Microeng* 13:S1.
78. Sanders, G. H. W., and A. Manz. 2000. Chip-based microsystems for genomic and proteomic analysis. *Trac-Trends Anal Chem* 19:364.
79. Pavlickova, P., E. M. Schneider, and H. Hug. 2004. Advances in recombinant antibody microarrays. *Clin Chim Acta* 343:17.
80. Bange, A., H. B. Halsall, and W. R. Heineman. 2005. Microfluidic immunosensor systems. *Biosens Bioelectron* 20:2488.
81. Henares, T. G., F. Mizutani, and H. Hisamoto. 2008. Current development in microfluidic immunosensing chip. *Anal Chim Acta* 611:17.
82. Hatch, A., et al. 2001. A rapid diffusion immunoassay in a t-sensor. *Nat Biotechnol* 19:461.
83. Choi, J. W., et al. 2002. An integrated microfluidic biochemical detection system for protein analysis with magnetic bead-based sampling capabilities. *Lab Chip* 2:27.
84. Wang, J. 2002. On-chip enzymatic assays. *Electrophoresis* 23:713.
85. Miyazaki, M., and H. Maeda. 2006. Microchannel enzyme reactors and their applications for processing. *Trends Biotechnol* 24:463.
86. Urban, P. L., D. M. Goodall, and N. C. Bruce. 2006. Enzymatic microreactors in chemical analysis and kinetic studies. *Biotechnol Adv* 24:42.
87. Cady, N. C., S. Stelick, and C. A. Batt. 2003. Nucleic acid purification using microfabricated silicon structures. *Biosens Bioelectron* 19:59–66.

8

Clinical Monitoring

8.1 Flow Cytometry

Cytometry is a measurement technique that captures the physical or chemical characteristics of single cells and uses these properties to sort them. *Flow cytometry* forces cells to flow in a single file past a detection region to facilitate the measurement of cell properties. These properties guide the electrical or mechanical sorting of cells into proper collection reservoirs. In fact, many biological and nonbiological entities can be analyzed and sorted using flow cytometry including mammalian cells, viruses, bacteria, particles, chromosomes, lipids, proteins, and ions. In medicine, flow cytometry is an indispensable tool for blood analysis, isolation of stem cells, detection of malignant cells, immunology, and genetic analysis. It is also an important technique in cell and molecular biology and environmental monitoring. A key feature of flow cytometry is that cells are sorted without the loss of their viability.

Early applications of flow cytometry date back to the 1940s, mainly in the area of cell counting. Modern-day conventional flow cytometry systems are capillary based but are still large and bulky. In addition, specialized personnel are required to operate such tools, and the process is still labor-intensive. Naturally, the exquisite fluid- and cell-handling capability of microfluidics was sought to address the limitations of current cytometric technology. Several excellent reviews exist on the topic, including general reviews [1–3], a focused review on single mammalian cells [4], a focused review on blood cells [5], and a review on cytometric analysis [6]. Here, we focus on the biomedical applications of cytometry. Chemical cytometry, which involves obtaining the chemical composition of single cells, is not covered here. Select topics for chemical cytometry such as mass spectrometry and capillary electrophoresis were covered in chapters 6 and 7. The interested reader is also referred to reference [7].

An illustration of a conventional flow cytometry system is shown in Figure 8.1. Cells contained within a sample stream are hydrodynamically focused using a surrounding sheath flow. By adjusting the pressure and the flow rates, the sample stream is squeezed to force the stained cells into a single file line. In a fluorescence-activated cell sorter (FACS), a cell crossing the detection region is exposed to an incident laser beam, which generates

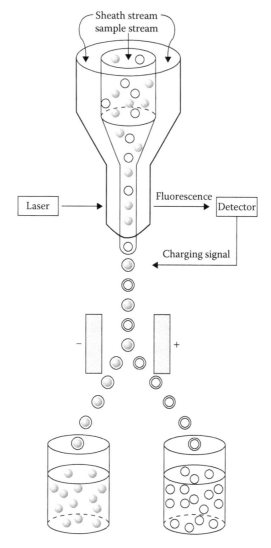

FIGURE 8.1
Conventional flow cytometry apparatus.

both light scattering and fluorescence emission. These signals are detected and processed to obtain relevant physical information on the interrogated cell. Then the analyzed cells exit the focusing system and are encapsulated in a droplet. Droplets are differentially charged and sorted in their respective collectors using electrostatic forces. In conventional systems, up to ~10,000 cells per second are processed. By utilizing microfluidics for hydrodynamic focusing, higher throughputs of ~25,000 cells per second were achieved [6]. In addition to higher throughput, miniaturizing flow cytometry enables on-site analysis and point of care (POC) applications.

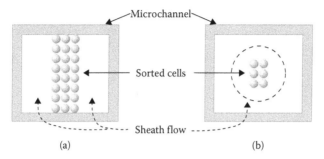

FIGURE 8.2
(a) Simple 2D microfluidic flow cytometry device and flow cross section and (b) 3D flow cytometry flow cross section.

A simple two-dimensional (2D) microfluidic flow cytometry device (Figure 8.2) can accurately position cells in the x-y plane (plane of the device) but may have difficulty with z-axis positioning. Accurate positioning in all three axes is required for detection. This is especially true of optical detection methods that rely on fluorescence and light scattering. Detection of a cell's presence, size, and dielectric properties is accomplished using electrical impedance measurements. This is the same principle applied in the decades-old Coulter counter. Advanced devices featuring three-dimensional (3D) hydrodynamic focusing have been developed [8,9]. In one device, SU-8 was patterned using a tilt-exposure process to obtain a complex flow through device that achieved 3D hydrodynamic focusing of labeled tanned sheep erythrocytes (red blood cells) by coaxial flow sheathing. Another simple method involved the use of casted silicone rubber for transverse and lateral hydrodynamic focusing. Two other device versions were investigated: high throughput and high resolution (Figure 8.3). The high-resolution device differs from standard cytometers in that the focused stream is a thin band (1 μm thick) designed to allow bright field and fluorescent imaging of cells and particles moving in the stream. Multiple channel heights (two- or three-level relief) were built up by using multiple layers of SU-8 on the master mold. Thus, only a single casted sheet of silicone rubber bonded to a glass substrate was required to form the microfluidic devices. Microbeads (0.75, 1.9, 2.5, and 10 μm) and multiple yeast strains (*Saccharomyces cerevisiae*) were processed and optically detected with these cytometers.

Most microfluidic flow cytometry devices still use 2D focusing because of the difficulty of fabricating devices capable of 3D focusing. However, 2D devices have demonstrated the potential of automated sorters that include precise fluid-handling capabilities on-chip. An integrated cell sorter was devised using multilayer soft lithography to include both valving and pumping on-chip [10]. Silicone rubber replicas containing the control lines (top layer) and microfluidic structures (bottom) layer were casted from the photoresist structures on a silicon substrate (Figure 8.4). The detection region was formed at a T-junction where the channels tapered down to 6 μm in width from the 30-μm wide main

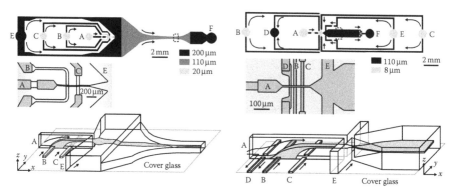

FIGURE 8.3
Left: High-throughput device with a top view illustration, top view photograph, and 3D illustration. Right: High-resolution device with a top view illustration, top view photograph, and 3D illustration. The detection region is indicated with a dashed line box, and gray scale codes the channel depths. In both parts, A and C provide top and bottom focusing, respectively, B is the sample inlet, E provides in-plane focusing, and F is the outlet. D provides auxiliary focusing. (From Simonnet, C., and A. Groisman. 2006. *Anal Chem* 78:5653. With permission.)

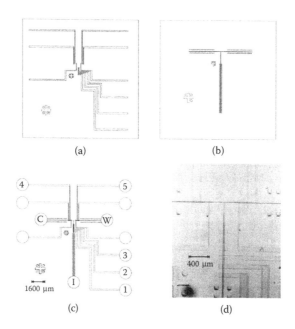

FIGURE 8.4
Integrated cell sorter fabricated from silicone rubber: (a) control layer, (b) microfluidic layer, (c) top view of the cell sorter, and (d) photograph of the cell sorter. The valves labeled 1–3 are sequentially arranged and combined to form a peristaltic pump. Numbers 4 and 5 are valves. I = input, C = collection, and W = waste. (From Fu, A. Y., et al. 2002. *Anal Chem* 74:2451. With permission.)

channel. Cell sorting and trapping was demonstrated on *E. coli*. Although the recovery and cell-processing speed were low compared to conventional systems, performance improvements are expected with optimization.

In addition to hydrodynamic methods, it is also possible to focus cells using magnetic methods such as magnetic activated cell sorting (MACS) or electrokinetic flows. For example, negative dielectrophoresis (DEP) is used for dielectrophoresis activated cell sorting (DACS) in which cells are repelled from higher electric field gradients near electrode edges [11,12]. An interesting application takes advantage of the differences in cell volume observed in mammalian cells in the different phases of their cell cycle (four distinct phases; Figure 8.5) [11]. Newborn cells are the smallest (G_1 phase), whereas cells just prior to mitosis are the largest (G_2 phase). This relationship is also true of other cells, such as bacteria, molds, and algae. The extreme size difference between the two phases was measured in human breast ductal carcinoma cells (MDA-MB-231; from 10 to 20 μm for G_1 and G_2, respectively). About 96% enrichment of these cells was achieved in a simple microfluidic device consisting of a microchannel and electrodes (Ti/Au) on a Pyrex borosilicate glass substrate. The channel walls were photolithographically defined in polyimide, and the open channel was sealed with a glass cap. A buffer and cells were introduced into the main channel, where cells were focused using electrodes arranged at 5° with respect to the flow to focus the cells along the *y*-axis to ensure that they traveled at an identical velocity before entering the separation area. Electrodes oriented at 10° with respect to the flow induced a DEP force on the cells such that the larger cells experienced a greater deflective force than the smaller cells. In the channel, the cells experienced both

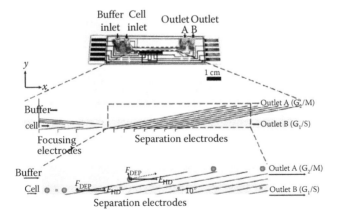

FIGURE 8.5
Overall geometry of the electrodes for focusing and separation. Dielectrophoresis (DEP) acting on cells having different radii results in their spatial separation and sorting into the two outlets in relation to their cell cycle. S = synthesis and M = mitosis. (From Kim, U., et al. 2007. *Proc Natl Acad Sci USA* 104:20708. With permission.)

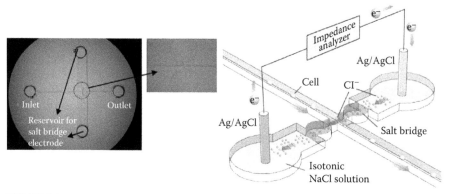

FIGURE 8.6
Left: Top view photograph of the electrical detector for flow cytometry, with a detail of the polyelectrolytic salt bridge electrode reservoirs in the inset. Right: Schematic showing layout of the detector and flow cytometry system. (From Chun, H. G., T. D. Chung, and H. C. Kim. 2005. *Anal Chem* 77:2490. With permission.)

DEP and hydrodynamic drag but the cubic dependence of DEP force and radii induced a spatial separation of the cells based on their size.

Many microfluidic flow cytometers use the process of optical detection. Electrical detection is an attractive alternative as it enables further miniaturization. However, a key challenge is that metal electrodes cannot be used at low frequencies or at direct currents (DC); measurements at these frequencies are desirable as impedance is inversely proportional to frequency (metal electrodes resemble a capacitor). At high frequencies, the electric double layer and Faradaic reactions minimize the impedance changes. A simple solution for this is to replace the metal electrodes with polyelectrolytic salt bridge electrodes (PSBEs) [13]. First, the flow cytometer channels (50 μm wide and 22 μm high) were fabricated from etched glass and thermally bonded to another glass piece to complete the system. The PSBEs were fabricated by introducing a prepolymer solution (65% diallyldimethlammonium chloride with 2% photoinitiator [2-hydroxy-4'-(2-hydroxyethoxy)-2-methylpropiophenone] and 2% cross-linker [N,N'-methylenebisacrylamide]) into the channel and selectively photopolymerizing the solution in ultraviolet light. The remaining prepolymer is then washed away. Silver/silver chloride (Ag/AgCl) electrodes are inserted into the sodium chloride (NaCl)-loaded reservoirs for impedance detection of cells (Figure 8.6). Pairs of PSBEs were used to extract the velocity of the passing particles and successfully classified them into red and white blood cells as a function of their size.

8.2 Microdialysis

The principle of dialysis was introduced earlier in Section 5.3.1.3. *Microdialysis* refers to miniaturized dialysis systems adapted for continuous sampling via mass transfer across a semipermeable membrane. For example, microdialysis for the continuous monitoring of glucose is of increasing importance in the treatment of patients with diabetes. A microdialysis probe placed in the tissue (abdomen or forearm for diabetes) is perfused with an isotonic fluid and acquires glucose from interstitial fluid through the concentration of gradient-driven diffusion. In addition to glucose detection, a great deal of microdialysis research has focused on understanding neurotransmitter release in the brain.

Microfabricated microdialysis systems are either chip or probe based. A chip-based system has been developed for continuous glucose monitoring [14]. A microdialysis sampler and on-chip impedance-based detection were integrated (Figure 8.7). Chromium/gold (Cr/Au) electrodes were patterned on a glass substrate, and then SU-8 channels for containing the perfusate were patterned (15 μm high). Following a short oxygen plasma pretreatment, a commercially available polycarbonate track-etched semipermeable membrane (15-nm pores in a 6-mm-thick membrane) was bonded to cap off the SU-8 channel. The upper silicone rubber channel was casted and bonded in the same manner. The Cr/Au electrodes were able to detect real-time conductivity changes in the dialyzed solution. Microdialysis and monitoring of calibration sample solutions of phosphate buffered saline in varying concentrations were demonstrated. One of the

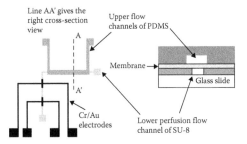

FIGURE 8.7
Top: Illustration showing the top and side view layout of the microdialysis chip. Bottom left: Close-up photograph showing the electrodes and channels. Bottom right: Photograph of the complete device. (From Hsieh, Y. C., and J. D. Zahn. 2007. *Biosens Bioelectron* 22:2422. With permission.)

authors also reported a needle-based microdialysis probe for continuous medical monitoring [15].

A probe-based device for continuous glucose monitoring was also reported by researchers; however, the probe was a conventional microdialysis needle [16]. An alternate approach to glucose detection was implemented. Amperometric glucose detection was performed in a sensor that was placed in line with the dialysis probe consisting of two fluidically connected cavities: one containing a working electrode (platinum [Pt]) in contact with immobilized glucose oxidase in a polyvinyl alcohol membrane and the other an Ag/AgCl reference electrode. Glucose detection in human serum was performed using a sequence of reactions starting with the glucose oxidase (an enzyme) catalyzed oxidation of glucose to hydrogen peroxide

$$glucose + O_2 \rightarrow gluconolactone + H_2O_2 \qquad (8.1)$$

Glucose is indirectly detected by the platinum working electrode, which detects the hydrogen peroxide through a second reaction as follows:

$$H_2O_2 \rightarrow O_2 + 2H^+ + 2e^- \qquad (8.2)$$

Other electrochemical glucose detection methods are reviewed in reference [17].

8.3 Catheter-Based Sensors

Catheters are well known for their use in cardiovascular applications involving local blood pressure measurement or injection of contrast medium in angiography (X-ray imaging of arteries and veins). These long, flexible, hollow tubes measure 0.3–3.0 mm in diameter, depending on the size of the blood vessels. Diagnostic catheters include one or more sensors at the tip for measuring quantities such as pressure, flow velocity, temperature, or oxygen (O_2) saturation. Catheters may also be used for interventions, as in the case of a balloon catheter for opening stenotic (constricted) arteries. To insert a catheter into the cardiovascular system, a solid guide wire is used. Typical catheterization starts with the insertion of the guide wire into a readily accessible vessel (in the femoral; leg, or brachial; arm artery; Figure 8.8). The wire is first manipulated into position and then exchanged with the catheter. Interestingly, catheterization dates back to 1711 (Stephen Hales performed the first catheterization on a horse) and has since spread from cardiovascular to other regions of the body where access is limited and minimally invasive methods are required.

The application of microtechnologies to improve catheters has been the topic of much research. Catheter-based imaging systems, actuators to steer

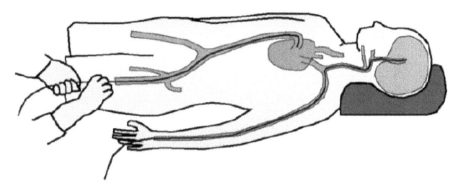

FIGURE 8.8
Typical cardiac catheterization through the femoral or brachial arteries. (From Haga, Y., and M. Esashi. 2004. *Proc IEEE* 92:98. With permission.)

catheters, and miniaturized catheter-based sensors are reviewed in references [18,19]. By miniaturizing ultrasonography, small catheter tools are possible for image-guided navigation of the catheterization process. This is typically done through an angiogram, which subjects the patient to X-ray exposure. Because repeated X-ray exposure poses a health risk, alternative navigation systems to avoid radiation doses have been developed either using ultrasonography or by integrating piezoelectric elements or electromagnetic positioning by integrating a magnetic sensor. In one study, magnetic sensors (Hall sensors or magnetoresistors) intended for integration at the tip of a guide wire were used with an externally generated pulsed magnetic field to monitor the tip position. By superimposing a previously obtained X-ray image, proper guidance of the wire to the desired location is possible [20]. The same group of researchers reported a multisensor chip suitable for integration into a catheter, featuring blood pressure, blood flow velocity, and oxygen saturation sensors (Figure 8.9). Piezoresistors on the edge of a membrane over a sealed cavity and connected together in a Wheatstone bridge could report blood pressure. The thermal flow sensor was configured in a calorimetric arrangement with two thermopiles placed around a central heater at ~2°C above ambient. The flow induced a temperature difference between the thermopiles, which could be correlated to the flow rate. Oxygen saturation was determined by photodiodes that compared light absorption between deoxygenated hemoglobin and oxygenated hemoglobin at two wavelengths (660 and 800 nm) [20,21].

Catheters can also be steered actively by integrating actuators at the tip to affect bending; multiple actuators offer more degrees of freedom. Flat, small form factor, shape memory alloy (SMA) actuators have been fabricated by electrochemical pulse etching of SMA sheets [22]. Serpentine actuators were etched from sheets (38 μm thick and ~30 μm wide wires) using a double-sided etching process, elongated, and attached to catheters (Figure 8.10). When activated, each serpentine actuator shrank causing the catheter to

326 Biomedical Microsystems

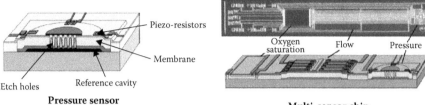

Pressure sensor — Piezo-resistors, Membrane, Etch holes, Reference cavity

Multi-sensor chip — Oxygen saturation, Flow, Pressure

FIGURE 8.9
Illustrations showing a detail of the pressure sensor (left) and the complete multi-sensor chip for catheter-based measurements (right). (From Tanase, D., et al. 2002. *Sens Actuators A Phys* 97–8:116. With permission.)

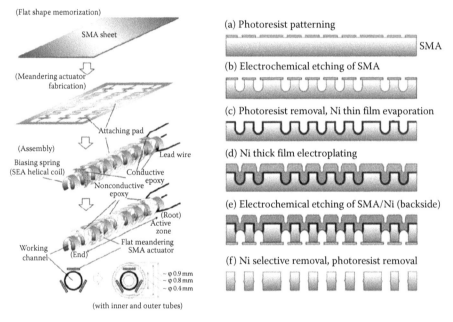

FIGURE 8.10
Left: Shape memory alloy (SMA)–actuated catheter concept. SEA = super elastic alloy. Right: Fabrication process for individual SMA actuators. (From Mineta, T., et al. 2001. *Sens Actuators A Phys* 88:112. With permission.)

bend (Figure 8.11). Active catheters with three SMA actuators (51% nickel–titanium super elastic alloy) measured 0.8 mm in diameter with an active length of 12.4 mm. A maximum bending angle of 50° and curvature of 8 mm with a 0.5-second response time were obtained.

Both capacitive and piezoresistive pressure sensors for catheters were described in Section 6.3.5 and in this chapter, respectively. However, most catheter-based pressure sensors use optical sensing methods, which allow further

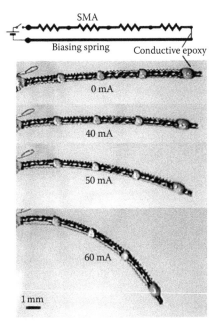

FIGURE 8.11
Current-controlled bending of the shape memory alloy (SMA) catheter. (From Mineta, T., et al. 2001. *Sens Actuators A Phys* 88:112. With permission.)

reduction of the form factor. Optical fibers form a low-loss optical conduction path for light, and can be instrumented with pressure sensitive cavities at fiber tips. In one study, a structured silicon dioxide membrane with an aluminum-coated central mesa was placed at the end of an optical fiber (125 µm) to form a Fabry–Perot interferometer [23]. The Fabry–Perot cavity was formed between two mirrors separated by a 2 µm polyimide spacer. The first mirror was the aluminum-coated mesa and the second was a half-mirror formed by coating 10 nm of chromium on the tip of the fiber (Figure 8.12). The flexible oxide diaphragm responded to local pressure variations resulting in a detectable change in the cavity length. A pressure sensitivity of −0.25 nm/mmHg and 4-mmHg resolution over the range −100 to 400 mmHg was obtained. The sensor obtained pressure measurements at multiple locations in goats.

In addition to blood flow and pressure, intravascular shear stress has also been measured using microfabricated sensors.

Hemodynamic forces, including shear stress, are thought to play a role in coronary artery disease, though they must be inferred from imaging data with low spatiotemporal resolution. In one study, metallic sensors were sandwiched in a protective Parylene film, and the whole sensor was freed from the supporting silicon substrate to allow mounting to a

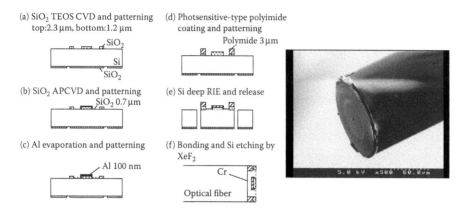

FIGURE 8.12
Left: Fabrication process for the fiber-optic pressure sensor. Right: Scanning electron microscope image of the sensor mounted on an optical fiber. CVD = chemical vapor deposition, APCVD = atmospheric CVD, and TEOS = tetraethyl orthosilicate. (From Totsu, K., Y. Haga, and M. Esashi. 2005. *J Micromech Microeng* 15:71. With permission.)

catheter [24]. Shear stress was indirectly determined from heat transfer between the heated thermal sensing element and the surrounding blood flow. Spatiotemporal variations in shear stress were obtained by catheterizing rabbit aortas.

In earlier examples, sensors were fabricated and then added to a catheter. This approach has difficulties during the packaging, assembly, and wiring of microdevices in the catheter format. In addition, the considerable bulk added by these methods may obstruct the catheter lumen. Recently, a new integrated approach was developed in which sensors were fabricated directly on a flexible strip that was later rolled into a hollow catheter tip [25]. The thin Kapton film (25 µm) supported multiple sensors, including temperature (resistance temperature detector), flow (hot film anemometer), and glucose (amperometric). This is shown in Figure 8.13. First, the Kapton (polyimide) film was supported on silicone rubber-coated silicon substrate to allow for planar fabrication of the sensing elements. Sensor electrodes and active elements were fabricated from titanium/gold patterns. For the glucose sensor, a reference electrode was formed on top of an existing electrode by plating with silver followed by reverse electroplating in potassium chloride (KCl) to obtain an Ag/AgCl surface. A glucose oxidase was immobilized in a polyacrylamide matrix on the electrochemical electrodes and was further secured under a screen-printed semipermeable membrane (2% w/v epoxy-polyurethane solution). The temperature and flow sensor elements were protected from blood using a Parylene coating. To form the catheter, a supporting strip of Kapton was rolled into a tube, silicone adhesive was applied, and then a second strip of Kapton containing the sensors was rolled over the first strip in the opposite direction such that the strips crossed for added strength (Figure 8.14).

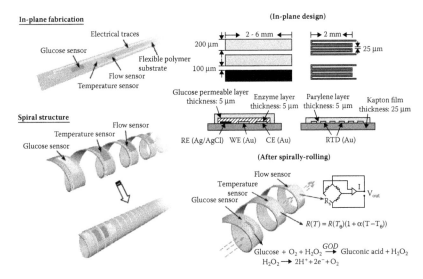

FIGURE 8.13
Left: Concept of the spiral catheter produced by winding a flexible sensor strip. Right: Illustration of the design of each sensor. (From Li, C. Y., et al. 2008. *Biomed Microdevices* 10:671. With permission.)

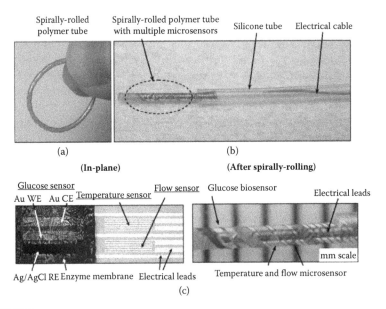

FIGURE 8.14
Photographs of the fabricated catheters: (a) rolled and curved Kapton tube, (b) assembled catheter that is 650 μm in outer diameter, (c) close-up in-plane and catheter images of the glucose, temperature, and flow sensors. (From Li, C. Y., et al. 2008. *Biomed Microdevices* 10:671. With permission.)

8.4 Endoscopy

Endoscopy is a minimally invasive procedure that allows a clinician to view the interior of the body through a flexible tube-based device. Its applications include photography, taking biopsies, retrieving foreign objects, and surgery. Normal clinical targets of endoscopy include the gastrointestinal tract, abdominal cavity, and respiratory tract although other organs or interior regions are also accessible with endoscopic tools. Endoscopes, like catheters, were first introduced hundreds of years ago. Philip Bozzini, a clinician, introduced the first endoscope in 1806; however, the invention was deemed inappropriate at that time and cost the young inventor his right to practice medicine. Although endoscopes are still large, they are now an established tool in modern medicine. Efforts to miniaturize imaging modules and active endoscope guidance actuators are reviewed in reference [18].

Imaging is usually performed in conventional endoscopes with fiber optic elements. Microfabricated approaches include ultrasound probes or optical scanning microscopy. Optical coherence tomography (OCT) is an optical scanning technique that acquires images using optical scattering in tissue with micrometer resolution. A thermally actuated micromirror for laser beam scanning was successfully demonstrated *in vivo* with a 5-mm-diameter endoscopic OCT system on a porcine urinary bladder [26]. The silicon micromirror (10–100 µm thick) was fabricated following a complementary metal oxide semiconductor (CMOS) process using only dry etching steps, and was performed without any photomasks (Figures 8.15 and 8.16). Actuation was accomplished using a thermal bimorph structure contained in the mesh hinge, an aluminum/silicon dioxide (Al/SiO_2) structure with polysilicon embedded in the oxide (1.8 µm thick). The as-fabricated mirrors exhibited a 17° tilt due to residual stress.

A new development in endoscopy eliminates the cable containing the power, video, and fiber optic lines. Small capsules containing a camera and wireless telemetry are swallowed, and endoscopy is performed without patient sedation and with minimal patient discomfort. The wide cables associated with standard fiber optic endoscopy not only result in discomfort, but also have difficulty reaching some gastrointestinal tract regions such as the largest part of the small bowel. Wireless endoscopic capsules were first introduced in the 1950s, and sensors were included later for the measurement of temperature, pressure, and pH [27]. These devices are reviewed in references [28–30]. Since then, there have been few contributions to the development of these capsular endoscopes, though microfabricated sensors may add a much-needed capability to these nascent devices. However, advances in microfabrication technology were in large part an enabler to realize these small capsular endoscopes.

The endoscopic capsule was first introduced in 2000 by Given Imaging Ltd. (Yoqneam, Israel). The pill measured 26 mm long, 11 mm in diameter,

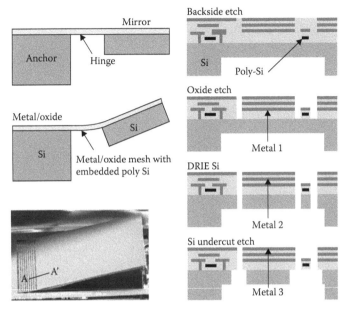

FIGURE 8.15
Left: Side view illustrations of the micromirror design and scanning electron microscope image of fabricated micromirror. Right: Fabrication process for the micromirror scanner. (From Xie, H. K., Y. T. Panc, and G. K. Fedder. 2003. *Sens Actuators A Phys* 103:237. With permission.)

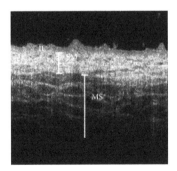

FIGURE 8.16
OCT image of the porcine bladder acquired with a CMOS-MEMS micromirror (U = urothelium, SM = submucosa, MS = muscularis layer). (From Xie, H. K., Y. T. Panc, and G. K. Fedder. 2003. *Sens Actuators A Phys* 103:237. With permission.)

and weighed 3.7 g. The capsules were swallowed, and they undertook an ~8-hour journey propelled naturally by peristalsis through the gastrointestinal tract. A wide field of view was provided by a short focal length lens, which focused images onto a CMOS image sensor (Figure 8.17). About 55,000 pictures were obtained while the capsule traveled through the body,

Nature Reviews | Drug Discovery

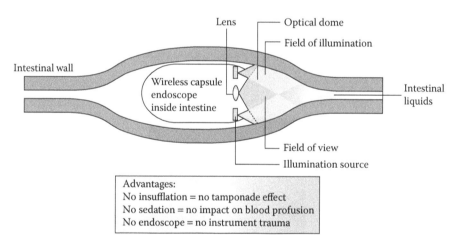

Nature Reviews | Drug Discovery

FIGURE 8.17
Top: M2A endoscopic capsule that includes (a) an optical dome, (b) a lens holder, (c) short-focal-length lens, (d) white light-emitting diodes, (e) CMOS chip camera, (f) a pair of silver oxide batteries, (g) a wireless transmitter, and (h) an antenna. Bottom: Concept of wireless endoscopy. (From Qureshi, W. A. 2004. *Nat Rev Drug Discov* 3:447. With permission.)

and were transmitted to a computing workstation. Hundreds of thousands of patients have benefited from these devices worldwide, but further improvements are necessary. The retention time of these devices is limited by peristaltic transport. Instead, active locomotion is planned in advanced devices.

8.5 Point of Care

POC is an emerging field that seeks to bring clinical diagnostics closer to the patient. While POC systems do exist today, most of the clinical tests in developed countries are still performed by skilled technicians using equipment available only in large testing laboratories. Thus, biological samples must be acquired and then transported to these facilities for processing. The transport of samples causes delay, and in addition, many tests still require labor-intensive processes (e.g., pipetting, incubating, temperature cycling, mixing) leading to significant delays between the initial clinic visit and the follow-up to review the laboratory results. In developing countries or in resource-poor settings, this approach is not practical for a number of reasons, including the inaccessibility of diagnostic techniques and trained personnel [32]. In fact, health diagnostics are severely lacking in developing nations and this particular area is often heralded as a large potential application area for microfluidics-based POC systems. Developing nations still struggle with infectious diseases such as human immunodeficiency virus/acquired immune deficiency syndrome (HIV/AIDS), malaria, and tuberculosis (TB). Early detection of diseases with POC systems may limit their spread and provide the ability to track the health status of populations in these areas.

The goal of POC systems is to measure health status, usually by assessing biomarkers. These may include proteins, hormones, and other biomolecules that are tracked through their levels, states, or activities that change during the course of a disease or medical treatment. Information obtained from these biomarkers may have diagnostic or prognostic implications and, whenever possible, should be chosen due to its link with the etiology (cause of the disease or abnormal condition) or pathobiology (anatomical and functional manifestation of the disease) of a particular disease [33]. For example, an increase in prostate-specific antigens points to an increased risk of prostate cancer. Ectopic pregnancy (occurring outside of the womb) is marked by a steep increase in the levels of beta-human chorionic gonadotropin (hCG) hormone [34].

POC is not a new technology. For example, the blood glucose meter home test is a well-established and successful example of a POC system. Many existing home-use POCs are based on lateral flow immunoassays in a "dipstick" format (e.g., the pregnancy, cardiovascular disease, and HIV-1 tests that are currently available). Other common POC immunoassays are in flow through or solid-phase format. Dipstick devices or lateral flow immunoassay devices (Figure 8.18) require no external pumping to drive the assay. Capillary flow drives the sample in contact with one end of the immunochromatographic strip across it. Interaction with a labeled immobilized antibody downstream produces a visible color change correlated to the result of the assay. While relatively inexpensive, the traditional dipstick POC provides only qualitative data with a "yes" or "no" type of output (Figure 8.19).

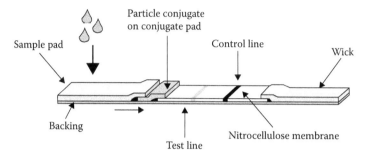

FIGURE 8.18
Example of an immunochromatographic strip test. (From Wong, R., and H. Tse, eds. 2008. *Lateral Flow Immunoassay*. 1st ed. New York: Springer. With permission.)

The lack of quantitative results and often the lack of sensitivity are the major drawbacks of current POC technology.

Given the limitations of current POC technologies, a great deal of interest exists in the application of microtechnologies toward advances in POC devices. Given the small format, expected niches for single-use, disposable, and inexpensive microPOC systems include health care in developing countries, home testing in developed countries, and diagnostics for first responder emergency situations [36]. Depending on the setting in which POC systems are used, they may use different technology, as shown in Figure 8.20. Regardless of the final format, POC systems must be extremely easy to use in low-resource settings [37]. Despite significant efforts both academically and commercially, to date, microPOCs have had limited success. Many technological barriers remain, including the difficulty of use, fouling of surfaces by nonspecific binding of biomolecules, and cost. Even so, the current POC market is estimated to be more than $10 billion [37] and will likely benefit from future advances in microPOCs. Many reviews dedicated to microPOCs exist, including those of a general nature [33,34] and single contributions focused on areas such as the use of microPOCs in developing countries [32], supporting microfluidics technology for POCs [36], optical detection miniaturization for POCs [38], cancer applications [39], and pathogen detection [40].

Micro POCs acquire measurements primarily from readily accessible biological samples, including whole blood [41], serum, plasma [41], saliva [42,43], urine [44,45], feces, sperm, tears, and sweat. Although noninvasive fluids (such as urine and saliva) are attractive for POC, these may include less protein than whole blood. Recently, a new format was introduced for microfluidic POCs using paper as a substrate instead of traditional materials such as silicon or glass [46]. Paper, which is primarily constructed from cellulose fibers, is also porous, enabling interesting possibilities for POC applications. A primary advantage of paper as an alternative substrate is the cost reduction in manufacturing gained by avoiding cleanroom fabrication processes, which are quite expensive. Other advantages are that paper is an abundant and sustainable resource and is disposable and easy to modify.

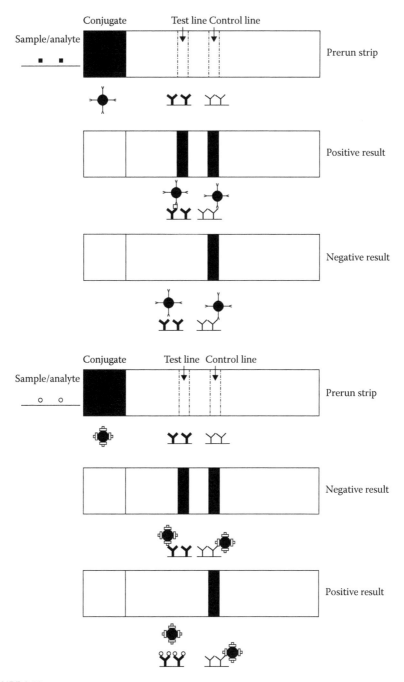

FIGURE 8.19
Two types of lateral flow immunoassays: (top) direct solid-phase immunoassay and (bottom) competitive solid-phase immunoassay. (From Wong, R., and H. Tse, eds. 2008. *Lateral Flow Immunoassay*. 1st ed. New York: Springer. With permission.)

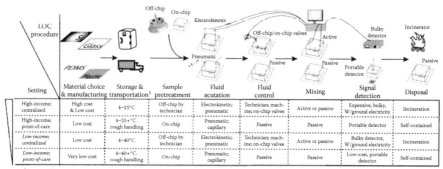

Setting	Material choice & manufacturing	Storage & transportation[1]	Sample pretreatment	Fluid acutation	Fluid control	Mixing	Signal detection	Disposal
High-income; centralized	High cost & Low cost	4–25°C	Off-chip by technician	Electrokinetic; pneumatic	Technician; mach-ine; on-chip valves	Active or passive	Expensive, bulky, W/ground electricity	Incineration
High-income; point-of-care	Low cost	4–25+°C, rough handling	On-chip	Pneumatic; capillary	Passive	Passive	Portable detector	Self-contained
Low-income; centralized	Low cost	4–40°C	Off-chip by technician	Electrokinetic; pneumatic	Technician; mach-ine; on-chip valves	Active or passive	Bulky detector, W/ground electricity	Incineration
Low-income; point-of-care	Very low cost	4–40+°C, rough handling	On-chip	Pneumatic; capillary	Passive	Passive	Low-cost, portable detector	Self-contained

[1]Our numbers are rough estimates for most transportation and storage conditions, excluding extremely hot and cold environments.

FIGURE 8.20
Some examples of lab-on-a-chip that may be used in different settings. (From Chin, C. D., V. Linder, and S. K. Sia. 2007. *Lab Chip* 7:41. With permission.)

In paper-based microfluidics, the paper itself becomes the channel. The channel walls are defined using photosensitive polymers that are patterned using photolithography. Fluid is further confined to distinct channels in the paper by taking advantage of its differential wetting properties. For example, hydrophilic chromatography paper was used as a substrate and channels were defined using hydrophobic photoresist impregnated on the paper [44]. The process by which this was done is illustrated in Figure 8.21. In this example, a chromatography paper disc was submerged into SU-8 2010 photoresist, then removed and spun (as in standard photolithography). From there, the process resembled standard photolithography. After development, the whole paper device was exposed to oxygen plasma to increase the hydrophilicity of the channels.

With this approach, millimeter-scale channels were produced and used for multiplex detection of biomarkers. Both glucose and protein were detected from 5-µL urine samples using colorimetric indication. Reagents were spotted and dried at the channel ends. For the glucose assay, potassium iodide and horseradish peroxidase/glucose oxidase were used. This chemical combination is clear initially and turns brown in the presence of glucose due to the oxidation of iodide to iodine. For the protein assay, citrate buffer and tetrabromophenol blue (TBPB) marked protein binding with a yellow-to-blue color change resulting from the ionization of TBPB were used. The results obtained with the paper-based device were comparable to those obtained with conventional dipsticks.

Recently, the ability to extend paper microfluidics technology to 3D was reported [45]. 2D paper microfluidics were made as described previously (with patterned ITW Technicloth TX 609 paper) and then bonded together using double-sided adhesive tape (carpet tape). Again, SU-8 2010 was used as the photoresist. The tape was cut into precise patterns by laser machining. The thickness of the tape created a gap between the tape and paper, which had to be filled with a cellulose powder/water mixture to allow controllable

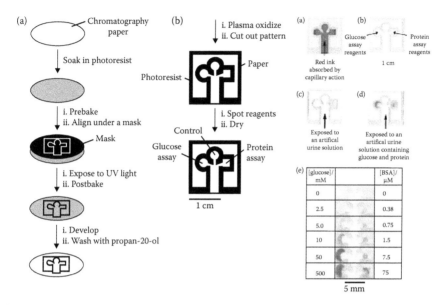

FIGURE 8.21

Left: Paper-based microfluidic patterning method showing (a) the photolithography process, which is followed by (b) the preparation of the multiplex bioassay. Right: (a) Dark (red) ink absorbed in the lithographically defined paper channel (gray regions contain photoresist and white regions do not), (b) paper-based assay with reagents spotted in left and right channel ends (circle for glucose and square for protein), (c) negative control experiment results, (d) positive results for a test solution of 550 mM glucose and 75 µM bovine serum albumin (BSA), and (e) results for varying concentrations of glucose and BSA. (From Martinez, A. W., et al. 2007. *Angew Chem Int Ed Engl* 46:1318. With permission.)

fluid flow in both horizontal and vertical directions (Figure 8.22). Multiple parallel glucose and protein assays using artificial urine were demonstrated (Figure 8.23). The projected cost for manufacturing such 3D paper micro-fluidic devices was only $0.003/cm². Although this exciting developing technology points to interesting possibilities in POC for developing nations, further fundamental research into fluid flow in paper, proper immobilization of sensing biomolecules, and their long-term stability is needed.

While blood contains a rich set of POC information, invasive procedures are required to obtain it and its complex composition often necessitates multiple complex enrichment processes to detect the target. One microfluidic POC device starts with whole blood samples and produces a "barcode"-style protein profile in less than 10 minutes [41]. Several parallel detection channels contained repeated "barcode" patterns of immobilized and labeled ssDNA oligomer–antibody conjugates chosen to target multiple proteins that were formed on a silicone rubber-glass chip format (Figure 8.24). These channels sampled plasma from a drop of whole blood obtained by a finger prick. To do this, the channels were carefully designed to take advantage of the Zweifach–Fung effect. A wide main channel was used to feed narrower

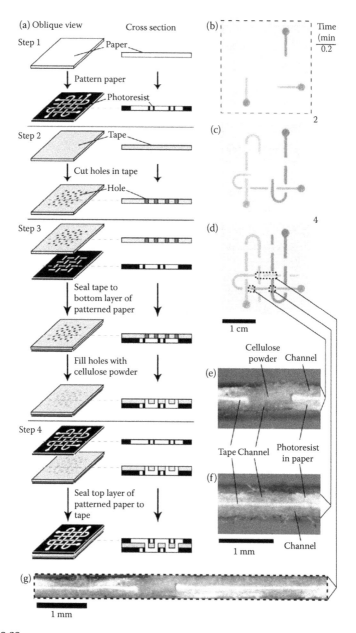

FIGURE 8.22
(a) Fabrication process for a four-channel paper-based microfluidic device. The channels (800 µm wide and ~5 cm long) are arranged in a basket-weave pattern. (b–d) Successive images taken over 4 minutes after adding dye (four colors: red, green, blue, yellow taken clockwise from the upper right corner) showing the progress of the dye front in each channel. (e–g) Cross-sectional images taken at the indicated locations showing the tape, channel, and photoresist-impregnated paper. (From Martinez, A. W., S. T. Phillips, and G. M. Whitesides. 2008. *Proc Natl Acad Sci USA* 105:19606. With permission.)

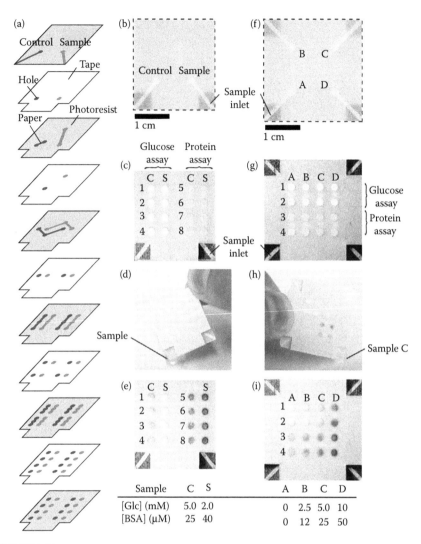

FIGURE 8.23
3D paper-based microfluidic device for parallel glucose and protein assays. (a) Detail of the layers used in a dual-assay device. Photograph of (b) front and (c) back of a dual-assay device. (d) Device dipped into artificial urine with 2 mM glucose and 40 μM BSA. (e) Results of the artificial urine assay with the control side-by-side. Photograph of the (f) front and (g) back of a four-assay device. (h) Four corners of the device dipped into different artificial urine samples. (i) Results of the assay. (From Martinez, A. W., et al. 2007. *Angew Chem Int Ed Engl* 46:1318. With permission.)

centimeter-long detection channels oriented perpendicularly. This resulted in a fluidic streamline near the detection channel entrances along the main channel sidewall. Since the streamline resides close to the side wall, blood cells and other larger species were rejected. Up to 15% of the plasma was pulled into the detection channels in this manner. Proteins in the plasma

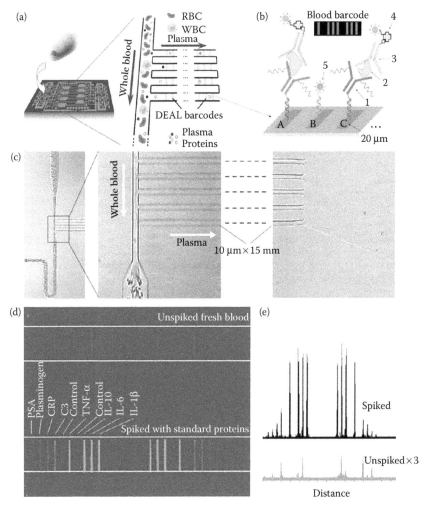

FIGURE 8.24
(a) Blood barcode chip concept and (b) illustration of the DNA barcodes printed in each plasma channel where A–C represent distinct oligomers. The numeric labels are (1) DNA-antibody conjugate, (2) plasma protein, (3) biotin-labeled capture antibody, (4) fluorescence probe, and (5) complementary reference probe. The inset shows a fluorescence protein barcode readout. Whole blood biomarker barcode chip images and results: (c) photograph of plasma separation from whole blood, (d) fluorescence images of barcode readouts in spiked and unspiked blood, and (e) corresponding fluorescence line scans of the barcodes. (From Fan, R., et al. 2008. *Nat Biotechnol* 26:1373. With permission.)

bound with the immobilized ssDNA–antibody conjugates (in 20 μm wide and 40 μm pitch barcode-like stripes) to produce a detectable fluorescent signal. Multiple detection channels allowed 8–12 multiprotein assays to be performed in parallel or sequentially.

8.6 Problems

1. Design your own microfluidic flow cytometry device. Draw the layout (top view) of your design, including all the major dimensions. Draw and briefly describe the fabrication and assembly process used to build the device. Finally, include a brief description of the intended use of your device, including any special features to optimize the performance.

2. Find a recent article on a microfabricated microdialysis device and answer the following questions: What is the intended application? What material was used for the construction of the semipermeable membrane? How was it integrated into the device? What is the target analyte (there may be more than one)? Provide the reference in IEEE format.

3. Search research literature and identify new microPOC technologies suitable for use in developing countries. What specific innovations were incorporated to address clinical monitoring needs in developing countries? Given the low-infrastructural environment, is this technology practical in this setting? Provide a critical review of at least one article and provide the reference in IEEE format. In the review, include the disease of interest, biomarker target(s), and requirements of the specific test.

4. Interview a medical professional or student who is clinically focused or research focused. Determine what current challenges they face in clinical monitoring and discuss what is required to improve the current state of the art. Is there a way in which a biomedical microdevice can be used to improve clinical monitoring applications?

References

1. Yi, C. Q., et al. 2006. Microfluidics technology for manipulation and analysis of biological cells. *Anal Chim Acta* 560:1.
2. Chung, T. D., and H. C. Kim. 2007. Recent advances in miniaturized microfluidic flow cytometry for clinical use. *Electrophoresis* 28:4511.
3. Ateya, D. A., et al. 2008. The good, the bad, and the tiny: A review of microflow cytometry. *Anal Bioanal Chem* 391:1485.
4. Sims, C. E., and N. L. Allbritton. 2007. Analysis of single mammalian cells on-chip. *Lab Chip* 7:423.
5. Toner, M., and D. Irimia. 2005. Blood-on-a-chip. *Annu Rev Biomed Eng* 7:77.

6. Huh, D., et al. 2005. Microfluidics for flow cytometric analysis of cells and particles. *Physiol Meas* 26:R73.
7. Dovichi, N. J., and S. Hu. 2003. Chemical cytometry. *Curr Opin Chem Biol* 7:603.
8. Yang, R., D. L. Feeback, and W. J. Wang. 2005. Microfabrication and test of a three-dimensional polymer hydro-focusing unit for flow cytometry applications. *Sens Actuators A Phys* 118:259.
9. Simonnet, C., and A. Groisman. 2006. High-throughput and high-resolution flow cytometry in molded microfluidic devices. *Anal Chem* 78:5653.
10. Fu, A. Y., et al. 2002. An integrated microfabricated cell sorter. *Anal Chem* 74:2451.
11. Kim, U., et al. 2007. Selection of mammalian cells based on their cell-cycle phase using dielectrophoresis. *Proc Natl Acad Sci U S A* 104:20708.
12. Kim, U., et al. 2008. Multitarget dielectrophoresis activated cell sorter. *Anal Chem* 80:8656.
13. Chun, H. G., T. D. Chung, and H. C. Kim. 2005. Cytometry and velocimetry on a microfluidic chip using polyelectrolytic salt bridges. *Anal Chem* 77:2490.
14. Hsieh, Y. C., and J. D. Zahn. 2007. On-chip microdialysis system with flow-through sensing components. *Biosens Bioelectron* 22:2422.
15. Zahn, J. D., D. Trebotich, and D. Liepmann. 2005. Microdialysis microneedles for continuous medical monitoring. *Biomed Microdevices* 7:59.
16. Steinkuhl, R., et al. 1996. Microdialysis system for continuous glucose monitoring. *Sens Actuators B Chem* 33:19.
17. Heller, A., and B. Feldman. 2008. Electrochemical glucose sensors and their applications in diabetes management. *Chem Rev* 108:2482.
18. Haga, Y., and M. Esashi. 2004. Biomedical microsystems for minimally invasive diagnosis and treatment. *Proc IEEE* 92:98.
19. Haga, Y., et al. 2006. Minimally invasive diagnostics and treatment using micro/nano machining. *Minim Invasive Ther Allied Technol* 15:218.
20. Tanase, D., et al. 2002. Multi-parameter sensor system with intravascular navigation for catheter/guide wire application. *Sens Actuators A Phys* 97–8:116.
21. Goosen, J. F. L., P. J. French, and P. M. Sarro. 2000. Pressure, flow and oxygen saturation sensors on one chip for use in catheters. In *Thirteenth Annual International Conference on Micro Electro Mechanical Systems*, 537. IEEE Piscataway, NJ.
22. Mineta, T., et al. 2001. Batch fabricated flat meandering shape memory alloy actuator for active catheter. *Sens Actuators A Phys* 88:112.
23. Totsu, K., Y. Haga, and M. Esashi. 2005. Ultra-miniature fiber-optic pressure sensor using white light interferometry. *J Micromech Microeng* 15:71.
24. Yu, H. Y., et al. 2008. Flexible polymer sensors for in vivo intravascular shear stress analysis. *J Microelectromech Syst* 17:1178.
25. Li, C. Y., et al. 2008. A flexible polymer tube lab-chip integrated with microsensors for smart microcatheter. *Biomed Microdevices* 10:671.
26. Xie, H. K., Y. T. Panc, and G. K. Fedder. 2003. Endoscopic optical coherence tomographic imaging with a CMOS-MEMS micromirror. *Sens Actuators A Phys* 103:237.
27. Iddan, G., et al. 2000. Wireless capsule endoscopy. *Nature* 405:417.
28. Moglia, A., et al. 2007. Wireless capsule endoscopy: From diagnostic devices to multipurpose robotic systems. *Biomed Microdevices* 9:235.

29. Schurr, M. O., et al. 2007. Microtechnologies in medicine: An overview. *Minim Invasive Ther Allied Technol* 16:76.
30. Moglia, A., et al. 2009. Capsule endoscopy: Progress update and challenges ahead. *Nat Rev Gastroenterol Hepatol* 6:353.
31. Qureshi, W. A. 2004. Current and future applications of the capsule camera. *Nat Rev Drug Discov* 3:447.
32. Chin, C. D., V. Linder, and S. K. Sia. 2007. Lab-on-a-chip devices for global health: Past studies and future opportunities. *Lab Chip* 7:41.
33. Sorger, P. K. 2008. Microfluidics closes in on point-of-care assays. *Nat Biotechnol* 26:1345.
34. Linder, V. 2007. Microfluidics at the crossroad with point-of-care diagnostics. *Analyst* 132:1186.
35. Wong, R., and H. Tse, eds. 2008. *Lateral Flow Immunoassay.* 1st ed. New York: Springer.
36. Weigl, B., et al. 2008. Towards non- and minimally instrumented, microfluidics-based diagnostic devices. *Lab Chip* 8:1999.
37. Sia, S. K., and L. J. Kricka. 2008. Microfluidics and point-of-care testing. *Lab Chip* 8:1982.
38. Myers, F. B., and L. P. Lee. 2008. Innovations in optical microfluidic technologies for point-of-care diagnostics. *Lab Chip* 8:2015.
39. Soper, S. A., et al. 2006. Point-of-care biosensor systems for cancer diagnostics/prognostics. *Biosens Bioelectron* 21:1932.
40. Mairhofer, J., K. Roppert, and P. Ertl. 2009. Microfluidic systems for pathogen sensing: A review. *Sensors* 9:4804.
41. Fan, R., et al. 2008. Integrated barcode chips for rapid, multiplexed analysis of proteins in microliter quantities of blood. *Nat Biotechnol* 26:1373.
42. Herr, A. E., et al. 2007. Integrated microfluidic platform for oral diagnostics. *Oral-Based Diagnostics* 1098:362.
43. Herr, A. E., et al. 2007. Microfluidic immunoassays as rapid saliva-based clinical diagnostics. *Proc Natl Acad Sci U S A* 104:5268.
44. Martinez, A. W., et al. 2007. Patterned paper as a platform for inexpensive, low-volume, portable bioassays. *Angew Chem Int Ed Engl* 46:1318.
45. Martinez, A. W., S. T. Phillips, and G. M. Whitesides. 2008. Three-dimensional microfluidic devices fabricated in layered paper and tape. *Proc Natl Acad Sci U S A* 105:19606.
46. Zhao, W. A., and A. van den Berg. 2008. Lab on paper. *Lab Chip* 8:1988.

9

MEMS Implants and Bioelectric Interfaces

9.1 Implantable MEMS

The small size and integration of functions featured in microelectromechanical systems (MEMS) provide new opportunities to create novel therapeutic and diagnostic devices for implant in the human body and to improve the existing ones. Much of the focus thus far has been on cardiovascular sensors, drug delivery devices, and microelectrodes (for use with the nervous system) that feature one or more MEMS components. Their extended contact with the corrosive environment within the body poses challenges to packaging technology, requiring that devices be housed or otherwise protected with ceramic, glass, or polymer cases or coatings. Reliability and safety are paramount and both contribute to the long road between the translation of the technology from conception to patients' use. Implantable bioMEMS are the subject of several reviews [1–4]. In this chapter, we explore key developments in implantable bioMEMS devices.

9.2 Microelectrodes and Neural Probes

A longstanding application of microfabrication technology has been to create electrical interfaces with which to communicate with neurons and other electrogenic (electrically active) cells. In neuroscience, electrical signals carry information between neurons and give rise to communication between them. Neurons consist of a cell body from which small processes (neurites) extend that are responsible for transmitting (axons) and receiving (dendrites) information from neighboring neurons (Figure 9.1). The cell bodies are ~10–50 µm in diameter and are packed at a density of 10^6 cells/mm^3 (although not all of these are neurons). Axons measure ~1–25 µm in diameter and ~0.001–1 m in length. However, neighboring cells are not directly connected to one another. A small gap called a "synapse" (~20 nm) exists between cell processes. Communication across the synapse occurs when a cell membrane (normally

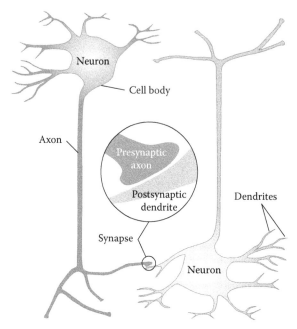

FIGURE 9.1
Basic structure of a neuron.

at a resting potential of ~−70 mV in humans) depolarizes as a result of an action potential and triggers the release of vesicle-bound neurotransmitters from the axon. These diffuse across the synapse and bind on a neighboring dendrite, where a chemical neurotransmitter signal is converted back into an electrical one (carried by ions and not by electrons as in a normal electrical circuit). This sequence of electrical and chemical events is in part due to the resting potential present in neurons; the selective permeability of the cell membrane to certain ionic species produces a measurable electrochemical gradient.

Neuroscientists have long had their own established tools for probing and manipulating the electrical communication between neurons. The earliest tools were simple wires. Later, glass micropipette electrodes enabled single-cell techniques known as the voltage-clamp (invented by Kenneth Cole in the 1940s) [5] and the patch-clamp (invented by Erwin Neher and Bert Sakmann in the 1970s) [6,7], which are still the gold standards of electrophysiological recordings in neuroscience. Glass pipettes for interfacing with neurons were formed by pulling on a heated glass capillary to form very fine hollow needle tips (~1 μm or higher in diameter), filling them with salt solution resembling the extracellular fluid, and adding a metal wire in contact with the solution for connecting external electronics. In the voltage-clamp technique, the pipette makes contact with intracellular contents for the study of the cell membrane potential. The patch-clamp technique is used for studying ion channels and requires a tight gigaohm ($G\Omega$) seal with the cell membrane for a proper ion channel interface. These tools can also be used for electrochemical recording. However, only a few cells can be accessed at a time due to the large size of the overall electrode. A highly skilled operator must carefully establish interfaces with single cells using precision micromanipulator setups. Each electrode is difficult to make and is not suitable for implantation in the human body. Currently, only short-term interfaces are practical.

In conventional neuroprosthetics, instead of glass micropipette electrodes, metallic electrodes are used to produce artificial electrical signals to manipulate the sensory or motor systems. While both recording and stimulation are possible with the same electrodes, neuroprosthetics primarily use stimulation to compensate for a neural deficit or disease. Currently, hundreds of thousands of patients worldwide use implantable neuroprosthetic devices, including cardiac pacemakers, implantable cardioverter defibrillators, cochlear implants, and deep brain stimulators. These devices possess only a few electrodes and are made using conventional technologies; however, they significantly improve the quality of life. Extension of these technologies to tackle complex neural deficits requires neuroprosthetics with a smaller form factor and a higher electrode density.

Planar patch-clamp devices were discussed in Chapter 7, but as mentioned, a high quality seal with the target cell can be difficult to achieve. Also, the size is still large and is not suitable for implantation. Microelectrode arrays (MEAs) on flat planar substrates were introduced in the 1970s and demonstrated that lithographically defined electrodes could be fabricated with features on the same scale as neurons [8–10]. These tools could acquire action potentials or provide stimulation to dissociated cell and tissue slice cultures [11,12]. More recently, finer control of microelectrode patterning has enabled multiple connections to single neurons [13]. A key limitation of MEAs is the inability to track individual neurons over long periods due to cell motility. To promote cell contact with electrodes, cell adhesion molecules or physical trapping structures such as wells [14] or cages [15] have been explored.

At about the same time that MEAs were introduced, electrodes patterned on a penetrating probe were also introduced (Figure 9.2) [16–18]. Single gold electrodes for biopotential recording were supported by etched silicon shanks and were insulated with silicon dioxide (SiO_2; Figure 9.3). Electrode tips of as little as 2 µm were possible with electrodes spacings of ~10 µm. These early probes recorded action potentials from single cells in cats. Later, integrated amplifier circuits were added to the base of the probes to allow both recording and stimulation (Figure 9.4).

Since this early work, many technological advances have been made, including high density single electrode probes (the so-called Utah probes were named after the University of Utah where they were developed) [19–22], multielectrode silicon-shank probes (the so-called Michigan probes, named after the University of Michigan, where they were developed; Figure 9.5) [23–25],

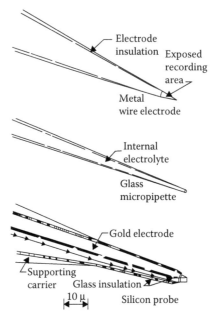

FIGURE 9.2
Comparison of the microwire electrode, micropipette electrode, and micromachined silicon-probe electrode. (From Wise, K. D., J. B. Angell, and A. Starr. 1970. *IEEE Trans Biomed Eng* BM 17:238. With permission.)

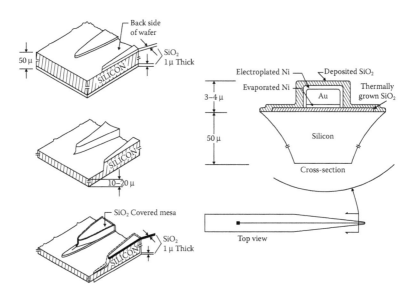

FIGURE 9.3
Left: Fabrication process for early silicon neuroprobe. Right: Cross-section and top view of the silicon neuroprobe. (From Wise, K. D., J. B. Angell, and A. Starr. 1970. *IEEE Trans Biomed Eng* BM 17:238. With permission.)

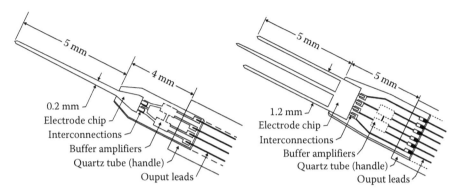

FIGURE 9.4
Left: An early single-electrode probe with integrated electronics for recording. Right: An array of probes for simultaneous recording and stimulation. (From Wise, K. D., and J. B. Angell. 1971. A microprobe with integrated amplifiers for neutrophysiology. In *Solid-State Circuits Conference*. Piscataway, NJ: IEEE. With permission.)

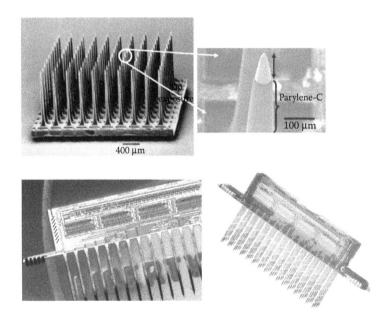

FIGURE 9.5
Silicon probe formats. Top: A 10 × 10 Utah electrode array with a single electrode at the tip of each silicon shank. The inset shows a single electrode at the tip of a shank insulated with Parylene C (from Bhandari, R., et al. 2009. *J Micromech Microeng* 19. With permission.) Bottom left: Michigan multielectrode probes with 16 shanks and 4 electrodes per shank. Electrodes are on 400 μm centers. Bottom right: Four probes in a silicon mounting platform forming a 256-site array (from Wise, K. D., et al. 2008. *Proc IEEE* 96:1184. With permission.)

wirelessly operated probes (Figure 9.6) [22,24,25,26], and electrode arrays on other substrates. In addition to silicon, ceramic- (e.g., aluminum oxide [Al$_2$O$_3$]) [27] and polymer-supported implantable electrode probes were developed. Polymers, in particular, have attracted a great deal of interest because they allow for flexible electrodes supported by polymers such as polyimide [26,28–30,], SU-8 [31] (Figure 9.7) or Parylene C [32,33]. Even cell-seeded probes have been developed (Figure 9.8) [34]. Electrode materials are typically platinum, gold, iridium, iridium oxide (IrO$_x$), silver, or carbon, and are selected for their charge transfer capacity, stability, and electrochemical properties. There are several excellent reviews of the recent advances in microfabricated central and peripheral nervous system (PNS) prosthetics [25,35–37].

Despite decades of development, neuroprosthetic microdevices still face many challenges, including damage to tissues, control of inflammatory immune responses, and the quality of long-term recording. Some researchers have chosen to address these issues by integrating microfluidic channels and components for local delivery of nutrients, neurotransmitters, growth factors, drugs, and toxins [25,32,38–40] (Figure 9.9).

Many neural deficits and diseases can be addressed with microfabricated neural prosthetic devices. A few will be explored in this chapter. Visual prostheses provide visual perceptions to the blind through electrical stimulation of (1) ganglion cells (epiretinal) [26,30,33,41,42], (2) the retinal network (subretinal) (Figure 9.10) [43–46], or (3) the visual cortex (cortical) [47]. Although this artificial vision is not a true analogue of vision achieved by light stimulation of photoreceptors in the retina, some level of useful vision has already been demonstrated in humans [48,49].

The epiretinal approach to visual prosthesis, in particular, has benefited from advances in microfabricated polymer microelectrodes. The retina is curved and is only ~250 μm thick. Thus, conventional rigid silicon-supported electrodes are insufficient to interface to the ganglion cells (about 1 million) in the retina. Diseases such as retinitis pigmentosa and age-related macular degeneration result in loss of photoreceptors of the retina. Photoreceptors convert light stimulus into electrical signals, which are processed by several cell layers in the retina and then carried by the optic nerve into the visual processing centers of the brain. The ganglion cell layer in patients with such ocular diseases is often intact and can be stimulated by MEAs to provide visual perceptions. As illustrated in Figure 9.10, visual information is first collected by an external camera and relayed wirelessly to an application specific integrated circuit that decodes the visual information converting it into a stimulus pattern that is applied to the ganglion cells through an implanted electrode array. Both polyimide- and Parylene-based flexible MEAs have been developed for this purpose (Figure 9.11). The polyimide-based device in Figure 9.11, top, shows the entire integrated system from the telemetry coil to the electronics and the electrode array. A special heat-forming process was used in the Parylene electrode array to curve the electrodes to allow a better

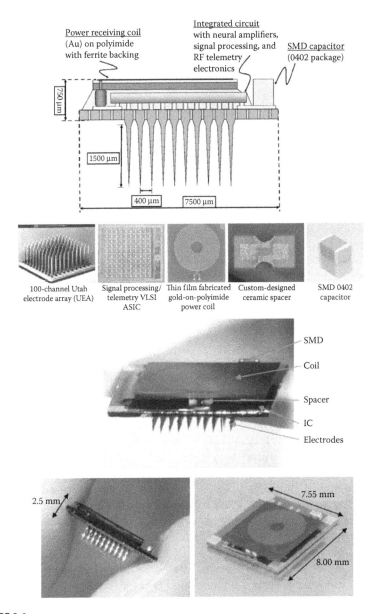

FIGURE 9.6
Wireless Utah probe format (RF = radio frequency; SMD = surface mount device; VLSI = very large scale integration; ASIC = application specific integrated circuit; IC = integrated circuit). Top: Schematic of system assembly showing the individual components. Bottom: Images of the assembled system. (From Kim, S., et al. 2009. *Biomed Microdevices* 11:453. With permission.)

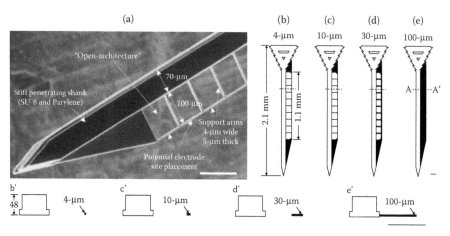

FIGURE 9.7
An SU-8 neuroprobe coated with Parylene C having an open architecture for reduction of tissue encapsulation: (a) scanning electron microscope image of the probe, (b–e) top views of different probe layouts, and (b′–e′) corresponding cross-sectional views of the layouts showing the thickness variation across the probe shank. The scale bar is 100 μm. (From Seymour, J. P., and D. R. Kipke. 2007. *Biomaterials* 28:3594. With permission.)

match to the curved retina and reduce the stimulating currents required to activate the ganglion cells (Figure 9.11, bottom).

Auditory prostheses date back to the pioneering work of William House and Jack Urban in the 1960s. An auditory prosthesis system consists of an external microphone, a speech encoder, and a wireless transmitter that directs the processed sound to a receiving unit (Figure 9.12). This provides an appropriate electrical stimulation pattern to implanted cochlear electrodes placed at various locations along the cochlea. Normally, sound is mechanically processed by hair cells (~16,000), which activate auditory neurons that begin the transmission of auditory information to the brain. In lieu of the hair cells, which are damaged during hearing loss, electrodes are used to directly stimulate the auditory neurons, which results in sound perception by the brain. Frequency is naturally encoded in a spatial manner so that electrodes near the base of the cochlea are stimulated by high frequency sound, and electrodes toward the apex are stimulated by low frequency auditory information. The amplitude of the stimulation current correlates to the loudness.

Current clinical cochlear implant systems consist of 16–22 electrodes arranged along a wire and are deficient in the areas of tonal languages, music, and speech in a crowded room. Thus, microfabricated electrodes are being considered to increase the number of stimulation sites in the limited space in the scala tympani of the ear (from 1 mm down to 200 μm diameter over its length) [50–52]. The electrodes must be placed near the inner wall of the scala tympani, which is closest to auditory nerves to minimize currents and improve frequency resolution [50]. An integrated system including

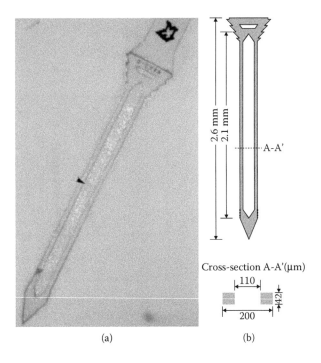

(a)   (b)

FIGURE 9.8
A neural stem-cell-seeded probe constructed from an SU-8 backbone and coated with Parylene C: (a) stained cells are indicated by the arrow and (b) the dimensions of the probe from the top and cross-sectional views. (From Purcell, E. K., et al. 2009. *J Neural Eng* 6. With permission.)

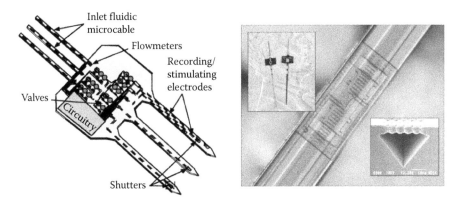

FIGURE 9.9
Left: Schematic of a neuroprobe with integrated microfluidics. Right: Photographs of a probe with a microfluidic channel and flow sensor. Insets show a photograph of two probes on a U.S. penny and a scanning electron microscope image of a channel cross-section. (From Wise, K. D., et al. 2008. *Proc IEEE* 96:1184. With permission.)

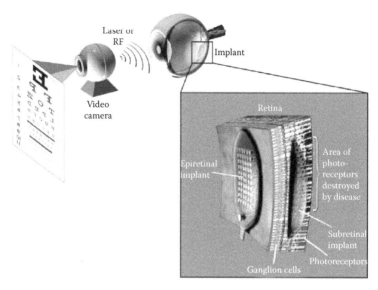

FIGURE 9.10
Illustration of epiretinal and subretinal visual prosthetic systems. An external camera acquires visual information that is wireless transmitted to the implant. An application-specific integrated circuit processes the information, and stimulation of the retina with a high-density multielectrode array on a flexible substrate results in a visual perception. (From Weiland, J. D., W. T. Liu, and M. S. Humayun. 2005. *Annu Rev Biomed Eng* 7:361. With permission.)

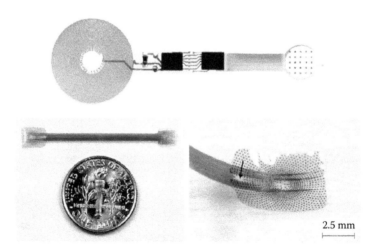

2.5 mm

FIGURE 9.11
Top: An epiretinal prosthesis in a polyimide substrate showing integration of a telemetry coil, electronics, and a stimulating electrode array (from Mokwa, W. 2004. *J Micromech Microeng* 14:S12. With permission.) Bottom: Photographs of (left) a high-density 1024 electrode array for an epiretinal prosthesis and (right) a close-up of a heat-formed curved array (from Rodger, D. C., et al. 2008. *Sens Actuators B Chem* 132:449. With permission.)

a 32-IrO electrode array, wireless interface, and microprocessor is shown in Figure 9.13. A Parylene ribbon cable connects the processor to the electrode array. The electrode arrays measure 6–8 μm thick, 8–16 mm long, and tapers from 600 down to 200 μm (Figure 9.14). Because positioning of these electrodes is directly related to the electrical stimulation parameters, the probe also integrates eight strain gauges and a tip wall-contact sensor.

Prostheses for neuromuscular stimulation and recording are often classified as PNS interfaces to distinguish from those related to the central nervous

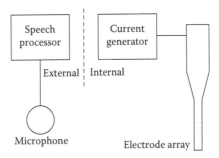

FIGURE 9.12
Diagram of an auditory prosthesis system.

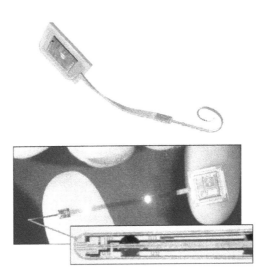

FIGURE 9.13
Top: Illustration of a cochlear implant with a microprocessor, data converter, wireless interface, ribbon cable, and electrode array. Bottom: Photograph of the system for use in guinea pigs with an inset showing a detail of the electrode array. The tip includes wall-contact and position-sensing strain gauges. (From Wise, K. D., et al. 2008. *Hear Res* 242:22. With permission.)

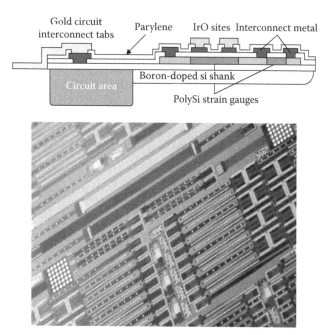

FIGURE 9.14
Top: Cross-section of the microfabricated cochlear implant. Bottom: View of a wafer containing several probes. (From Wise, K. D., et al. 2008. *Hear Res* 242:22. With permission.)

system (CNS). In fact, both invasive and noninvasive (surface) electrodes can be used to stimulate peripheral motor nerves or muscles for patients suffering from spinal cord injury [35]. In the PNS, neural cell bodies are located in the spinal cord and extend their axons to throughout the body as peripheral nerves bundled into fascicles. Intact peripheral neural pathways can be stimulated to control the diaphragm for ventilatory assistance (phrenic nerve), the sphincter muscles for bladder emptying (sacral roots), and limbs for walking and grasping. Approaches to invasive PNS electrodes include surface, muscle, extraneural (adjacent or around), and intraneural (penetrating or sieve). Few examples of microfabricated surface electrodes exist. Muscle electrodes allow stimulation of muscle into which an electrode is implanted. Surgical placement of extraneural or intraneural electrodes can be difficult, and precautions must be taken to avoid damage to the nerve fibers. However, if they are placed appropriately, localized PNS electrodes can reduce the stimulation intensity and can even be used to selectively stimulate nerve fascicles.

A wireless microstimulator capsule was devised for implantation into muscle by injection through a 12-G hypodermic needle [53]. The capsule, measuring only $2 \times 2 \times 10$ mm^3, consisted of a silicon substrate that supported the stimulating electrodes and electronics contained within a hermetically sealed glass chamber (Figure 9.15). The electrodes, exposed at the ends of the capsule, were arranged in a waffle format with many small interconnected electrodes instead of a single continuous electrode. This arrangement avoided the concentration of current at the perimeter of continuous electrodes and results in a more uniform current distribution across the electrode area. Also, IrO_x was selected as the electrode material for its high charge injection capacity when compared to platinum and stainless steel. The electronics were protected in a hermetically sealed glass-silicon chamber by using a modified electrostatic anodic bonding process between Pyrex 7740 and fine-grain polysilicon at a reduced temperature of 320°C (~2000 V) [54]. To connect the electronics within the package to the electrodes, electrical feedthroughs using phosphosilicate glass-insulated conductive polysilicon wires were used. With this packaging scheme, a projected device lifetime of 116 years at body temperature (37°C) was obtained from extrapolated mean time to failure (MTTF) data from accelerated lifetime soak testing performed at elevated temperatures.

Extraneural PNS interfaces allow simultaneous interfaces with many axons in a nerve fascicle. An example is the popular cuff electrode that wraps around a nerve. The cuff consists of a sheath containing one or more electrodes exposed to the nerve. Because the sheath must fit around the nerve without damaging it, the sheath is usually split or spirals around the nerve.

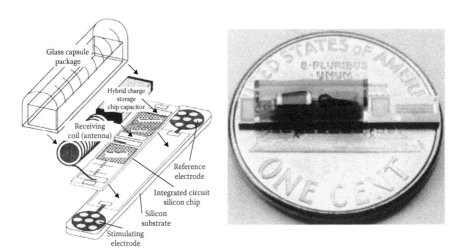

FIGURE 9.15
Left: Schematic showing the construction of the microstimulator capsule with the supporting silicon substrate, integrated circuitry, discrete electronic components, and glass lid (from Ziaie, B., et al. 1996. *J Microelectromech Syst* 5:166. With permission.) Right: Photograph of a microstimulator on a U.S. penny (from Ziaie, B., et al. 1997. *IEEE Trans Biomed Eng* 44:909. With permission.)

However, the cuff's presence itself can damage the nerve, and appropriate sizing is difficult. Polyimide and polyimide-silicone rubber cuffs (10 mm long, 4–8 mm in diameter) were developed with up to 18 platinum electrodes (500 μm in diameter) [55] and integrated with electronics [36]. The flat cuffs were rolled and held in place (~1.4 mm overlap), following a thermal tempering process (340°C, 2 hours; Figure 9.16). Some were then encapsulated in silicone rubber.

Intraneural electrodes permit contacts with small groups of neurons and further enhance selectivity of activated nerves. These include electrode probes that penetrate fascicles and/or regenerative interfaces that consist of a sieve through which fibers must grow through for repair. Penetrating intraneural electrodes have a similar layout as probes for the (CNS) [56,57]. In the sieve electrode approach, the nerve bundle must be initially severed. The purpose of these implants is to interface to peripheral nerve fibers as they regenerate across the implant through perforations in the substrate supporting the electrodes [36,58–60]. A guidance tube is used to support the regenerating nerve and the electrode. In an earlier work, regeneration of peripheral nerves was demonstrated in multiple animals for both silicon and polymer devices [35]. A regeneration sieve electrode for the rat sciatic nerve

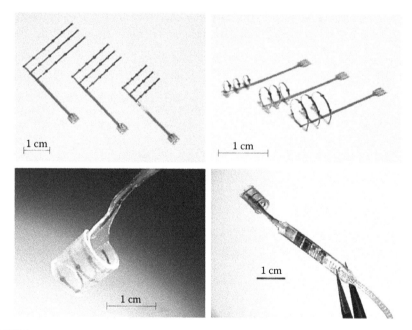

FIGURE 9.16
Cuff electrodes. Top left: Polyimide cuffs prior to assembly. Top right: Assembled polyimide cuffs in 4, 6, or 8 mm diameters. Bottom left: A single 6-mm-diameter polyimide-silicone rubber cuff. Bottom right: Polyimide-silicone rubber cuff with an integrated stimulator or multiplexer unit. (From Stieglitz, T., M. Schuettler, and K. P. Koch. 2005. *IEEE Eng Med Biol Mag* 24:58. With permission.)

was demonstrated. A supporting silicone rubber tube (1.8 mm inner diameter and 0.5 mm wall thickness) was flanked on both sides of a polyimide sieve electrode (Figure 9.17). The area measured 1.8 mm in diameter and consisted of 27 ring-shaped electrodes distributed on both sides of the substrate. Each electrode was arranged around one of 571 holes measuring 40 µm in diameter and arranged hexagonally with a pitch of 70 µm. Additional electrodes were positioned away from the central sieve and were used as reference electrodes for recording or counter electrodes during simulation. All electrodes were made of platinum. Both nerve regeneration across the sieve and the recording of neural signals in rats were achieved.

9.3 Implantable Sensors

The largest class of microfabricated implantable sensors is pressure sensors, which are briefly discussed in Chapter 6. Here, a few examples of implantable pressure sensors used in different body cavities are discussed.

Implantable pressure sensors for continuous monitoring of intraocular pressure (IOP) date back to the 1960s [61]. Elevated IOP (ocular hypertension) is a significant risk factor for glaucoma, an incurable disease that affects the

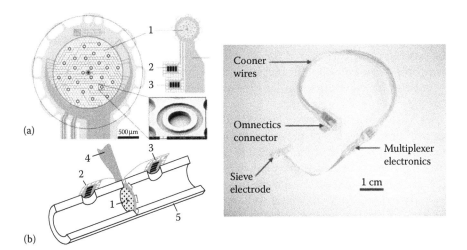

FIGURE 9.17
Left: Schematic showing the layout of the (a) sieve electrode and (b) nerve-supporting silicone rubber tubing: (1) sieve area, (2) counter electrode in front of the sieve, (3) counter electrode in back of the sieve, (4) polyimide substrate, (5) silicone rubber tubing. Right: Photograph of the assembled system including the multiplexer and external wiring. (From Ramachandran, A., et al. 2006. *J Neural Eng* 3:114. With permission.)

optic nerve. If left untreated, glaucoma causes patients to lose their peripheral vision. However, at present, monitoring of IOP is only performed periodically in clinics using tonometry. This technique uses the force required to
flatten a predetermined area of the cornea to infer IOP. Because the physical
contact with the eye is uncomfortable, frequent measurements are impractical. In patients with severe glaucoma, continuous monitoring of IOP with the
prescribed treatment to preserve vision is necessary.

IOP sensors must measure pressures in the normal IOP range of 10–22
mmHg as well as at elevated IOPs above 22 mmHg. Most approaches to
measure IOP take into account the small size of the eye and adopt wireless
transmission of pressure data to prevent the use of wires crossing the eye
wall, which are potential sites of infection. Also, the placement of the sensor
should not impede vision. Wireless methods use inductance-capacitance (LC)
circuits with varying resonant frequency as a function of pressure; either the
inductance or the capacitance or both can be pressure responsive [41,61–64].
No internal power source or storage device is required. Instead, the sensor is
excited wirelessly through inductive coupling. One such device is described
in Chapter 6.

A completely mechanical approach to detect pressure changes was
recently introduced [65]. A hollow Archimedean spiral tube with an internal
reference pressure and anchored at the center unwound in response to an
ambient pressure decrease, whereas an increase caused the spiral tube to
coil more tightly (Figure 9.18). This principle is based on the classic Bourdon
tube. An arm placed at the end of the tube was calibrated to indicate a pressure change in millimeters of mercury (mmHg). This was read out using
a standard clinical microscope. Sensors measuring $3 \times 0.6 \times 0.5$ mm^3 were
implanted through a 19-G hypodermic needle. The tiny feet at the bottom of
the sensing platform secured the device onto the textured iris. These devices
were able to measure IOP in rabbits with 3.6 mmHg resolution.

Catheter-based cardiovascular pressure sensors are discussed in Chapters
6 and 8.

An early pressure sensor used bulk micromachined diaphragms for
capacitive monitoring of pressure. In one early work, two pressure sensors
were implanted in the left ventricle at different elevations to obtain the
intramyocardial pressure gradient, which is linked to hemodynamic function. Using the data acquired from these sensors, the heart rate was tracked,
and arrhythmias were detected. When used with a pacemaker or defibrilla-
tor, rate-responsive pacing and arrhythmia treatment, respectively, were possible [66]. Another approach avoided complex surgical procedures and used
a catheter to deploy a telemetric pressure sensing capsule to the target artery.
Upon release from the catheter, the levers in the capsule would expand and
lock the sensors into place without requiring any sutures (Figure 9.19). Prior
to deployment, the capsule maintained its cylindrical shape (2.6 mm diameter and 20 mm long) within the catheter. The capsule included a surface

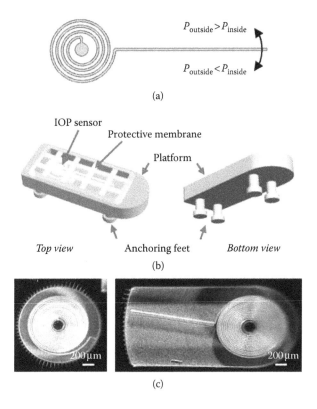

FIGURE 9.18
(a) Schematic showing the spiral tube design with a long indicator arm. The arrows indicate the direction of arm movement in response to differential pressure between the interior and exterior of the tube. (b) 3D illustration showing the overall design of the IOP sensing platform. (c) Photographs of two different IOP sensors with varying arm length. The indicator arm points to tick marks that were used to determine IOP values (IOP = intraocular pressure). (From Chen, P. J., et al. 2007. *J Micromech Microeng* 17:1931. With permission.)

micromachined capacitive pressure sensor fabricated using a complementary metal oxide semiconductor (CMOS) process and telemetry unit for wireless monitoring [67].

Stents play an important role in maintaining arteries that are affected by cardiovascular diseases, and may also be used in other ducts or flow paths in the body. By integrating pressure sensors on the stents, the intraluminal pressure can be monitored [68]. A stent was fabricated from electrodischarge machining of thin stainless steel foils (50 μm thick, 200 mm long, and 3.5 mm diameter; Figure 9.20). This element also performed as an inductor or wireless antenna. On either side, capacitive pressure sensor chips were integrated to form an LC circuit for wireless pressure measurement. Having two sensors allows average pressure measurement or differential pressure measurement to determine the flow rate. Stents were deployed using standard angioplasty

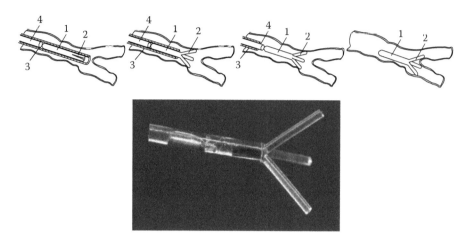

FIGURE 9.19
Top: Sequence of images illustrating the process by which the pressure sensing capsule is placed in the artery: (1) capsule, (2) expanding levers, (3) catheter, (4) push rod. Bottom: Wireless blood pressure monitoring device. (From Schnakenberg, U., et al. 2004. *Sens Actuators A Phys* 110:61. With permission.)

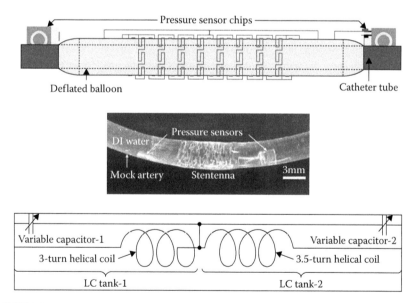

FIGURE 9.20
Top: Illustration showing the inductive stent with capacitive pressure sensor chips at either end. The relationship of the angioplasty balloon to the stent prior to deployment is indicated. Middle: Device deployed on an artificial artery after removal of the balloon. Bottom: Equivalent circuit of deployed sensor. (From Takahata, K., Y. B. Gianchandani, and K. D. Wise. 2006. *J Microelectromech Syst* 15:1289. With permission.)

balloons. Once the balloons were inflated, the stents took on their final permanent helical configuration.

Implantable pressure sensors have also been developed for measuring intracranial pressure (ICP) in the cavity between the brain and skull. For healthy adults, ICP is usually between 0 and 10 mmHg, with 20 mmHg being abnormal. Elevated ICP can point to one of a number of severe medical conditions including hemorrhage, tumors, encephalitis, hydrocephalus, abscesses, meningitis, and stroke. Capacitive pressure sensors with telemetric readout have been suggested as an alternative to conventional approaches. Surface [69,70] and bulk [71] micromachined pressure-sensitive diaphragms have been used. Although there are implantable sensors for other parameters such as pH and analytes, the lack of long-term sensor stability following implantation has limited their success to date.

9.4 Drug Delivery

Medical treatment often entails drug administration, which is still a very active research area and an ongoing challenge in modern medicine. The goal of drug delivery is to provide the drug within a therapeutic window over the prescribed duration of the treatment. Microfabricated devices, either invasive or noninvasive, assist this process by controlling the time and location of drug release to achieve the desired therapeutic concentrations.

Conventional drug administration by oral route or injection (e.g., intravenous, intramuscular, or subcutaneous) results in an initial peak in drug concentration followed by a slow decline related to the half-life of the drug and metabolic uptake (Figure 9.21). Other routes (e.g., topical, inhalation, ophthalmic, rectal, and vaginal) also exist if oral or injections routes are unsuitable. Readministration repeats this process and is necessary to maintain the concentration within the effective window. The initial peak is of some concern, as the concentration may briefly surpass the toxicity level. The drug administration route and method impact the kinetics of drug distribution and elimination, which ultimately impact the effectiveness of the treatment. Controlled, sustained, and targeted release of the drug is possible using devices that provide optimal methods of delivery. Microfabricated devices for drug delivery are the subject of many reviews [72–78]. In general, these devices may be divided into microparticles, biocapsules, needles, and implantable systems. Although the focus of this chapter is on implantable devices, examples of externally applied drug delivery systems will also be covered.

Conventional oral drug delivery prohibits the effective delivery of peptides and proteins, which are degraded into smaller molecules in the gastrointestinal tract. To enhance the bioavailability of these drugs, micromachined

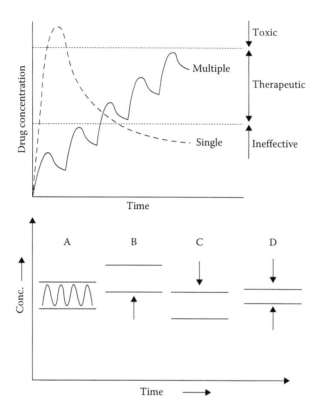

FIGURE 9.21

Top: Drug delivery profiles resulting from a single large dose or multiple small doses. The target therapeutic drug concentration is indicated by the dotted line boundaries. Bottom: Therapeutic window in an adult defined by the upper toxic limit and lower threshold limit for effective drug administration (A: desired case with the concentration fluctuations within the window, B: upward shift of the window as a result of high resistance or antagonistic drug interactions, C: downward shift as a result of hypersensitivity or synergistic drug interactions, and D: narrower window for a child).

particles containing drug-filled wells are ingested, and they adhere to the intestines and release their payload locally (Figure 9.22) [79]. Silicon dioxide square wells (0.5–1 μm deep and 35–100 μm long on a side) containing 80–150 μm square particles (~2 μm deep) can be fabricated on top of a polysilicon sacrificial layer using a low temperature oxide process. The actual wells and particles are defined by either a wet chemical or dry reactive ion etching (RIE) process and then released. Surfaces are decorated with lectins by an avidin-biotin conjugation method; lectins allow selective binding to the intestinal mucosa.

Biocapsules with nanoporous membranes can be used for the immunoisolation of transplanted pancreatic islet cells (Figure 9.23). Fabrication of these nanoporous membranes is described in Section 5.3.1.2; this process uses thin sacrificial layers to define nanometer pores [80]. Two such membranes can be

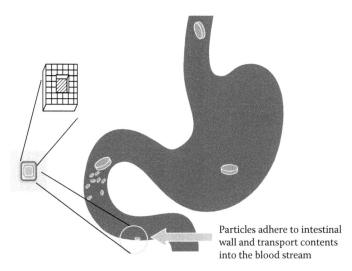

FIGURE 9.22
Microparticle-based oral drug delivery system. The insets show a photograph and illustration of a micromachined particle containing a well for storing a drug. (From Ahmed, A., C. Bonner, and T. A. Desai. 2002. *J Control Release* 81:291. With permission.)

joined together to allow insulin secreted by trapped islet cells and glucose to permeate but prevent the passage of immunologic cells (e.g., macrophages) and antibodies that can disrupt or destroy the transplanted cells. Protection of islet cells *in vitro* [81] and *in vivo* (rats) was demonstrated in reference [82]. Diffusion of immunoglobulin was severely impeded by the nanoporous membranes in a size-dependent manner.

Orally administered drugs may undergo undesirable gastrointestinal degradation. Transdermal modes of drug delivery avoid this but must contend with the low permeability of the skin. To bypass the skin, delivery via conventional hypodermic needles is possible, but may cause undesirable pain to the patient. These conventionally machined needles can be made as small as ~300 μm in diameter. With micromachining technology, shorter and smaller needles are possible for more painless transdermal drug delivery. This technology is reviewed in reference [83].

Skin is composed of two major layers—the epidermis and dermis. Pain receptors are located within the dermis. For painless transdermal delivery, microneedles must penetrate the outer layer of dead tissue in the epidermis, the stratum corneum (10–20 μm thick), to reach the viable epidermis (50–100 μm thick), which is living tissue with few nerves. In a study by Henry et al., solid silicon needle arrays were used to penetrate the stratum corneum, which is the primary barrier to transdermal delivery, to expose the viable epidermis [84]. The needle tracks left behind greatly increased the permeability of human skin samples to a model drug, calcein. Needles were created by deep reactive ion etching (DRIE) to first create an array of posts (150 μm

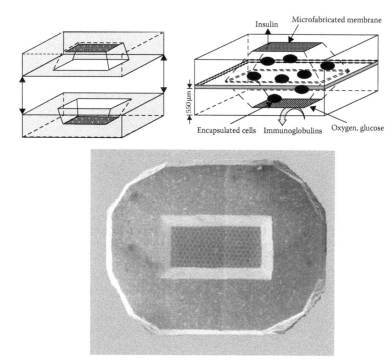

FIGURE 9.23
Top: Illustration of biocapsule assembly process and an assembled biocapsule containing cells (from Desai, T. A., D. Hansford, and M. Ferrari. 1999. *J Memb Sci* 159:221. With permission.) Bottom: Photograph of a biocapsule with 24.5 nm pores (from Tao, S. L., and T. A. Desai. 2003. *Adv Drug Deliv Rev* 55:315. With permission.)

long). Second, an RIE was performed to undercut the metal mask used to define the posts in the first step. In this manner, high aspect ratio microneedle arrays were made (Figure 9.24). These needles were easily inserted into the forearm or hand of human subjects without pain.

Several hollow needles for transdermal delivery have also been investigated, including needles with openings on the top [85] or side [86,87] of the shank. However, similar to conventional flat-tipped noncoring needles, silicon needles with centered bores on flat tips cored the skin during puncture, which may later result in clogging. In Figure 9.25, a needle having a tapered tip with a bore positioned off-center that prevents coring is shown [85]. A combination of dry and wet etching techniques was used to create a tall, tapered needle tip. The tapered pore corresponds to the {111} silicon plane and was exposed by anisotropic wet etching.

Alternatively, microneedles with side openings also avoid unwanted coring. Needles with side openings have a low flow resistance [86]. These intricate structures were produced through a selected combination of isotropic and anisotropic plasma etches (Figure 9.26). The bore was etched in a separate backside process. Later, a newer version of these needles was developed and integrated with a dispensing unit (Figure 9.27). The dispensing actuator uses thermally induced, irreversible expansion of a silicone-expanding microsphere composite material to push the drug through the needle outlets. Heating was performed using a resistive element patterned on a printed circuit board. Needles were fabricated using a similar process in which the bore was first etched from the backside and the final needle shape was produced using a combination of isotropic and anisotropic dry etches. The needle and the dispensing units were joined to create an entire transdermal drug delivery system. Insulin delivery using this system was demonstrated in diabetic rats.

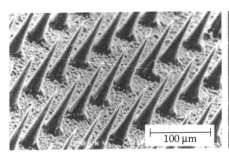 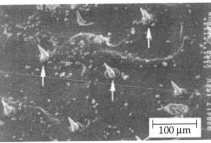

FIGURE 9.24
Left: Scanning electron microscope image of microneedles. Right: Scanning electron microscope image showing microneedle tips exiting from the underside of the epidermis. Arrows indicate needle tips. (From Henry, S., et al. 1998. *J Pharm Sci* 87:922. With permission.)

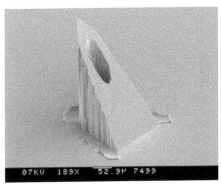 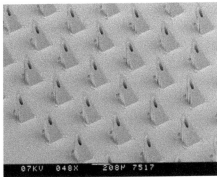

FIGURE 9.25
Left: Scanning electron microscope image of a microneedle (350 μm tall, 250 μm wide at the base, and elliptical flow channel with 70-μm major axis). Right: Scanning electron microscope image of a microneedle array (555-μm pitch). (From Gardeniers, H., et al. 2003. *J Microelectromech Syst* 12:855. With permission.)

FIGURE 9.26
Left: Scanning electron microscope image of a microneedle with side openings measuring 210 µm tall. Right: Scanning electron microscope image of a microneedle array. (From Griss, P., and G. Stemme. 2003. *J Microelectromech Syst* 12:296. With permission.)

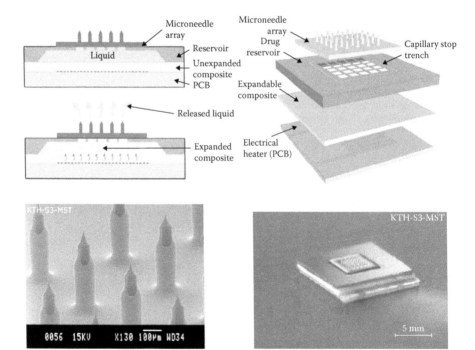

FIGURE 9.27
Top left: Illustration showing the operation of the transdermal drug delivery system. Top right: Illustration of transdermal drug delivery system. Bottom left: Scanning electron microscope image of needles with side openings (400 µm tall). Bottom right: Photograph of an assembled transdermal drug delivery system (PCB = printed circuit board). (From Roxhed, N., et al. 2008. *IEEE Trans Biomed Eng* 55:1063. With permission.)

One method of achieving the controlled release of therapeutics is to use implantable drug reservoirs. By using lithography, reservoirs of precise dimensions can be defined. Many reservoir-based drug delivery formats have been devised and are reviewed in reference [88]. Anisotropically wet-etched silicon reservoirs gated with a thin gold film can be filled with a solid, liquid, or gel-based drug. The drug was sealed using a plastic film applied with epoxy. The 0.3-μm-thick gold films covering the reservoirs were connected to the anode in an electrochemical circuit. Each reservoir was individually addressable and the contents were selectively exposed by the electrochemical dissolution of the gold film [89]. Thus, a pulsatile release profile was obtained as single or multiple reservoirs were activated. A total of 34 reservoirs (25 nL) were patterned onto a $17 \times 17 \times 0.3$ mm^3 die (Figure 9.28). This reservoir-based drug delivery scheme was commercialized by MicroChips Inc., but a faster electrothermal process to activate each reservoir, as opposed to the original slower electrochemical one, was used [90]. Thin films covering one side of the reservoirs were selectively opened by applying current, which induced membrane failure and thus the release of the drug. The chips measured $15 \times 15 \times 1$ mm^3 and contained 100 individually addressable reservoirs, each holding 300 nL. The drug reservoir chip and electronics were mounted on PCBs and then packaged in a welded titanium case, similar to those used to package pacemakers (Figure 9.29). Drug delivery in canines was demonstrated with a model drug. Other reservoirs constructed of polymers have also been described in reference [91].

Both external and internal drug pumps using MEMS technology have been developed. An inexpensive pump for the delivery of insulin to patients

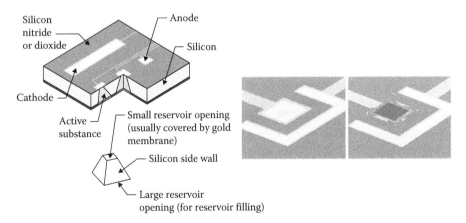

FIGURE 9.28
Left: Illustration showing an array of drug delivery reservoirs selectively opened by electrochemical dissolution of the anode. The inset shows the geometry of a single reservoir. Right: Sequence of scanning electron microscope images showing the reservoir before and after anode dissolution. (From Santini, J. T., M. J. Cima, and R. Langer. 1999. *Nature* 397:335. With permission.)

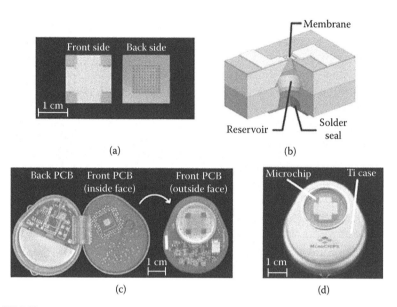

FIGURE 9.29
A microreservoir drug delivery system: (a) photographs of the reservoir chip, (b) illustration of a single reservoir, (c) PCB containing electronics to control the chip, (d) the fully packaged drug delivery system (PCB = printed circuit board). (From Prescott, J. H., et al. 2006. *Nat Biotechnol* 24:437. With permission.)

with diabetes has been developed [92]. Compared to an earlier version, this fabrication was simplified to reduce the cost to achieve a practical disposable pump. The micromachined structure is modular in that the actuator is not included and can be attached afterward. This allows a number of different actuation methods to be used; piezoelectric bimorph discs, piezoelectric bimorph cantilevers, and pneumatic actuation were demonstrated with this device. The diaphragm pump consisted of a pump chamber and two check valves assembled from a stack of 3 anodically bonded substrates: two Pyrex dies that sandwich a silicon die (Figure 9.30). The diaphragm precisely delivered 160 nL per stroke using a double limiter scheme. A mesa structure and chamber walls limit the maximum displacement of the diaphragm to 20 μm. By overdriving the diaphragm with the actuator, the precise stroke volume was obtained.

An implantable drug delivery pump was developed for treating incurable ocular diseases and drug delivery was demonstrated acutely in rabbits [93]. Intraocular diseases present many challenges to conventional drug administration methods because drugs must penetrate many tissue barriers before reaching the target tissues. Both intraocular injections and controlled-release implants have a limited duration of therapeutic benefit and require multiple invasive clinical interventions for lifetime patient treatment. A miniaturized

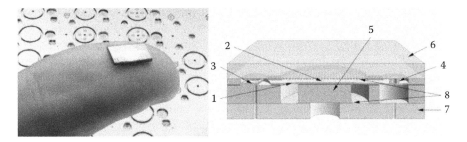

FIGURE 9.30
Left: Photograph of the disposable drug delivery pump (6×10 mm^2). Right: Cross-sectional view of the pump without the actuator: (1) pump diaphragm, (2) pump chamber, (3) inlet check valve, (4) outlet check valve, (5) stroke-limiting mesa, (6) upper glass plate, (7): lower glass plate, (8) thin film surface treatment. (From Maillefer, D., et al. 2001. A high-performance silicon micropump for disposable drug delivery systems. In *MEMS 2001: 14th IEEE International Conference on Micro Electro Mechanical Systems*. Piscataway, NJ: IEEE. With permission.)

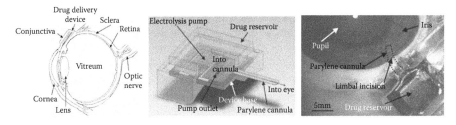

FIGURE 9.31
Left: Schematic showing drug pump implantation in the eye. Middle: 3D illustration of the ocular drug delivery device. Right: Drug delivery device implanted in a rabbit eye. (From Li, P. Y., et al. 2008. *Sens Actuators A Phys* 143:41. With permission.)

pump uses the electrolysis of water to push the drug into the eye through an integrated polymer cannula, limiting the diffusion distance to the intended target. The pump also contains a refillable reservoir to maintain the long-term therapeutic benefit. The pump is implanted in the eyewall, and the cannula is directed into the anterior chamber of the eye (Figure 9.31).

9.5 Tissue Engineering

Cells in tissue are surrounded by a matrix possessing both micro- and nano-scale features. Using microtechnologies for tissue supporting scaffold design and fabrication, functional tissue replacements for damaged, diseased, or aged tissues may be possible. These 3D tissue constructs may also be used as *in vitro* models for the development of drugs and physiological research. Tissue engineering is a complex endeavor that requires the appropriate cell-material interactions for cell survival and for producing the desired tissue construct. Scaffold construction incorporates design parameters such as composition, architecture, porosity, surface chemistry, strength, shape, and compliance. In some instances, a scaffold need not be permanent and may be constructed of biodegradable materials. The structure may also incorporate extracellular matrix proteins and other relevant stimuli (e.g., electrical and mechanical signals) to emulate the natural cell surroundings. At a cell culture level, microfabrication has already demonstrated control of cell-material interactions such as cell alignment, selective adhesion, and guidance. Tissue engineering at the cell and tissue level using microtechnologies are reviewed in references [94–96].

Biodegradable scaffolds are attractive because they are not permanent and exist long enough to establish an appropriate structure for the formation of the intended tissue construct. Popular biodegradable polymers for tissue engineering include polyglycolic acid, polylactic acid, poly(DL-lactic-*co*-glycolic acid), polycaprolactone, and poly(glycerol-sebacate; PGS). Microfabrication of some these biodegradable polymers are reviewed in reference [97]. In addition to the presence of the scaffold to establish a geometrical relationship between seeded cells, perfusion is necessary to provide nutrients, dispose of metabolic waste products, or provide other stimuli. A PGS tissue-engineering scaffold was constructed for creating both vascular and hepatocyte constructs [98]. The scaffold consisted of peg structures separated by microfluidic channels possessing geometries specific to each application. For the vascular scaffold, an overall microfluidic reactor was designed to apply a constant maximum shear stress in each channel so as to mimic natural conditions. The hepatocyte device possessed a different geometry for ensuring high perfusion rates and to promote cell attachment. Each microfluidic reactor was produced by molding PGS using etched silicon substrates (Figure 9.32). Sucrose was applied to the silicon substrate as a release layer for the PGS. Separated layers could be bonded by covalent cross-linking into multilayer devices (up to five layers demonstrated). The PGS channels were coated with collagen prior to cell seeding to promote adhesion. The survival of Hep G2 (human hepatocarcinoma) cells seeded and perfused with these devices was demonstrated for a period of 1 week.

Adhesion of cells to scaffolds is an important factor in cell survival and controlling cell-material and cell-cell interactions. A permanent structured silicone rubber scaffold was used to control the cell attachment, orientation, and morphology in neonatal rat primary cardiac myocytes [99]. Primary cells are obtained directly from animals and have a limited lifespan, whereas cell lines are modified to proliferate indefinitely. The topographical features (pegs and trenches) molded into the silicone rubber possessed sizes (5–10 µm

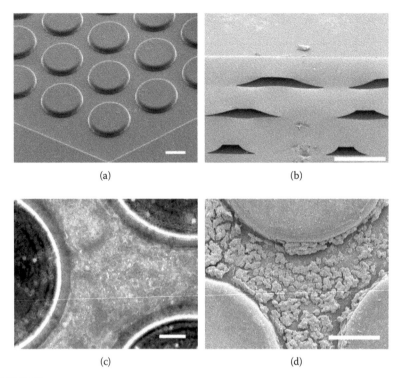

FIGURE 9.32
Poly(glycerol sebacate) biodegradable microfluidic tissue engineering devices. (a) Scanning electron microscope image showing structure of a hepatocyte device (scale bar = 200 μm). (b) Scanning electron microscope image of cross-section of a three-layer device (scale bar = 200 μm). (c) Hep G2 cells photographed after 1 week (scale bar = 50 μm). (d) Fixed Hep G2 cells in the hepatocyte device imaged by scanning electron microscope after 1 week in culture (scale bar = 100 μm). (From Bettinger, C. J., et al. 2006. *Adv Mater* 18:165. With permission.)

features) at the same scale as the myocytes. Substrates were initially coated with laminin but later altered with peptide linkages to further promote adhesion [100].

Primary cells often lose their function in standard cell cultures. This is true of hepatocytes that perform metabolic processes in the liver. Microfabricated scaffolds were investigated to promote morphogenesis (organ shaping arising from biological processes such as differentiation) in rat primary hepatocyte cultures [101]. The scaffolds featured a silicon wafer with channels produced by DRIE and sandwiched in a 3D bioreactor for tissue culture. A filter membrane placed below the silicon scaffold provided support and retained cells while allowing continuous perfusion (Figure 9.33). This cross-flow perfusion method was adopted to provide homogenous flow and mass transfers for tissue maintenance. Each of the channels formed in the silicon

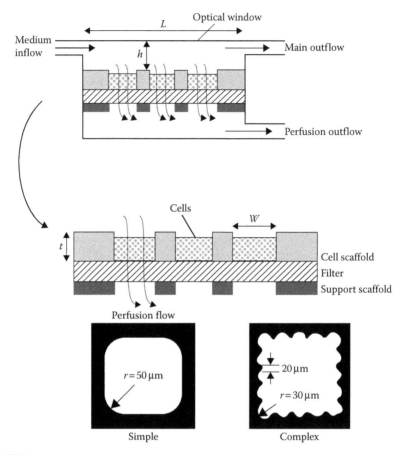

FIGURE 9.33

Top: Illustration of hepatocyte scaffold and reactor design ($L = 400$ μm, $h = 805$ μm, $t = 235$ μm, $w = 300$ μm). Bottom: Two different channel geometries etched into silicon scaffolds. (From Powers, M. J., et al. 2002. *Biotechnol Bioeng* 78:257. With permission.)

substrate was modified to promote cell-cell and cell-substrate adhesion. Silicon and silicon-fluoropolymer surfaces were evaluated with and without collagen I (Figure 9.34). Collagen I resulted in the desired cell spreading and adhesion. Along with the overall reactor structure, hepatocytes were able to rearrange to form structures resembling tissue and survive for up to 2 weeks in a culture.

The successful transplant of cells bound to scaffolds was demonstrated in rat retina. Certain ocular diseases, such as retinitis pigmentosa and age-related macular degeneration, exhibit loss of photoreceptors, which results in blindness. A promising method to replace these cells is the bolus injection of retinal

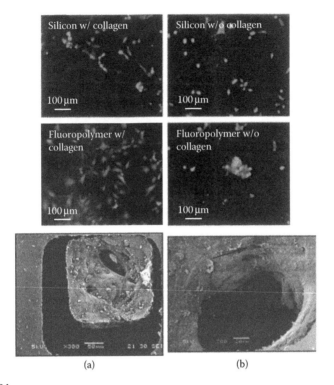

FIGURE 9.34
Top: Four fluorescent images showing the difference in cell morphology and density for different surface treatments. Bottom: Scanning electron microscope images of tissue morphology obtained after 2 days. Tissue shrank by ~30% during the sample preparation process. The left figure (a) is a low magnification view showing the channel and tissue construct. The right (b) is a high magnification view in which vessel-like structures and cells with cobblestone vessel endothelium morphology are apparent (arrow). (From Powers, M. J., et al. 2002. *Biotechnol Bioeng* 78:257. With permission.)

progenitor cells (RPCs). However, injection is not optimal; it is associated with poor cell survival and other complications. RPCs were cultured on thin polymethylmethacrylate scaffolds (6 μm thick) with and without pores (11 μm diameter, 63 μm apart) [102]. Scaffolds were produced by spinning a prepolymer onto supporting substrates, etching pores by RIE, and then releasing the thin films. Following surface treatment with poly-L-lysine and laminin coating, the RPCs obtained from postnatal day 1 transgenic mice were cultured and transplanted into the subretinal space of mice. Smooth, nonporous scaffolds exhibited poor retention of transplanted RPCs, whereas porous scaffolds enabled greater process outgrowth and cell migration into the host retina (Figure 9.35). Transplanted tissue on scaffolds resulted in an increased cell survival and localization compared to bolus injection.

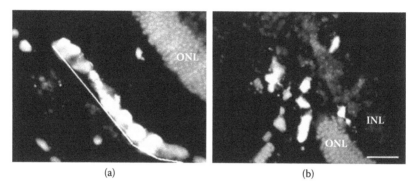

(a)   (b)

FIGURE 9.35
Fluorescent microscopy images of retina sections showing (a) the adhesion of retinal progenitor cells to a porous polymethylmethacrylate scaffold and (b) the migration of retinal progenitor cells into the host retina (ONL = outer nuclear layer; INL = inner nuclear layer). The line in the left figure shows the location of the scaffold. The scale bar is 50 µm. (From Tao, S., et al. 2007. *Lab Chip* 7:695. With permission.)

9.6 Problems

1. Implanted neural recording electrodes are known to suffer from signal degradation over time, partly due to the growth of an encapsulating sac induced by the body's natural immune response to a foreign object. The sac distances recording sites from the target neurons. What strategies have been devised by researchers to lengthen the lifespan of these devices? Search the research literature and provide at least three examples. Each example should explicitly state what the specific strategy is and the exact materials and/or methods to achieve the goal. Cite any references in IEEE format.

2. Many types of microfabricated drug delivery devices have been devised. Applications requiring implantable pumps face additional challenges compared to passive, diffusion-based delivery devices. Likewise, few conventional implantable pumps exist. Perhaps, the most notable is the Medtronic insulin pump. Describe the construction and operational principle behind this pump. Why do you think certain design choices were made? How might these decisions extend to microfabricated drug delivery pumps?

3. The application of microsystems technology is one approach to tissue engineering. Many other "conventional" approaches have also been investigated. Discover two additional tissue engineering approaches by searching the research literature. What advantages for their particular applications do these approaches offer over the methods described in the text? When do you think it is appropriate to apply

microfabricated solutions to tissue engineering over these other methods? Cite any references in IEEE format.

4. Despite many efforts to commercialize implantable microsystems, few devices are available for patients today. What are some of the technological and regulatory hurdles that may play a role in the long path to the market?

References

1. Grayson, A. C. R., et al. 2004. A bioMEMS review: MEMS technology for physiologically integrated devices. *Proc IEEE* 92:6.
2. Mokwa, W. 2007. Medical implants based on microsystems. *Meas Sci Technol* 18:R47.
3. Receveur, R. A. M., F. W. Lindemans, and N. F. de Rooij. 2007. Microsystem technologies for implantable applications. *J Micromech Microeng* 17:R50.
4. Schurr, M. O., et al. 2007. Microtechnologies in medicine: An overview. *Minim Invasive Ther Allied Technol* 16:76.
5. Kandel, E. R., J. H. Schwartz, and T. M. Jessell. 2000. *Principles of Neural Science.* 4th ed. New York: McGraw-Hill, Health Professions Division.
6. Neher, E., and B. Sakmann. 1976. Single-channel currents recorded from membrane of denervated frog muscle-fibers. *Nature* 260:799.
7. Hamill, O. P., et al. 1981. Improved patch-clamp techniques for high-resolution current recording from cells and cell-free membrane patches. *Pflugers Arch* 391:85.
8. Thomas, C. A., et al. 1972. Miniature microelectrode array to monitor bioelectric activity of cultured cells. *Exp Cell Res* 74:61.
9. Pine, J. 1980. Recording action-potentials from cultured neurons with extracellular micro-circuit electrodes. *J Neurosci Methods* 2:19.
10. Gross, G. W., W. Y. Wen, and J. W. Lin. 1985. Transparent indium tin oxide electrode patterns for extracellular, multisite recording in neuronal cultures. *J Neurosci Methods* 15:243.
11. Boppart, S. A., B. C. Wheeler, and C. S. Wallace. 1992. A flexible perforated microelectrode array for extended neural recordings. *IEEE Trans Biomed Eng* 39:37.
12. Rutten, W. L. C. 2002. Selective electrical interfaces with the nervous system. *Annu Rev Biomed Eng* 4:407.
13. Smith, S. L., J. W. Judy, and T. S. Otis. 2004. An ultra small array of electrodes for stimulating multiple inputs into a single neuron. *J Neurosci Methods* 133:109.
14. Maher, M. P., et al. 1999. The neurochip: A new multielectrode device for stimulating and recording from cultured neurons. *J Neurosci Methods* 87:45.
15. Tooker, A., et al. 2005. Biocompatible parylene neurocages. *IEEE Eng Med Biol Mag* 24:30.
16. Wise, K. D., J. B. Angell, and A. Starr. 1970. An integrated-circuit approach to extracellular microelectrodes. *IEEE Trans Biomed Eng* BM17:238.
17. Wise, K. D., and J. B. Angell. 1971. A microprobe with integrated amplifiers for neutrophysiology. In *Solid-State Circuits Conference,* Piscataway, NJ: IEEE, 100–1.

18. Wise, K. D., and J. B. Angell. 1975. Low-capacitance multielectrode probe for use in extracellular neurophysiology. *IEEE Trans Biomed Eng* BM22:212.
19. Campbell, P. K., et al. 1991. A silicon-based, 3-dimensional neural interface - manufacturing processes for an intracortical electrode array. *IEEE Trans Biomed Eng* 38:758.
20. Branner, A., R. B. Stein, and R. A. Normann. 2001. Selective stimulation of cat sciatic nerve using an array of varying-length microelectrodes. *J Neurophysiol* 85:1585.
21. Bhandari, R., et al. 2009. A novel masking method for high aspect ratio penetrating microelectrode arrays. *J Micromech Microeng* 19.
22. Kim, S., et al. 2009. Integrated wireless neural interface based on the Utah electrode array. *Biomed Microdevices* 11:453.
23. Vetter, R. J., et al. 2004. Chronic neural recording using silicon-substrate microelectrode arrays implanted in cerebral cortex. *IEEE Trans Biomed Eng* 51:896.
24. Wise, K. D., et al. 2004. Wireless implantable microsystems: High-density electronic interfaces to the nervous system. *Proc IEEE* 92:76.
25. Wise, K. D., et al. 2008. Microelectrodes, microelectronics, and implantable neural microsystems. *Proc IEEE* 96:1184.
26. Walter, P., et al. 2005. Cortical activation via an implanted wireless retinal prosthesis. Invest *Ophthalmol Vis Sci* 46:1780.
27. Burmeister, J. J., K. Moxon, and G. A. Gerhardt. 2000. Ceramic-based multisite microelectrodes for electrochemical recordings. *Anal Chem* 72:187.
28. Rousche, P. J., et al. 2001. Flexible polyimide-based intracortical electrode arrays with bioactive capability. *IEEE Trans Biomed Eng* 48:361.
29. Lee, K., et al. 2004. Polyimide based neural implants with stiffness improvement. *Sens Actuators B Chem* 102:67.
30. Mokwa, W. 2004. MEMS technologies for epiretinal stimulation of the retina. *J Micromech Microeng* 14:S12.
31. Seymour, J. P., and D. R. Kipke. 2007. Neural probe design for reduced tissue encapsulation in CNS. *Biomaterials* 28:3594.
32. Takeuchi, S., et al. 2005. Parylene flexible neural probes integrated with microfluidic channels. *Lab Chip* 5:519.
33. Rodger, D. C., et al. 2008. Flexible parylene-based multielectrode array technology for high-density neural stimulation and recording. *Sens Actuators B Chem* 132:449.
34. Purcell, E. K., et al. 2009. In vivo evaluation of a neural stem cell-seeded prosthesis. *J Neural Eng* 6.
35. Navarro, X., et al. 2005. A critical review of interfaces with the peripheral nervous system for the control of neuroprostheses and hybrid bionic systems. *J Peripher Nerv Syst* 10:229.
36. Stieglitz, T., M. Schuettler, and K. P. Koch. 2005. Implantable biomedical microsystems for neural prostheses. *IEEE Eng Med Biol Mag* 24:58.
37. Kipke, D. R., et al. 2008. Advanced neurotechnologies for chronic neural interfaces: New horizons and clinical opportunities. *J Neurosci* 28:11830.
38. Chen, J. K., et al. 1997. A multichannel neural probe for selective chemical delivery at the cellular level. *IEEE Trans Biomed Eng* 44:760.
39. Papageorgiou, D., et al. 2001. A shuttered probe with in-line flowmeters for chronic in-vivo drug delivery. *Technical Digest. MEMS 2001. 14th IEEE International Conference on Micro Electro Mechanical Systems, Interlaken, Switzerland, Jan. 21–25, 2001*, Piscataway, NJ: IEEE, 212–15.

40. Rathnasingham, R., et al. 2004. Characterization of implantable microfabricated fluid delivery devices. *IEEE Trans Biomed Eng* 51:138.
41. Mokwa, W., and U. Schnakenberg. 2001. Micro-transponder systems for medical applications. *IEEE Trans Instrum Meas* 50:1551.
42. Rizzo, J. F., et al. 2003. Methods and perceptual thresholds for short-term electrical stimulation of human retina with microelectrode arrays. *Invest Ophthalmol Vis Sci* 44:5355.
43. Zrenner, E., et al. 1999. Can subretinal microphotodiodes successfully replace degenerated photoreceptors? *Vision Res* 39:2555.
44. Hammerle, H., et al. 2002. Biostability of micro-photodiode arrays for subretinal implantation. *Biomaterials* 23:797.
45. Bauerdick, S., et al. 2003. Substrate-integrated microelectrodes with improved charge-transfer capacity by 3-dimensional micro-fabrication. *Biomed Microdevices* 5:93.
46. Chow, A. Y., et al. 2004. The artificial silicon retina microchip for the treatment of vision loss from retinitis pigmentosa. *Arch Ophthal* 122:460.
47. Normann, R. A., et al. 2001. High-resolution spatiotemporal mapping of visual pathways using multi-electrode arrays. *Vision Res* 41:1261.
48. Weiland, J. D., W. T. Liu, and M. S. Humayun. 2005. Retinal prosthesis. *Annu Rev Biomed Eng* 7:361.
49. Weiland, J. D., and M. S. Humayun. 2008. Visual prosthesis. *Proc IEEE* 96:1076.
50. Bell, T. E., K. D. Wise, and D. J. Anderson. 1998. A flexible micromachined electrode array for a cochlear prosthesis. *Sens Actuators A Phys* 66:63.
51. Wise, K. D., et al. 2008. High-density cochlear implants with position sensing and control. *Hear Res* 242:22.
52. Wang, J. B., and K. D. Wise. 2009. A thin-film cochlear electrode array with integrated position sensing. *J Microelectromech Syst* 18:385.
53. Ziaie, B., et al. 1997. A single-channel implantable microstimulator for functional neuromuscular stimulation. *IEEE Trans Biomed Eng* 44:909.
54. Ziaie, B., et al. 1996. A hermetic glass-silicon micropackage with high-density on-chip feedthroughs for sensors and actuators. *J Microelectromech Syst* 5:166.
55. Rodriguez, F. J., et al. 2000. Polyimide cuff electrodes for peripheral nerve stimulation. *J Neurosci Methods* 98:105.
56. Rutten, W. L. C., H. J. Vanwier, and J. H. M. Put. 1991. Sensitivity and selectivity of intraneural stimulation using a silicon electrode array. *IEEE Trans Biomed Eng* 38:192.
57. Rutten, W. L. C., et al. 1995. 3D neuro-electronic interface devices for neuromuscular control - design studies and realization steps. *Biosens Bioelectron* 10:141.
58. Kovacs, G. T. A., C. W. Storment, and J. M. Rosen. 1992. Regeneration microelectrode array for peripheral-nerve recording and stimulation. *IEEE Trans Biomed Eng* 39:893.
59. Akin, T., et al. 1994. A micromachined silicon sieve electrode for nerve regeneration applications. *IEEE Trans Biomed Eng* 41:305.
60. Ramachandran, A., et al. 2006. Design, in vitro and in vivo assessment of a multi-channel sieve electrode with integrated multiplexer. *J Neural Eng* 3:114.
61. Collins, C. C. 1967. Miniature passive pressure transensor for implanting in eye. *IEEE Trans Biomed Eng* BM14:74.
62. Backlund, Y., et al. 1990. Passive silicon transensor intended for biomedical, remote pressure monitoring. *Sens Actuators A Phys* 21:58.

63. Puers, R., G. Vandevoorde, and D. De Bruyker. 2000. Electrodeposited copper inductors for intraocular pressure telemetry. *J Micromech Microeng* 10:124.
64. Chen, P. J., et al. 2008. Microfabricated implantable parylene-based wireless passive intraocular pressure sensors. *J Microelectromech Syst* 17:1342.
65. Chen, P. J., et al. 2007. Implantable micromechanical parylene-based pressure sensors for unpowered intraocular pressure sensing. *J Micromech Microeng* 17:1931.
66. Puers, B., et al. 1990. An implantable pressure sensor for use in cardiology. *Sens Actuators A Phys* 23:944.
67. Schnakenberg, U., et al. 2004. Intravascular pressure monitoring system. *Sens Actuators A Phys* 110:61.
68. Takahata, K., Y. B. Gianchandani, and K. D. Wise, 2006. Micromachined antenna stents and cuffs for monitoring intraluminal pressure and flow. *J Microelectromech Syst* 15:1289.
69. Hierold, C., et al. 1999. Low power integrated pressure sensor system for medical applications. *Sens Actuators A Phys* 73:58.
70. Eggers, T., et al. 2000. Advanced hybrid integrated low-power telemetric pressure monitoring system for biomedical applications. In *MEMS 2000. 13th IEEE International Conference on Micro Electro Mechanical Systems, Miyazaki, Japan, Jan. 23–27, 2000*, Piscataway, NJ: IEEE, 329–34.
71. Flick, B. B., and R. Orglmeister. 2000. A portable microsystem-based telemetric pressure and temperature measurement unit. *IEEE Trans Biomed Eng* 47:12.
72. Razzacki, S. Z., et al. 2004. Integrated microsystems for controlled drug delivery. *Adv Drug Deliv Rev* 56:185.
73. Deo, S., et al. 2003. Responsive drug delivery systems. *Anal Chem* 75:207A.
74. Tao, S. L., and T. A. Desai. 2003. Microfabricated drug delivery systems: From particles to pores. *Adv Drug Deliv Rev* 55:315.
75. Grayson, A. C. R., et al. 2004. Electronic MEMS for triggered delivery. *Adv Drug Deliv Rev* 56:173.
76. Ziaie, B., et al. 2004. Hard and soft micromachining for bioMEMS: Review of techniques and examples of applications in microfluidics and drug delivery. *Adv Drug Deliv Rev* 56:145.
77. Tsai, N. C., and C. Y. Sue. 2007. Review of MEMS-based drug delivery and dosing systems. *Sens Actuators A Phys* 134:555.
78. Nuxoll, E. E., and R. A. Siegel. 2009. BioMEMS devices for drug delivery improved therapy by design. *IEEE Eng Med Biol Mag* 28:31.
79. Ahmed, A., C. Bonner, and T. A. Desai. 2002. Bioadhesive microdevices with multiple reservoirs: A new platform for oral drug delivery. *J Control Release* 81:291.
80. Desai, T. A., et al. 1999. Nanopore technology for biomedical applications. *Biomedical Microdevices* 2:11.
81. Desai, T. A., D. Hansford, and M. Ferrari. 1999. Characterization of micromachined silicon membranes for immunoisolation and bioseparation applications. *J Memb Sci* 159:221.
82. Leoni, L., A. Boiarski, and T. A. Desai. 2002. Characterization of nanoporous membranes for immunoisolation: Diffusion properties and tissue effects. *Biomed Microdevices* 4:131.
83. McAllister, D. V., M. G. Allen, and M. R. Prausnitz. 2000. Microfabricated microneedles for gene and drug delivery. *Annu Rev Biomed Eng* 2:289.

84. Henry, S., et al. 1998. Microfabricated microneedles: A novel approach to transdermal drug delivery. *J Pharm Sci* 87:922.
85. Gardeniers, H., et al. 2003. Silicon micromachined hollow microneedles for transdermal liquid transport. *J Microelectromech Syst* 12:855.
86. Griss, P., and G. Stemme. 2003. Side-opened out-of-plane microneedles for microfluidic transdermal liquid transfer. *J Microelectromech Syst* 12:296.
87. Roxhed, N., et al. 2008. Painless drug delivery through microneedle-based transdermal patches featuring active infusion. *IEEE Trans Biomed Eng* 55:1063.
88. Randall, C. L., et al. 2007. 3d lithographically fabricated nanoliter containers for drug delivery. *Adv Drug Deliv Rev* 59:1547.
89. Santini, J. T., M. J. Cima, and R. Langer. 1999. A controlled-release microchip. *Nature* 397:335.
90. Prescott, J. H., et al. 2006. Chronic, programmed polypeptide delivery from an implanted, multireservoir microchip device. *Nat Biotechnol* 24:437.
91. Elman, N. M., et al. 2009. The next generation of drug-delivery microdevices. *Clin Pharmacol Ther* 85:544.
92. Maillefer, D., et al. 2001. A high-performance silicon micropump for disposable drug delivery systems. In *MEMS 2001: 14th IEEE International Conference on Micro Electro Mechanical Systems.* Piscataway, NJ: IEEE, 413–17.
93. Li, P. Y., et al. 2008. An electrochemical intraocular drug delivery device. *Sens Actuators A Phys* 143:41.
94. Pancrazio, J. J., F. Wang, and C. A. Kelley. 2007. Enabling tools for tissue engineering. *Biosens Bioelectron* 22:2803.
95. Puleo, C. M., H. C. Yeh, and T. H. Wang. 2007. Applications of MEMS technologies in tissue engineering. *Tissue Eng* 13:2839.
96. Ainslie, K. M., and T. A. Desai. 2008. Microfabricated implants for applications in therapeutic delivery, tissue engineering, and biosensing. *Lab Chip* 8:1864.
97. Lu, Y., and S. C. Chen. 2004. Micro and nano-fabrication of biodegradable polymers for drug delivery. *Adv Drug Deliv Rev* 56:1621.
98. Bettinger, C. J., et al. 2006. Three-dimensional microfluidic tissue-engineering scaffolds using a flexible biodegradable polymer. *Adv Mater* 18:165.
99. Deutsch, J., et al. 2000. Fabrication of microtextured membranes for cardiac myocyte attachment and orientation. *J Biomed Mater Res* 53:267.
100. Boateng, S., et al. 2002. Peptides bound to silicone membranes and 3d microfabrication for cardiac cell culture. *Adv Mater* 14:461.
101. Powers, M. J., et al. 2002. A microfabricated array bioreactor for perfused 3d liver culture. *Biotechnol Bioeng* 78:257.
102. Tao, S., et al. 2007. Survival, migration and differentiation of retinal progenitor cells transplanted on micro-machined poly(methyl methacrylate) scaffolds to the subretinal space. *Lab Chip* 7:695.

Index

A

Absolute viscosity, 139
Absorbance method, 248–249
Active pumps
  mechanical, 174–175
  nonmechanical, 175–177
Active valves, 166
  and actuation methods, 168–172
Adhesion of thin films, 37
Adhesive bonding, 118–119
Affinity chromatography, 212, 214
Amorphous materials, 23
Amperometry, 239–241
Amplification, PCR, 294
Animal Welfare Act (1985), 55
Anisotropic etching, 81–82, 103–105
Anodic bonding, 117
Antigen–antibody binding
  interaction, 307
APCVD, *see* Atmospheric pressure CVD
Arrhenius law, 105
Assembly, 120–121
Atmospheric pressure CVD
  (APCVD), 93–95
Atomic force microscopes (AFMs)
  probes, 113
Auditory prostheses, 352, 355
Axial filtration, 191

B

Bias, definition of, 82
Bimetallic valve, 173
Biocapsule assembly process, 364, 366
Biocompatibility, 52, 55–57
Biodegradable polymers, 49
  for tissue engineering, 372
Biological polymers, 44
Biology, microanalytical systems in,
  187–188
Biomaterials, 48–50
Biomedical applications, material
  selection in, 49–50

BioMEMS, applications, 5
Biorecognition, 300
Biosensors, 238
Blood barcode chip, 340
Body-centered cubic cell, 24
Boltzmann distribution, 155
Bonding methods, 116
Boron dipyrromethane
  (BODIPY), 199
Bourdon tube, 360
Brownian motion, 151, 162
Bulk micromachining, 102
Bull's eye effect, 84

C

CAE, *see* Capillary array
  electrophoresis
Capacitive pressure sensor, 360
Capacitive sensors, 231
Capillarity, 144–145
Capillary array electrophoresis (CAE),
  216, 300, 302, 303
Capillary electrophoresis (CE), 160,
  216–217
Capillary gel electrophoresis (CGE),
  216–217
Capillary pumping, 145
Cardiac catheterization, 325
Casting lithography, 110–111
Cell-based sensors, 289–292
Cells
  cell adhesion, 271–274
  culture, 269–271
  electroporation, 283–287
  lysis, 284, 288–289
  manipulation, 281–283
  polymer matrices and, 280–281
  retention, 275
Cell sorting, 321
Center for Devices and Radiological
  Health (CDRH), 50
Ceramics, 49
Chaotic advection, 178

For Product Safety Concerns and Information please contact our EU
representative GPSR@taylorandfrancis.com
Taylor & Francis Verlag GmbH, Kaufingerstraße 24, 80331 München, Germany

www.ingramcontent.com/pod-product-compliance
Ingram Content Group UK Ltd.
Pitfield, Milton Keynes, MK11 3LW, UK
UKHW021115180425
457613UK00005B/94

9 7 8 1 4 2 0 0 5 1 2 2 3